AIRCRAFT FLIGHT DYNAMICS AND CONTROL

Aerospace Series List

Aircraft Flight Dynamics and Control	Durham	August 2013
Civil Avionics Systems, Second Edition	Moir, Seabridge and Jukes	August 2013
Modelling and Managing Airport Performance	Zografos	July 2013
Advanced Aircraft Design: Conceptual Design, Analysis and Optimization of Subsonic Civil Airplanes	Torenbeek	June 2013
Design and Analysis of Composite Structures: With applications to aerospace Structures, Second Edition	Kassapoglou	April 2013
Aircraft Systems Integration of Air-Launched Weapons	Rigby	April 2013
Design and Development of Aircraft Systems, Second Edition	Moir and Seabridge	November 2012
Understanding Aerodynamics: Arguing from the Real Physics	McLean	November 2012
Aircraft Design: A Systems Engineering Approach	Sadraey	October 2012
Introduction to UAV Systems, Fourth Edition	Fahlstrom and Gleason	August 2012
Theory of Lift: Introductory Computational Aerodynamics with MATLAB and Octave	McBain	August 2012
Sense and Avoid in UAS: Research and Applications	Angelov	April 2012
Morphing Aerospace Vehicles and Structures	Valasek	April 2012
Gas Turbine Propulsion Systems	MacIsaac and Langton	July 2011
Basic Helicopter Aerodynamics, Third Edition	Seddon and Newman	July 2011
Advanced Control of Aircraft, Spacecraft and Rockets	Tewari	July 2011
Cooperative Path Planning of Unmanned Aerial Vehicles	Tsourdos et al	November 2010
Principles of Flight for Pilots	Swatton	October 2010
Air Travel and Health: A Systems Perspective	Seabridge et al	September 2010
Unmanned Aircraft Systems: UAVS Design, Development and Deployment	Austin	April 2010
Introduction to Antenna Placement & Installations	Macnamara	April 2010
Principles of Flight Simulation	Allerton	October 2009
Aircraft Fuel Systems	Langton et al	May 2009
The Global Airline Industry	Belobaba	April 2009
Computational Modelling and Simulation of Aircraft and the Environment: Volume 1 – Platform Kinematics and Synthetic Environment	Diston	April 2009
Handbook of Space Technology	Ley, Wittmann Hallmann	April 2009
Aircraft Performance Theory and Practice for Pilots	Swatton	August 2008
Aircraft Systems, Third Edition	Moir & Seabridge	March 2008
Introduction to Aircraft Aeroelasticity And Loads	Wright & Cooper	December 2007
Stability and Control of Aircraft Systems	Langton	September 2006
Military Avionics Systems	Moir & Seabridge	February 2006
Design and Development of Aircraft Systems	Moir & Seabridge	June 2004
Aircraft Loading and Structural Layout	Howe	May 2004
Aircraft Display Systems	Jukes	December 2003
Civil Avionics Systems	Moir & Seabridge	December 2002

AIRCRAFT FLIGHT DYNAMICS AND CONTROL

Wayne Durham
Virginia Polytechnic Institute and State University, USA

This edition first published in 2013
© 2013, John Wiley & Sons, Ltd

Registered office
John Wiley & Sons Ltd, The Atrium, Southern Gate, Chichester, West Sussex, PO19 8SQ, United Kingdom

For details of our global editorial offices, for customer services and for information about how to apply for permission to reuse the copyright material in this book please see our website at www.wiley.com.

The right of the author to be identified as the author of this work has been asserted in accordance with the Copyright, Designs and Patents Act 1988.

Reprinted with corrections March 2014.

All rights reserved. No part of this publication may be reproduced, stored in a retrieval system, or transmitted, in any form or by any means, electronic, mechanical, photocopying, recording or otherwise, except as permitted by the UK Copyright, Designs and Patents Act 1988, without the prior permission of the publisher.

Wiley also publishes its books in a variety of electronic formats. Some content that appears in print may not be available in electronic books.

Designations used by companies to distinguish their products are often claimed as trademarks. All brand names and product names used in this book are trade names, service marks, trademarks or registered trademarks of their respective owners. The publisher is not associated with any product or vendor mentioned in this book

Limit of Liability/Disclaimer of Warranty: While the publisher and author have used their best efforts in preparing this book, they make no representations or warranties with respect to the accuracy or completeness of the contents of this book and specifically disclaim any implied warranties of merchantability or fitness for a particular purpose. It is sold on the understanding that the publisher is not engaged in rendering professional services and neither the publisher nor the author shall be liable for damages arising herefrom. If professional advice or other expert assistance is required, the services of a competent professional should be sought.

MATLAB® is a trademark of The MathWorks, Inc. and is used with permission. The MathWorks does not warrant the accuracy of the text or exercises in this book. This book's use or discussion of MATLAB® software or related products does not constitute endorsement or sponsorship by The MathWorks of a particular pedagogical approach or particular use of the MATLAB® software.

For MATLAB® and Simulink® product information, please contact:

The MathWorks, Inc.
3 Apple Hill Drive
Natick, MA, 01760-2098 USA
Tel: 508-647-7000
Fax: 508-647-7001
E-mail: info@mathworks.com
Web: mathworks.com

Library of Congress Cataloging-in-Publication Data

Durham, Wayne, 1941-
  Aircraft flight dynamics and control / by Wayne Durham.
    1 online resource.
  Includes bibliographical references and index.
  Description based on print version record and CIP data provided by publisher; resource not viewed.
  ISBN 978-1-118-64678-6 (MobiPocket) – ISBN 978-1-118-64679-3 – ISBN 978-1-118-64680-9 (ePub) – ISBN 978-1-118-64681-6 (cloth) 1. Aerodynamics. 2. Flight. 3. Flight control. 4. Airplanes–Performance. I. Title.
  TL570
  629.132'3–dc23
                                                                                              2013020974

A catalogue record for this book is available from the British Library.

ISBN: 978-1-118-64681-6

Set in 10/12pt Times by Laserwords Private Limited, Chennai, India

1 2013

For Fred Lutze. If I got anything wrong here it's because I didn't listen to him closely enough.
For Hank Kelley. He was right. Sometimes you have to stare at the problem for a very long time before you see it. Sitzfleisch.

Contents

Series Preface   xiii

Glossary   xv

1 Introduction   1
1.1 Background   1
1.2 Overview   2
1.3 Customs and Conventions   6
References   6

2 Coordinate Systems   7
2.1 Background   7
2.2 The Coordinate Systems   7
    2.2.1 *The inertial reference frame, F_I*   7
    2.2.2 *The earth-centered reference frame, F_{EC}*   8
    2.2.3 *The earth-fixed reference frame, F_E*   8
    2.2.4 *The local-horizontal reference frame, F_H*   8
    2.2.5 *Body-fixed reference frames, F_B*   10
    2.2.6 *Wind-axis system, F_W*   12
    2.2.7 *Atmospheric reference frame*   12
2.3 Vector Notation   13
2.4 Customs and Conventions   14
    2.4.1 *Latitude and longitude*   14
    2.4.2 *Body axes*   14
    2.4.3 *'The' body-axis system*   14
    2.4.4 *Aerodynamic angles*   15
Problems   16
References   16

3 Coordinate System Transformations   17
3.1 Problem Statement   17
3.2 Transformations   18
    3.2.1 *Definitions*   18
    3.2.2 *Direction cosines*   18

	3.2.3	Euler angles	21
	3.2.4	Euler parameters	25
3.3	Transformations of Systems of Equations		26
3.4	Customs and Conventions		27
	3.4.1	Names of Euler angles	27
	3.4.2	Principal values of Euler angles	27
	Problems		27
	Reference		29
4	**Rotating Coordinate Systems**		**31**
4.1	General		31
4.2	Direction Cosines		34
4.3	Euler Angles		34
4.4	Euler Parameters		36
4.5	Customs and Conventions		38
	4.5.1	Angular velocity components	38
	Problems		38
5	**Inertial Accelerations**		**43**
5.1	General		43
5.2	Inertial Acceleration of a Point		43
	5.2.1	Arbitrary moving reference frame	43
	5.2.2	Earth-centered moving reference frame	46
	5.2.3	Earth-fixed moving reference frame	46
5.3	Inertial Acceleration of a Mass		47
	5.3.1	Linear acceleration	48
	5.3.2	Rotational acceleration	49
5.4	States		53
5.5	Customs and Conventions		53
	5.5.1	Linear velocity components	53
	5.5.2	Angular velocity components	54
	5.5.3	Forces	54
	5.5.4	Moments	56
	5.5.5	Groupings	56
	Problems		57
6	**Forces and Moments**		**59**
6.1	General		59
	6.1.1	Assumptions	59
	6.1.2	State variables	60
	6.1.3	State rates	60
	6.1.4	Flight controls	60
	6.1.5	Independent variables	62
6.2	Non-Dimensionalization		62
6.3	Non-Dimensional Coefficient Dependencies		63

		6.3.1	General	63
		6.3.2	Altitude dependencies	64
		6.3.3	Velocity dependencies	64
		6.3.4	Angle-of-attack dependencies	64
		6.3.5	Sideslip dependencies	66
		6.3.6	Angular velocity dependencies	68
		6.3.7	Control dependencies	69
		6.3.8	Summary of dependencies	70
	6.4	The Linear Assumption		71
	6.5	Tabular Data		71
	6.6	Customs and Conventions		72
		Problems		73
7		**Equations of Motion**		**75**
7.1		General		75
7.2		Body-Axis Equations		75
		7.2.1	Body-axis force equations	75
		7.2.2	Body-axis moment equations	76
		7.2.3	Body-axis orientation equations (kinematic equations)	77
		7.2.4	Body-axis navigation equations	77
7.3		Wind-Axis Equations		78
		7.3.1	Wind-axis force equations	78
		7.3.2	Wind-axis orientation equations (kinematic equations)	80
		7.3.3	Wind-axis navigation equations	81
7.4		Steady-State Solutions		81
		7.4.1	General	81
		7.4.2	Special cases	83
		7.4.3	The trim problem	88
		Problems		89
		Reference		91
8		**Linearization**		**93**
8.1		General		93
8.2		Taylor Series		94
8.3		Nonlinear Ordinary Differential Equations		95
8.4		Systems of Equations		95
8.5		Examples		97
		8.5.1	General	97
		8.5.2	A kinematic equation	99
		8.5.3	A moment equation	100
		8.5.4	A force equation	103
8.6		Customs and Conventions		105
		8.6.1	Omission of Δ	105
		8.6.2	Dimensional derivatives	105
		8.6.3	Added mass	105

8.7	The Linear Equations	106
	8.7.1 Linear equations	106
	8.7.2 Matrix forms of the linear equations	108
	Problems	111
	References	112

9   Solutions to the Linear Equations — 113
9.1	Scalar Equations	113
9.2	Matrix Equations	114
9.3	Initial Condition Response	115
	9.3.1 Modal analysis	115
9.4	Mode Sensitivity and Approximations	120
	9.4.1 Mode sensitivity	120
	9.4.2 Approximations	123
9.5	Forced Response	124
	9.5.1 Transfer functions	124
	9.5.2 Steady-state response	125
	Problems	125

10   Aircraft Flight Dynamics — 127
10.1	Example: Longitudinal Dynamics	127
	10.1.1 System matrices	127
	10.1.2 State transition matrix and eigenvalues	127
	10.1.3 Eigenvector analysis	129
	10.1.4 Longitudinal mode sensitivity and approximations	132
	10.1.5 Forced response	137
10.2	Example: Lateral–Directional Dynamics	140
	10.2.1 System matrices	140
	10.2.2 State transition matrix and eigenvalues	140
	10.2.3 Eigenvector analysis	142
	10.2.4 Lateral–directional mode sensitivity and approximations	144
	10.2.5 Forced response	148
	Problems	149
	References	150

11   Flying Qualities — 151
11.1	General	151
	11.1.1 Method	152
	11.1.2 Specifications and standards	155
11.2	MIL-F-8785C Requirements	156
	11.2.1 General	156
	11.2.2 Longitudinal flying qualities	157
	11.2.3 Lateral–directional flying qualitities	158
	Problems	166
	References	166

12 Automatic Flight Control — 169

12.1 Simple Feedback Systems — 170
    12.1.1 First-order systems — 170
    12.1.2 Second-order systems — 172
    12.1.3 A general representation — 177

12.2 Example Feedback Control Applications — 178
    12.2.1 Roll mode — 178
    12.2.2 Short-period mode — 184
    12.2.3 Phugoid — 188
    12.2.4 Coupled roll–spiral oscillation — 198

Problems — 206
References — 207

13 Trends in Automatic Flight Control — 209

13.1 Overview — 209

13.2 Dynamic Inversion — 210
    13.2.1 The controlled equations — 212
    13.2.2 The kinematic equations — 215
    13.2.3 The complementary equations — 221

13.3 Control Allocation — 224
    13.3.1 Background — 224
    13.3.2 Problem statement — 225
    13.3.3 Optimality — 231
    13.3.4 Sub-optimal solutions — 232
    13.3.5 Optimal solutions — 235
    13.3.6 Near-optimal solutions — 241

Problems — 243
References — 244

A Example Aircraft — 247
Reference — 253

B Linearization — 255
B.1 Derivation of Frequently Used Derivatives — 255
B.2 Non-dimensionalization of the Rolling Moment Equation — 257
B.3 Body Axis Z-Force and Thrust Derivatives — 258
B.4 Non-dimensionalization of the Z-Force Equation — 260

C Derivation of Euler Parameters — 263

D Fedeeva's Algorithm — 269
Reference — 272

E MATLAB® Commands Used in the Text — 273
E.1 Using MATLAB® — 273
E.2 Eigenvalues and Eigenvectors — 274

E.3	State-Space Representation	274
E.4	Transfer Function Representation	275
E.5	Root Locus	277
E.6	MATLAB® Functions (m-files)	277
	E.6.1 Example aircraft	278
	E.6.2 Mode sensitivity matrix	278
	E.6.3 Cut-and-try root locus gains	278
E.7	Miscellaneous Applications and Notes	280
	E.7.1 Matrices	280
	E.7.2 Commands used to create Figures 10.2 and 10.3	281

Index   **283**

Series Preface

The Aerospace Series covers a wide range of aerospace vehicles and their systems, comprehensively covering aspects of structural and system design in theoretical and practical terms. This book offers a clear and systematic treatment of flight dynamics and control which complements other books in the Series, especially books by McClean, Swatton and Diston.

The subject of flight dynamics and control has always been of importance in the design and operation of any aircraft, much of it learned by trial and error in the development of very early aircraft. It developed as an engineering science throughout succeeding generations of aircraft to support increasing demands of aircraft stability and control and it now has a major role to play in the design of modern aircraft to ensure efficient, comfortable and safe flight. The emergence of a need for unstable and highly manoeuvrable combat aircraft, and the dependence on full authority fly-by-wire software based control systems for both military and commercial aircraft together with a demand for economic automatic operation has ensured that the understanding of flight dynamics is essential for all designers of integrated flight systems. Growing trends towards unmanned air vehicles will serve to strengthen this dependency. Modern on-board sensors and computing in integrated systems offers the opportunity to sense aircraft motions and rates and to include aircraft models in the control systems to further improve aircraft performance. Engineers with an interest in these aspects will find this book essential reading.

The book has been built up from a combination of practical flying experience, the evolutionary improvement of a mentor's text and a desire that students should understand the basic concepts underlying modern modelling practices before applying them – an excellent way to evolve a text book to provide a real teaching experience. Much of the content has been validated by use in a teaching environment over a period of years.

This is a book for all those working in the field of flight control systems and aircraft performance for both manned and unmanned flight control as well as auto-flight control for real time applications in aircraft and high fidelity simulation.

Peter Belobaba, Jonathan Cooper and Allan Seabridge

Glossary

Greek symbols

α Angle of attack. The aerodynamic angle between the projection of the relative wind onto the airplane's plane of symmetry and a suitably defined body fixed x-axis.

n/α The change in load factor n resulting from a change in angle-of-attack α, or more properly the partial derivative of the former with respect to the latter.
A parameter used in the determination of short-period frequency requirements in flying qualities specifications, often called the 'control anticipation parameter'.

β Sideslip angle. The aerodynamic angle between the velocity vector and the airplane's plane of symmetry.

$\boldsymbol{\omega}, \omega$ As a vector (**bold**), usually signifies angular velocity. As a scalar, often subscripted, a component of such a vector.

χ Tracking angle. One of three angles that define a 321 rotation from inertial to the wind reference frames.

δ_ℓ A generic control effector that generates rolling moments L. It is often taken to be the ailerons, δ_a.

δ_a The ailerons, positive with the right aileron trailing-edge down and left aileron trailing-edge up.

δ_e The elevator, positive with trailing-edge down.

δ_m A generic control effector that generates pitching moments M. It is often taken to be the elevator, δ_e, or horizontal tail, δ_{HT}.

δ_n A generic control effector that generates yawing moments N. It is often taken to be the rudder, δ_r.

δ_r The rudder, positive with trailing-edge left.

δ_T Thrust, or throttle control.

Δ Indicates a change from reference conditions of the quantity it precedes. Often omitted when implied by context.

γ Flight-path angle. One of three angles that define a 321 rotation from inertial to the wind reference frames.

λ An eigenvalue, units s^{-1}.

λ Latitude on the earth.

Λ A diagonal matrix of a system's eigenvalues.

μ Longitude on the earth.

μ	Wind-axis bank angle. One of three angles that define a 321 rotation from inertial to the wind reference frames.
ω_d	Damped frequency of an oscillatory mode.
ω_n	Natural frequency of an oscillatory mode.
Ω	Every combination of control effector deflections that are admissible, i.e., that are within the limits of travel or deflection.
ϕ	Bank attitude. One of three angles that define a 321 rotation from inertial to body-fixed reference frames.
Φ	The effects, usually body-axis moments, of every combination of control effector deflections in Ω. Sometimes called the Attainable Moment Subset.
ψ	Heading angle. One of three angles that define a 321 rotation from inertial to body-fixed reference frames.
ρ	Density (property of the atmosphere).
θ	Pitch attitude. One of three angles that define a 321 rotation from inertial to body-fixed reference frames.
ζ	Damping ratio of an oscillatory mode.

Acronyms, abbreviations, and other terms

$\cdot$	Placed above a symbol of a time-varying entity, differentiation with respect to time.
$\hat{}$	Placed above a symbol to indicate that it is a non-dimensional quantity.
$\{v_a^b\}_c$	A vector **v** that is some feature of a (position, velocitiy, etc.) relative to b and represented in the coordinate system of c.
f	A vector of scalar functions, or a function of a vector.
F	A vector usually signifying force. See X, Y, Z and L, C, D.
h	A vector usually signifying angular momentum.
M	A vector usually signifying body-axis moments. See L, M, N.
q, q	As a vector (**bold**), usually signifies the transformed states of a system, such transformation serving to uncouple the dynamics. As a scalar, a component of such a vector.
r, r	As a vector (**bold**), usually signifies position. As a scalar, often subscripted, a component of such a vector.
T	A vector usually signifying thrust.
u	Vector of control effector variables.
v, v	As a vector (**bold**), usually signifies linear velocity. As a scalar, often subscripted, a component of such a vector.
W	A vector usually signifying weight.
x	Vector of state variables.
$\mathcal{A}$	Aspect ratio.
$\mathcal{L}$	LaPlace transform operator.
$\sim$	Placed above a symbol to indicate that it is an approximation or an approximate quantity.
A, B	Matrices of the linearized equations of motion, as in $\dot{x} = Ax + Bu$. A is the system matrix, B is control-effectiveness matrix.
C_{xy}	The non-dimensional stability or control derivative of x with respect to y. It is the non-dimensional form of X_y, q.v.

Glossary

c_i, r_j	The ith column, jth row of a matrix.
Comp	Complementary. A superscript to certain dynamic responses.
Cont	Controllable. A superscript to certain dynamic responses.
$D(\cdot)$	Non-dimensional differentiation.
$d(s)$	The characteristic polynomial of a system. The roots of the characteristic equation, $d(s) = 0$, are the systems eigenvalues.
d	Desired. A subscript to a dynamical response.
d	Subscript identifying the Dutch roll response mode.
DR	Subscript identifying the Dutch roll response mode. In flying qualities specifications the subscript is d.
F_B	Body-fixed reference frames.
F_E	Earth-fixed reference frame.
F_{EC}	Earth-centered reference frame.
F_H	Local-horizontal reference frame.
F_I	Inertial reference frame.
F_P	Principal axes.
F_S	Stability-axis system.
F_W	Wind-axis system.
F_Z	Zero-lift body-axis system.
$G(s)$	A matrix of transfer functions.
g	Acceleration of gravity. As a non-dimensional quantity g is the load factor n, q.v.
I	Identity matrix.
I	With subscripts, moment of inertia.
j	Imaginary number, $j = \sqrt{-1}$. Preference for j rather than i often stems from a background in electrical engineering, where i is electrical current.
Kine	Kinematic. A superscript to certain dynamic responses.
L, C, D	Lift, side force, and drag. Wind-axis forces in the $-x$-, $-y$- and $-z$-directions, respectively.
L, M, N	Body-axis rolling, pitching, and yawing moments, respectively.
L	Lift, or rolling moment, depending on context.
LD	Lateral–directional. Sometimes Lat–Dir.
Long	Longitudinal.
M	A matrix whose columns are the eigenvectors of a system.
M	Mach number.
m	Mass.
$N_{1/2}, N_2$	Number of cycles to half or double amplitude.
n	Load factor, the ratio of lift to weight, $n = L/W$. Measured in gs.
p_W, q_W, r_W	Wind-axis roll rate, pitch rate, and yaw rate, respectively.
p, q, r	Body-axis roll rate, pitch rate, and yaw rate, respectively.
P	A pseudo-inverse of a matrix B. $BPB = B$ and $PBP = P$, with appropriate dimensions.
Ph	Subscript identifying the phugoid response mode.
$q_0 \ldots q_3$	Euler parameters.
$q, \bar{q}$	The pitch rate is q. The dynamic pressure is $\bar{q}$, Kevin.
R	Subscript identifying the roll subsidence response mode.

Ref	Subscript, 'evaluated in reference conditions'.
RS	Subscript identifying the coupled roll–spiral response mode.
$S, \bar{c}, b$	Wing area, chord, and span, respectively.
s	Complex variable in LaPlace transformations.
S	Subscript identifying the spiral mode.
SP	Subscript identifying the short-period response mode.
ss	Subscript signifying steady state.
$t_{1/2}, t_2$	Time to half or double amplitude, seconds.
$T_{a,b}$	A transformation matrix that transforms vectors in coordinate system *b* to their representation in system *a*.
T	The period of an oscillatory response, seconds.
t	Time, seconds.
V_C	Magnitude of the velocity of the center of mass.
x_W, y_W, z_W	Names of wind axes.
X_y	Where *X* is a force or moment and *y* is a state or control, a dimensional derivative, $\partial X/\partial y$. It is the dimensional form of C_{x_y}, q.v. Note that the definition does *not* include division by mass or moment of inertia in this book.
X, Y, Z	Body-axis forces in the *x*-, *y*- and *z*-directions, respectively.
x, y, z	Names of axes. With no subscripts usually taken to be body axes.
8785C	Short for MIL-F-8785C, 'Military Specification, Flying Qualities of Piloted Airplanes'.
ACTIVE	Advanced Control Technology for Integrated Vehicles. A research F-15 with differential canards, axisymmetric thrust vectoring, and other novel features.
AMS	Attainable Moment Subset. See Φ.
ARI	Aileron–Rudder Interconnect. Normally used to reduce adverse yaw due to aileron deflection.
BIUG	Background Information and User's Guide, companion to Military Specifications for Flying Qualities.
CAS	Control Augmentation System.
CHR	Cooper–Harper Rating; sometimes HQR.
Control effector	The devices that directly effect control by changing forces or moments, such as ailerons or rudders. When we say 'controls' with no qualification, we usually mean the control effectors. The sign convention for conventional flapping control effectors follows a right-hand rule, with the thumb along the axis the effector is designed to generate moments, and the curled fingers denoting the positive deflection of the trailing edge.
Control inceptor	Cockpit devices that control, through direct linkage or a flight-control system or computer, the control effectors. Positive control inceptor deflections correspond to positive deflections of the effectors they are connected to, barring such things as aileron–rudder interconnects (ARI, q.v.).
E	The capital letter in Euler's name, not lowercase. Like the capital V in the Victorian era. 'Euler' is pronounced 'Oh e ler' by Swiss Germans, or 'Oiler' by many English speakers, but never 'Yuler'.

FBW	Fly By Wire. The pilot flies the computer, the computer flies the airplane.
Ganged	Said of mechanical devices linked so that they move in fixed relation to each other, such as ailerons and the rudder.
HARV	High Angle-of-Attack Research Vehicle.
HQR	Handling Qualities Rating.
Kt	Abbreviation for knot, a nautical mile per hour.
Lat–Dir	Lateral–directional.
OBM	On-Board Model. A set of aerodynamic data for an aircraft stored in a computer in the aircraft's flight control computer.
PA	Powered Approach. One of several flight phases defined in flying qualities specifications. See Section 11.2 for a complete list.
PIO	Pilot-Induced Oscillation. There's a more politically correct term that removes the onus from the pilot.
PR	Pilot Rating; sometimes HQR.
SAS	Stability Augmentation System.
SSSLF	Steady, Straight, Symmetric, Level Flight.
SVD	Singular-Value Decomposition.
TEU, TED, TEL, TER	Trailing-Edge Up, Down, Left, Right. Terms used to describe the deflection of flapping control surfaces.

1

Introduction

1.1 Background

This book grew out of several years of teaching a flight dynamics course at *The Virginia Polytechnic Institute & State University,* more commonly known as *Virginia Tech,* in Blacksburg, Virginia, USA. That course was initially based on Bernard Etkin's excellent graduate level text *Dynamics of Atmospheric Flight* (Etkin, 1972). There is a newer edition than that cited, but the author prefers his copy, as it can be relied on to fall open to the desired pages.

The author was taken on at Virginia Tech after a full career in the U.S. Navy as a fighter pilot and engineering test pilot. They taught an old dog new tricks, awarded him his PhD, and put him to work.

The author's background crept into the course presentation and Etkin's treatment became more and more modified as different approaches were taken to explaining things.

A sheaf of hand-written notes from a mentor who had actually designed flight control systems at Northrop; course material and flight experience from two different test-pilot schools; the experience of thousands of hours of flight in aircraft at the leading edge of the technology of their time; the precise and clear-minded approach to the analysis of flight dynamics problems that Fred Lutze demanded: all these things and more overlaid the tone and style of the course.

Then, one day, the author's course notes were so different from Etkin's work that it made no sense to continue using that book, and this book was born.

The course as taught at Virginia Tech was intended for first-year graduate students in aerospace engineering. The students all had previous course work in engineering mathematics. For purposes of the current treatment, multi-variable calculus, and a good understanding of ordinary differential equations and their solutions in the time domain and using LaPlace transforms are needed.

The undergraduate preparation at Virginia Tech also included a sound course in aircraft performance as in, for example, Anderson's excellent text (Anderson, 1989). Our undergraduates also had an award-winning sequence of courses in aircraft design taught by Bill Mason. While that course undoubtedly gave the students a better feel for what makes airplanes fly, such background is in no way essential to the understanding of this book.

The undergraduates had also studied intoductory flight stability and control, most often using another of Etkin's books (Etkin and Reid, 1995). Once again, previous exposure to this subject matter is by no means essential to mastering the material in this book. The author has seen mechanical engineering students who had *no* previous course-work involving airplanes or *any* airplane experience stand at the top of their classes. These students often had small models of airplanes that they brought with them to class.

The chapters on automatic flight control were not part of the course as originally taught. The major thrust of the book is airplane flight dynamics, but it was felt that some discussion of control was desirable to motivate future study. It seemed unlikely that anything as comprehensive as, for example, Stevens and Lewis (1992) could be included. Therefore just a basic introduction to feedback control is presented, but with some examples that are probably not often found in flight control design.

The last chapter was motivated by the author's pride in the accomplishments of many of his past students in real-world applications of flight dynamics and control.

The method of choice in current flight control system design appears to have settled on to dynamic inversion, and the associated problem of control allocation, and so a brief introduction to these disciplines is offered. Enough material is presented that the reader will be comfortable in the midst of modern flight control system engineers, and may even know something they do not.

Finally, almost all references to MATLAB® will be new to previous students. The author's approach to flight dynamics and control has always been to learn the basics, then adopt the modern tools and software to implement the basics. It is not expected that any reader will often use Fedeeva's algorithm in his work, but understanding it does afford one a singular look inside the minds of the men and women who solved these problems with pencil and paper, and who later went on to develop the algorithms that underlie the simple looking MATLAB® commands. But the dimension of the problems kept getting bigger and bigger, so some MATLAB® tools are now used.

1.2 Overview

The study of aircraft flight dynamics boils down to the determination of the position and velocity of an aircraft at some arbitrary time. This determination will be developed as the *equations of motion* of the airplane. The equations of motion consist of nonlinear ordinary differential equations in which the independent variables are the *states* of the airplane–the variables that fully describe the position and velocity.

The results of the analysis of flight dynamics–the equations of motion–are important in several related studies and disciplines. Chief among these are:

- Aircraft performance. Items of flight performance typically include stall speeds, level flight performance (range, endurance, etc.), excess power and acceleration characteristics, turn performance and agility, climb performance, descent performance, and takeoff and landing performance. Each of these items is governed by the equations of motion. The equations of motion are analyzed to determine the relevant parameters of the performance item, and these parameters are used to devise flight-test techniques to measure performance, or alternatively, to modify aircraft design to improve a particular area of performance.
- Aircraft control. Aircraft control is a very broad discipline, with primary sub-disciplines of manned control, automatic control, and optimal control.

- Manned control refers to a human operator, manipulating control inceptors in the cockpit to drive external control effectors to modify the state and change the trajectory of the aircraft. The relative ease with which a pilot can control the aircraft is described as the aircraft's *flying qualities*. Here, the equations of motion are analyzed to determine the factors that influence the pilot's workload in controlling the aircraft. These factors drive flight-test techniques and provide design guidance to improve flying qualities.
- Automatic control ranges from stability augmentation systems, which modify the aircraft's response to manned control, to autopilots, to full fly-by-wire systems. In fly-by-wire systems, the pilot effectively flies a computer, and the computer decides how to drive the external control effectors to best satisfy what its program thinks the pilot wants to do. Most forms of automatic control are typically based on a special, linearized, form of the equations of motion. The linearized equations of motion are used to design feedback systems to achieve the desired aircraft responses.
- Optimal control is often analyzed open-loop, that is, with no human operator. Representative objectives of optimal control are to minimize (or maximize) some performance measure, such as minimum time to climb, maximum altitude, or minimum tracking error in the presence of external disturbances. The complexity of the solution techniques used to determine optimal control usually means that some simplification of the equations of motion is required. Some ways in which the equations of motion are simplified include treating the aircraft as a point-mass with no rotational dynamics, or reduction in order (reducing the number of states) of the equations of motion.
- Flight simulation. In flight simulation the full nonlinear equations of motion, as ordinary differential equations, are programmed into a computer that integrates the equations (often using high-speed computers in real-time) using numerical integration algorithms. The forces that act upon the airplane are often provided as data in tabular form. The data may be aerodynamic (from wind tunnel or flight tests), or any other description of external forces (thrust, landing gear reactions, etc.)

The information that describes an airplane's position and velocity is usually expressed as relative to some external reference frame, for instance:

- Position and velocity relative to the earth, required for navigation. A pilot navigating from city to city needs to maintain altitude, track, and airspeed within parameters dictated by air-traffic control, and aircraft performance requirements. During terminal flight phases, especially approach and landing, the position and velocity must be maintained within very close tolerances.
- Position and velocity relative to another aircraft, required for rendezvous, formation flying, in-flight refueling, air-to-air combat, and so on. Military aircraft often fly within a few feet of one another; only the relative position and speed are important to the wingman, so that position and velocity relative to the earth are relatively unimportant.
- Position and velocity relative to the atmosphere determines the aerodynamic forces and moments. Students of aircraft performance immediately recognize the relationship of an aircraft's angle-of-attack to its lift coefficient, and of lift coefficient and dynamic pressure to the lift force itself. These simple relationships are the tip of the iceberg:

angular velocities, as well as other aerodynamic relationships, are needed for the most basic understanding of the forces and moments acting on a maneuvering aircraft.
- The last example offered is less intuitive than the others. We consider the position and velocity of an aircraft relative to some reference condition. If the position tends back to the reference position, and the change in velocity tends to zero, then we have described a stable flight condition. The changes in position and velocity might be induced by some external influence, such as a gust or turbulence. An aircraft that is stable in response to such disturbances will be easier to fly. On the other hand, the changes in relative position and velocity might be created by the pilot redefining the reference condition, say by changing from straight, level flight to turning flight. In this case the response of the aircraft determines the ease with which pilot is able to control it.

The approach taken in this book is essentially an application of Newton's Second Law,

$$F = ma \tag{1.1}$$

Simplistically, we assume some knowledge of the mass m and the applied forces F, solve for the acceleration $a = F/m$, and integrate the result with respect to time twice to yield velocity and position, respectively. There are, however, several considerations that will make our analysis non-trivial:

- Newton's Second Law correctly stated is that the external forces acting on a particle of infinitesimal mass are proportional to the time rate of change of the inertial momentum of the particle. We will show that the formula $F = ma$ can be applied to the aircraft as a whole, so long as we are talking about the acceleration of the instantaneous center of mass. The problem of finding an inertial reference frame in which to measure the accelerations will be dealt with by determining a reference frame that is almost inertial, and hopefully showing that the approximation does not introduce too great an error.
- The preceding mention of Newton's Second Law dealt with linear accelerations and can be expected to yield inertial position and linear velocity. An aircraft's angular position and velocity is also of great interest and must be determined through the extension of Newton's Law that relates externally applied moments (torque) to the time rate of change of angular momentum. This will yield the angular accelerations, and twice integrating we obtain angular position and velocity. One major problem with this formulation is that the various reference frames of interest generally rotate with respect to one another, which will require consideration of the various rotational accelerations.
- The mass of what we consider to be the aircraft may become redistributed and change, for example as fuel is burned or external stores are released. The redistribution of mass within the aircraft can arise from many sources and may be easy to consider (rotating machinery) or very difficult to formulate (aeroelastic flutter). It is impractical and undesirable to keep track of the various particles of mass that are moving around or are no longer attached to the aircraft, so we will look for reasonable approximations that allow us to neglect them.
- The various motions, forces, and moments we will consider all have coordinate systems in which they are most naturally characterized. For example, in a suitably defined coordinate system gravity always points 'down' or in the z-direction. For another, our

study of airfoils yields characterizations in which lift is perpendicular to the mean airflow and drag is parallel to it, suggesting a coordinate system in which some axis points in the direction of the relative wind. We will therefore define several coordinate systems that typically are rotating relative to one another. The problem then becomes one of describing the orientation of one system with respect to another, and determining how the orientation varies with time.

- The externally applied forces and moments may be hard to calculate accurately. Only for the force of gravity do we have a reasonably accurate approximation. Propulsive forces depend on a variety of factors that are difficult to predict, such as propeller efficiency or duct losses. Aerodynamic forces and moments in particular create difficulties because they are dependent in complex ways upon the various quantities we are trying to determine. Our understanding of those dependencies relies on empirical and analytical studies, the most extensive of which capture only the most salient relationships. Our study of flight dynamics will typically ignore the uncertainties in aerodynamic and propulsive data, marking these uncertainties for future consideration.

- Assuming we can find tools and reasonable approximations to deal with the aforementioned difficulties and can formulate $a = F/m$ (and its rotational counterpart) then we have the problem of solving these equations for positions and velocities. While it is easy to state that the velocity is the first integral of acceleration, and that position is its second, it is quite another thing to actually solve these equations. The complex relationships among the variables involved guarantee that we will have to find solutions to coupled, nonlinear ordinary differential equations for which analytical solutions are generally unavailable. We will overcome this difficulty by first considering unaccelerated motion of the aircraft, and then asking the question: how does the aircraft behave following small disturbances from this motion, or in response to control inputs? The result of this analysis will be systems of coupled, linear ordinary differential equations for which solutions are available.

- The solutions to the systems of linear ordinary differential equations (equations of motion) will tell us a good deal about how the aircraft behaves as a dynamical system. Unanswered is the question of what the behavior should be in order for the aircraft to be a 'good' aircraft. The problem is that most aircraft are piloted, and the human pilot does not want to spend all the time making corrections to keep the aircraft pointing in the right direction. The response and behavior of a piloted aircraft is called flying qualities (sometimes, handling qualities). The difficulty here is that how a pilot would like an aircraft to handle varies from pilot to pilot, and depends on what the pilot is trying to get the aircraft to do. Our study of flying qualities will be based on statistical analyses of a range of pilots' opinions that give rise to criteria and guidance for the design of aircraft.

- In the event an aircraft does not have inherently good flying qualities, it may be necessary to modify certain dynamical response characteristics. Structural modifications to an aircraft late in the design or in its operational phase are costly. It is relatively inexpensive to incorporate electronic control systems that can be tuned to provide the desired characteristics. The approach we will first investigate utilizes classical feedback control theory. Then we will briefly examine a recent development in automatic control called *dynamic inversion,* and examine its application to interesting control problems.

1.3 Customs and Conventions

We will attempt to proceed from the general to the specific in addressing the various aspects of flight dynamics and control. We will define certain generic quantities and operations needed to address the problem. In applying these results to the study of aircraft, custom and convention dictate that certain variables get their own names. The definitions used here are largely those of Etkin (1972), generally held in the United States to be authoritative. One should always be careful when encountering any terminology to be sure the exact definitions are understood.

References

Anderson, J. D. J. (1989) *Introduction to Flight,* McGraw-Hill

Etkin, B. (1972) *Dynamics of Atmospheric Flight*, 1st edn, John Wiley & Sons, Inc.

Etkin, B. and Reid, L.D. (1995) *Dynamics of Flight: Stability and Control,* 3rd edn, John Wiley & Sons, Inc.

Stevens, B. L. and Lewis, F. L. (1992) *Aircraft Control and Simulation,* 1st edn, John Wiley & Sons, Inc. pp. 255–259.

2

Coordinate Systems

2.1 Background

The need to define appropriate coordinate systems arises from two considerations. First, there may be some particular coordinate system in which the position and velocity of the aircraft 'make sense'. For navigation we are concerned with position and velocity with respect to the Earth, whereas for aircraft performance we need position and velocity with respect to the atmosphere. Second, there are coordinate systems in which the phenomena of interest are most naturally expressed. The direction of a jet engine's propulsive force may often be considered fixed with respect to the body of the aircraft.

All coordinate systems will be right-handed and orthogonal. Coordinate systems will be designated by the symbol F with a subscript intended to be a mnemonic for the name of the system, such as F_I for the inertial reference frame. The origin of the system will be denoted by O and a subscript (e.g., O_I). If we speak of where a coordinate system is we mean where its origin is. Axes of the system are labled x, y, and z with the appropriate subscript. Unit vectors along x, y, and z will be denoted **i, j,** and **k,** respectively, and subscripted appropriately.

It is customary in flight dynamics to omit subscripts when speaking of certain body-fixed coordinate systems. If this is not the case then the lack of subscripts will be taken to mean a generic system.

The definition of a coordinate system must state the location of its origin and the means of determining at least two of its axes, the third axis being determined by completing the right-hand system. The location of the origin and orientation of the axes may be arbitrary within certain restrictions, but once selected may not be changed. Following are the main coordinate systems of interest.

2.2 The Coordinate Systems

2.2.1 The inertial reference frame, F_I

The location of the origin may be any point that is completely unaccelerated (inertial), and the orientation of the axes is usually irrelevant to most problems so long as they too are fixed with respect to inertial space. For the purposes of this book the origin is at the Great Galactic Center.

2.2.2 The earth-centered reference frame, F_{EC}

As its name suggests this coordinate system has its origin at the center of the Earth (Figure 2.1). Its axes may be arbitrarily selected with respect to fixed positions on the surface of the Earth. We will take x_{EC} pointing from O_{EC} to the point of zero latitude and zero longitude on the Earth's surface, and z_{EC} in the direction of the spin vector of the Earth. This coordinate system obviously rotates with the Earth.

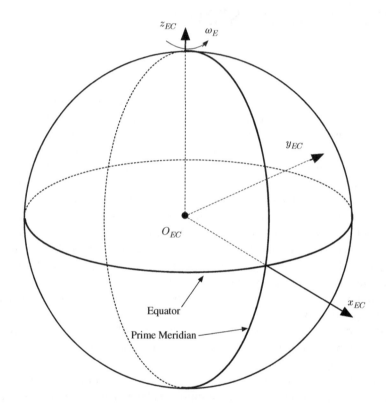

Figure 2.1 Earth-centered reference frame.

2.2.3 The earth-fixed reference frame, F_E

This coordinate system (Figure 2.2) has its origin fixed to an arbitrary point on the surface of the Earth (assumed to be a uniform sphere). x_E points due north, y_E points due east, and z_E points toward the center of the Earth.

2.2.4 The local-horizontal reference frame, F_H

This coordinate system (Figure 2.3) has its origin fixed to any arbitrary point that may be free to move relative to the Earth (assumed to be a uniform sphere). For example, the origin may be fixed to the center of gravity (CG) of an aircraft and move with the CG. x_H points due north, y_H points due east, and z_H points toward the center of the Earth.

Coordinate Systems

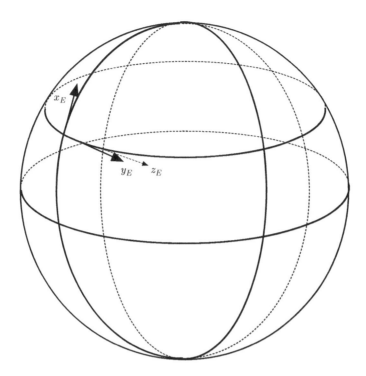

Figure 2.2 Earth-fixed reference frame, F_E.

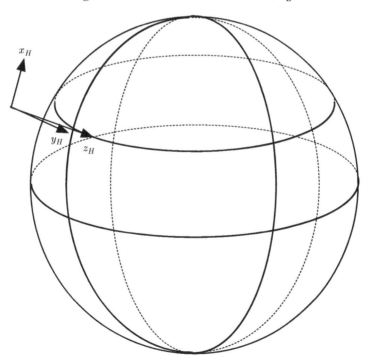

Figure 2.3 Local-horizontal reference frame, F_H.

2.2.5 Body-fixed reference frames, F_B

Body-fixed means the origin and axes of the coordinate system are fixed with respect to the (nominal) geometry of the aircraft. This must be distinguished from body-carried systems in which the origin is fixed with respect to the body but the axes are free to rotate relative to it. In flight dynamics body-fixed reference frames usually have their origin at the *CG*. In some applications in which the *CG* will change appreciably (e.g., flight simulation) the origin may be at some fixed fuselage reference point. (In ship stability and control, body-fixed coordinate systems are usually at the ship's center of buoyancy.) Determination of the orientation of the axes is as follows (see Figure 2.4). If the aircraft has a plane of symmetry (and we will assume in this book that they all do) then x_B and z_B lie in that plane of symmetry. x_B is chosen to point forward and z_B is chosen to point downward. There are some obvious difficulties with this specification: forward may mean 'toward the pointy end' or 'in the direction of flight'. In high angle-of-attack flight this gets confusing, since the direction of flight may be more toward the underside of the aircraft than toward the nose. There are clearly an infinite number of reference frames that satisfy the definition of a body-fixed coordinate system. Figure 2.5 shows two possibilities.

Figure 2.4 Body-fixed reference frame. x_B and z_B are in the aircraft plane of symmetry. The axes intersect at the airplane center of gravity. x_B has been chosen here to point through the aircraft's pitot tube, mounted at the front of the radar dome. The author sits astraddle the x_B-axis. Source: US Navy.

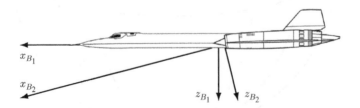

Figure 2.5 Two different body-fixed reference frames. y_B is common, and the plane of the page is the aircraft plane of symmetry. Source: NASA.

Coordinate Systems

Some of the more important body-fixed coordinate systems are as follows.

Principal axes, F_P

For every rigid body an orthogonal coordinate system may be found in which cross-products of inertia are zero. By the assumption of a plane of symmetry (geometric and mass symmetry) two of these axes will lie in the plane of symmetry. These are named x_P and z_P, and the typical longish nature of aircraft will permit one of these axes to be selected to be toward the nose, and this is x_P.

Zero-lift body-axis system, F_Z

Assuming the relative wind lies in the plane of symmetry (no sideslip) then there is a direction of the wind in this plane for which the net contribution of all surfaces of the aircraft toward creating lift is zero. The x_Z axis is selected to be into the relative wind when lift is zero. This is usually toward the nose of the aircraft, permitting the other axes to be chosen as described.

Stability-axis system, F_S

This is a special system defined as follows (see Figure 2.6). We consider the aircraft in some reference flight condition, usually steady flight so that the relative wind is seen from a constant direction by the aircraft. The x_S axis is taken as the projection of the velocity vector of the aircraft relative to the air mass into the aircraft plane of symmetry. Normally this is toward the nose of the aircraft, permitting the other axes to be chosen as for other

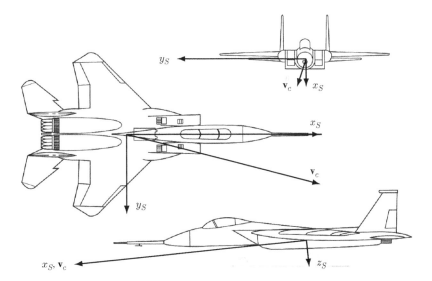

Figure 2.6 Stability-axis reference frame. x_S is the orthogonal projection of $\mathbf{v}_c$ onto the aircraft plane of symmetry. $\mathbf{v}_c$ is normally chosen in some steady flight condition and taken as the reference for subsequent analysis. Source: NASA.

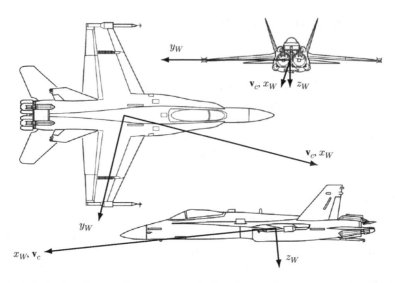

Figure 2.7 Wind-axis reference frame. z_W lies in the aircraft plane of symmetry, but in general x_W does not. Source: NASA.

body-axis systems. It is important to remember that this is a true body-fixed system, and that once defined the orientation of the axes relative to the aircraft is fixed, even if the direction of the relative wind changes.

This definition of the stability-axis system is that of Etkin (1972), which is also used in later work (Etkin, and Reid, 1995). Other authors, Stevans and Lewis (1992) for instance, define the stability-axes as a body-carried (not fixed) coordinate system in which the x-axis is the projection of the time-varying velocity vector into the plane of symmetry. By this definition the stability axes are mid-way between the wind axes (below, Section 2.2.6) and some body-fixed axis system.

2.2.6 Wind-axis system, F_W

Shown in Figure 2.7, F_W is a body-carried (origin fixed to the body, normally the CG) coordinate system in which the x_W axis is in the direction of the velocity vector of the aircraft relative to the air mass. The z_W axis is chosen to lie in the plane of symmetry of the aircraft and the y_W axis to the right of the plane of symmetry. Note that x_W need not lie in the plane of symmetry. If the relative wind changes, the orientation of the wind axes changes too, but z_W always lies in the plane of symmetry as defined.

2.2.7 Atmospheric reference frame

It is hard to define a reference frame that characterizes the motion of the atmosphere. What we usually think we know is the motion of the atmosphere relative to the Earth's surface, so a local horizontal reference frame might seem appropriate. However, the atmosphere is not a rigid body with a meaningful center of gravity to which to affix the origin of such a system, and at any rate such a system would have to be characterized with rotational properties so that the atmosphere did not fly tangentially off into space. In flight dynamics problems (as opposed to navigation problems) it is the instantaneous interaction of the airframe with the air mass that is of interest. For these problems the atmosphere is typically thought of as a separate earth-fixed system to which appropriate extra components are added to account for winds, gusts, and turbulence.

2.3 Vector Notation

Vectors are denoted by **bold** symbols. Vectors are often defined as some measurement (position, linear velocity, angular velocity, linear acceleration, angular acceleration, etc.) of one point or body relative to another. These points and bodies will be referred to by some physical name (or a symbol such as c for CG) or by the name of a coordinate system whose origin is fixed to the point or body. We will use a subscript to denote the first such point and a superscript for the other. Omission of a superscript may usually be taken to mean we are referring to an inertial reference frame. Commonly used symbols are $\mathbf{r}$ for position, $\mathbf{v}$ for linear velocity, $\mathbf{a}$ for linear acceleration, $\boldsymbol{\omega}$ for angular velocity, and $\boldsymbol{\alpha}$ for angular acceleration. Examples:

$\mathbf{r}_p^E$:   Position vector of the point p relative to (the origin of) some F_E.
$\boldsymbol{\omega}_E^{EC}$:   Rotation of some F_E relative to the Earth-centered coordinate system.
$\mathbf{a}_c$:   Linear acceleration of the point c relative to inertial space.

Vectors can exist notionally as just described, but to quantify them they must be *represented* in some coordinate system. Once defined, the notional vector can be represented in any coordinate system by placing the vector at the origin of the coordinate system and finding its components in the x, y, and z directions. In general these components will be different in two different coordinate systems, unless the two coordinate systems have parallel axes. We will denote the coordinate system in which a vector is represented by affixing a second subscript. In order to avoid multiple subscripts, curly braces $\{\cdot\}$ will be used to enclose the vector, and the subscript appended to the right brace will indicate the system in which the vector is represented. Examples:

$\{\mathbf{r}_p^E\}_E$:   The vector is $\mathbf{r}_p^E$, the position vector of the point p relative to (the origin of) some F_E. The vector is represented in F_E, as $\{\mathbf{r}_p^E\}_E = r_x \mathbf{i}_E + r_y \mathbf{j}_E + r_z \mathbf{k}_E$ or, in vector notation

$$\{\mathbf{r}_p^E\}_E = \begin{Bmatrix} r_x \\ r_y \\ r_z \end{Bmatrix}_E$$

$\{\mathbf{v}_c^E\}_B$:   The vector is $\mathbf{v}_c^E$, the velocity of point c relative to the origin of an Earth-fixed reference frame. The components of the vector are as projected into a body-fixed coordinate system F_B.

$\{\boldsymbol{\omega}_B\}_B$:   The vector is $\boldsymbol{\omega}_B$, the inertial angular rotation rate of a body-fixed coordinate system. The omission of a superscript suggests the inertial reference frame. The vector is represented in a body-fixed coordinate system.

Once defined it may be convenient to omit the subscripts and superscripts of a vector, especially if the vector is referred to repeatedly, or if the notation becomes cumbersome.

2.4 Customs and Conventions

2.4.1 Latitude and longitude

Position on the Earth is measured by *latitude* and *longitude*. Denote latitude by the symbol λ and longitude by the symbol μ. We will measure latitude positive north and negative south of the equator, $-90 \deg \leq \lambda \leq +90 \deg$; and longitude positive east and negative west of zero longitude, $-180 \deg < \mu \leq +180 \deg$. Latitude and longitude may refer to either an Earth-fixed coordinate system λ_E, μ_E or a local horizontal cooridinate system λ_H, μ_H.

2.4.2 Body axes

The names of the axes in a body-axis system are often not subscripted, and appear simply as x, y, and z.

2.4.3 'The' body-axis system

This ill-defined term is frequently used at all levels of discussion of aircraft flight dynamics. Its usage often seems to imply that there is only one body-axis system, when in fact there are an infinite number of them. If the meaning is unclear, then the user of such an expression should be asked to specify how the system was chosen.

'The' body-axis system often refers to a fuselage reference system described by Liming (1945). Liming's system, or something very much like it, is used in aircraft fabrication and assembly. The system will often survive the shop floor and live on in the operational life of the aircraft.

In Liming's system the y-axis is positive aft (the negative of our x_B-axis) and the x-axis is positive out of the left wing (the negative of our y_B-axis). This leaves the z-axis positive up (the negative of our z_B-axis). Liming described the $x - y$ plane as the *waterline plane*, taken as horizontal when the aircraft is in the 'rigged' position. He was likely thinking of the aircraft on the shop floor during assembly, or later when performing weight and balance measurements. Waterlines are measured along the z-axis from this plane. The $x - z$ plane is called the *fuselage station plane*. Fuselage stations (coordinates along the y-axis) are numbered beginning at the nose of the aircraft, not the origin. Thus, a component may be mounted on a bulkhead at Fuselage Station (FS) 300, meaning the bulkhead is 300 inches back from the nose. The $y - z$ plane is the plane of symmetry, and coordinates along the x-axis (spanwise) are called buttock, or butt lines.

Liming describes the origin of the system as simply the intersection of the plane of symmetry (which is pretty well defined) and the other two planes (which are not very well defined). The origin of the system is not generally the center of gravity of the aircraft, but is often close to it.

2.4.4 Aerodynamic angles

The angle between the velocity vector and the plane of symmetry, measured in the plane $x_W - y_W$, is called sideslip and is denoted by the symbol β. Sideslip is positive when the relative wind is from the right of the plane of symmetry, as shown in Figure 2.8.

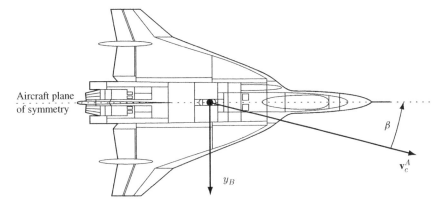

Figure 2.8 Sideslip angle, β. The axis y_B and velocity vector $\mathbf{v}_c^A$ lie in the plane of the page. Source: NASA.

Consider the projection of the velocity vector $\mathbf{v}_c^A$ into the plane of symmetry, and assume some body-fixed coordinate system has been defined. The angle between this projection and the x_B axis is called the angle-of-attack and is given the symbol α. It is positive when the relative wind is from below the x_B axis as shown in Figure 2.9. Since the definition of α depends on the choice of x_B it may be necessary to subscript α, such as α_S in the stability-axis system.

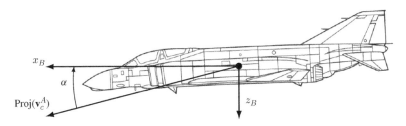

Figure 2.9 Angle-of-attack, α. Plane of the page is the aircraft plane of symmetry, $x_B - z_B$. $\text{Proj}(\mathbf{v}_c^A)$ is the projection of $\mathbf{v}_c^A$ onto the plane of symmetry. Source: NASA.

Problems

1. Where does y_{EC} intersect the surface of the Earth?
2. Assume the Earth's center is in uniform, unaccelerated motion with respect to inertial space, and that the Earth rotates at a fixed rate of $\omega_E = 360$ deg/day. Find $\{\omega_{EC}\}_{EC}$, the inertial rotation of the Earth as represented in F_{EC}. That is, express the inertial rotation of the Earth in terms of $\mathbf{i}_{EC}$, $\mathbf{j}_{EC}$, and $\mathbf{k}_{EC}$ (rad/s).
3. Find $\{\mathbf{g}_P^E\}_E$, the acceleration of gravity at a point p relative to the Earth as represented in F_E (ft/s²). Use $|\mathbf{g}| \equiv g = 32.174$ ft/s². The gravity vector usually does not have subscripts or superscripts as these will be suggested by context. In this problem it would simply be $\{\mathbf{g}\}_E$.
4. Use the same assumptions as in Problem 2. Consider an Earth-fixed reference frame whose origin is 90 deg east longitude and on the equator. Find $\{\boldsymbol{\omega}_E\}_E$.
5. Consider an aircraft flying due east along the equator. Define an Earth-fixed coordinate system F_E at zero longitude and zero latitude, and a local horizontal coordinate system F_H fixed to the center-of-gravity (CG) of the aircraft. Describe the relationship between the axes of F_E and F_H at the instant the aircraft is at zero longitude and zero latitude, and again at the instant it is at 180 deg east longitude and zero latitude.
6. Assume the Earth (uniform sphere) has a diameter of 6875 NM (nautical miles, NM). The aircraft in Problem 5 is flying at 600 knots relative to an assumed stationary atmosphere, at a constant altitude of 2 NM. Find $\{\omega_H^E\}_H$ and $\{\omega_H^{EC}\}_H$.
7. Answer true or false, and explain:
   (a) All body-fixed axis systems have a common x-axis.
   (b) All body-fixed axis systems have a common y-axis.
   (c) All body-fixed axis systems have a common z-axis.
8. Answer true or false, and explain:
   (a) All body-carried axis systems have a common x-axis.
   (b) All body-carried axis systems have a common y-axis.
   (c) All body-carried axis systems have a common z-axis.
9. An arbitrary body-fixed system F_B is related to F_Z by a positive rotation of 10 deg about their common y-axis. What is the angle of the relative wind to x_z when the net lift on the aircraft is zero (no sideslip)?
10. An aircraft is flying with velocity relative to the air mass of 500 ft/s. Find $\{\mathbf{v}_C^A\}_W$.
11. The figure that was used to define sideslip β shows y_B in the plane of $x_W - y_W$. Is this always the case? Explain.

References

Etkin, B. (1972) *Dynamics of Atmospheric Flight*, 1st edn, John Wiley & Sons, Inc.
Etkin, B. and Reid, L.D. (1995) *Dynamics of Flight: Stability and Control*, 3rd edn, John Wiley & Sons, Inc.
Liming, R.A. (1945) *Practical Analytic Geometry with Applications to Aircraft*, The MacMillan Company.
Stevens, B. L. and Lewis, F. L. (1992) *Aircraft Control and Simulation*, 1st edn, John Wiley & Sons, Inc.

3

Coordinate System Transformations

3.1 Problem Statement

We have met the various coordinate systems that will be of primary interest in the study of flight dynamics. Now we address the subject of how these coordinate systems are related to one another. For instance, when we begin to sum the external forces acting on the aircraft, we will have to relate all these forces to a common reference frame. If we take some body-fixed system to be the common frame, then we need to be able to take the gravity vector (weight) from the local horizontal reference frame, the thrust from some other body-axis frame, and the aerodynamic forces from the wind-axis frame, and represent all these forces in the body-fixed frame.

Some of these relationships are easy because they are fixed: two body-fixed systems, once defined, are always related by a single rotation around their common y-axis. Others are much more complicated because they vary with time: the orientation of a given body-fixed system with respect to the wind axes determines certain aerodynamic forces and moments which change that orientation.

First we must characterize the relationship between two coordinate systems at some frozen instant in time. The instantaneous relationship between two coordinate systems will be addressed by determining a *transformation* that will take the representation of an arbitrary vector in one system and convert it to its representation in the other.

Three approaches to finding the needed transformations will be presented. The first is called *Direction Cosines*, a sort of brute-force approach to the problem. Next we will discuss *Euler angles*, by far the most common approach but one with a potentially serious problem. Finally we will examine *Euler parameters*, an elegant solution to the problem found with Euler angles.

Aircraft Flight Dynamics and Control, First Edition. Wayne Durham.
© 2013 John Wiley & Sons, Ltd. Published 2013 by John Wiley & Sons, Ltd.

3.2 Transformations

3.2.1 Definitions

Consider two reference frames, F_1 and F_2, and a vector **v** whose components are known in F_1, represented as $\{\mathbf{v}\}_1$:

$$\{\mathbf{v}\}_1 = \begin{Bmatrix} v_{x_1} \\ v_{y_1} \\ v_{z_1} \end{Bmatrix}$$

We wish to determine the representation of the same vector in F_2, or $\{\mathbf{v}\}_2$:

$$\{\mathbf{v}\}_2 = \begin{Bmatrix} v_{x_2} \\ v_{y_2} \\ v_{z_2} \end{Bmatrix}$$

These are linear spaces, so the transformation of the vector from F_1 to F_2 is simply a matrix multiplication which we will denote $T_{2,1}$, such that $\{\mathbf{v}\}_2 = T_{2,1}\{\mathbf{v}\}_1$. Transformations such as $T_{2,1}$ are called similarity transformations. The transformations involved in simple rotations of orthogonal reference frames have many special properties that will be shown.

The order of subscripts of $T_{2,1}$ is such that the left subscript goes with the system of the vector on the left side of the equation and the right subscript with the vector on the right. For the matrix multiplication to be conformal $T_{2,1}$ must be a 3x3 matrix:

$$T_{2,1} = \begin{bmatrix} t_{11} & t_{12} & t_{13} \\ t_{21} & t_{22} & t_{23} \\ t_{31} & t_{32} & t_{33} \end{bmatrix}$$

(The t_{ij} probably need more subscripting to distinguish them from those in other transformation matrices, but this gets cumbersome.)

The whole point of the three approaches to be presented is to figure out how to evaluate the numbers t_{ij}. It is important to note that for a fixed orientation between two coordinate systems, the numbers t_{ij} are the same quantities no matter how they are evaluated.

3.2.2 Direction cosines

Derivation

Consider our two reference frames F_1 and F_2, and the vector **v**, shown in Figure 3.1. We claim to know the representation of **v** in F_1. The vector **v** is the vector sum of the three components $v_{x_1}\mathbf{i}_1$, $v_{y_1}\mathbf{j}_1$, and $v_{z_1}\mathbf{k}_1$ so we may replace the vector by those three components, shown in Figure 3.2.

Now, the projection of **v** onto x_2 is the same as the vector sum of the projections of each of its components $v_{x_1}\mathbf{i}_1$, $v_{y_1}\mathbf{j}_1$, and $v_{z_1}\mathbf{k}_1$ onto x_2, and similarly for y_2 and z_2. Define the angle generated in going from x_2 to x_1 as $\theta_{x_2 x_1}$. The projection of $v_{x_1}\mathbf{i}_1$ onto x_2 therefore has magnitude $v_{x_1} \cos \theta_{x_2 x_1}$ and direction $\mathbf{i}_2$, as shown if Figure 3.3.

Coordinate System Transformations

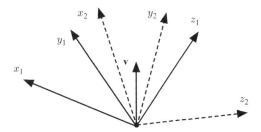

Figure 3.1 Two coordinate systems with co-located origins. The vector **v** may be represented in either reference frame.

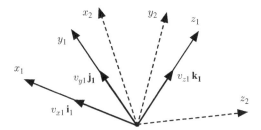

Figure 3.2 Vector in component form, components in F_1.

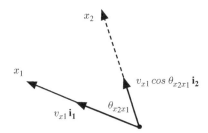

Figure 3.3 One component of the vector.

To the vector $v_{x_1} \cos\theta_{x_2 x_1} \mathbf{i_2}$ must be added the other two projections, $v_{y_1} \cos\theta_{x_2 y_1} \mathbf{i_2}$ and $v_{z_1} \cos\theta_{x_2 z_1} \mathbf{i_2}$. Similar projections onto y_2 and z_2 result in:

$$v_{x_2} = v_{x_1} \cos\theta_{x_2 x_1} + v_{y_1} \cos\theta_{x_2 y_1} + v_{z_1} \cos\theta_{x_2 z_1}$$

$$v_{y_2} = v_{x_1} \cos\theta_{y_2 x_1} + v_{y_1} \cos\theta_{y_2 y_1} + v_{z_1} \cos\theta_{y_2 z_1}$$

$$v_{z_2} = v_{x_1} \cos\theta_{z_2 x_1} + v_{y_1} \cos\theta_{z_2 y_1} + v_{z_1} \cos\theta_{z_2 z_1}$$

In vector-matrix notation, this may be written

$$\begin{Bmatrix} v_{x_2} \\ v_{y_2} \\ v_{z_2} \end{Bmatrix} = \begin{bmatrix} \cos\theta_{x_2 x_1} & \cos\theta_{x_2 y_1} & \cos\theta_{x_2 z_1} \\ \cos\theta_{y_2 x_1} & \cos\theta_{y_2 y_1} & \cos\theta_{y_2 z_1} \\ \cos\theta_{z_2 x_1} & \cos\theta_{z_2 y_1} & \cos\theta_{z_2 z_1} \end{bmatrix} \begin{Bmatrix} v_{x_1} \\ v_{y_1} \\ v_{z_1} \end{Bmatrix}$$

Clearly this is $\{v\}_2 = T_{2,1}\{v\}_1$, so we must have $t_{ij} = \cos\theta_{(axis)_2(axis)_1}$ in which i or j is 1 if the corresponding $(axis)$ is x, 2 if $(axis)$ is y, and 3 if $(axis)$ is z.

$$T_{2,1} = \begin{bmatrix} \cos\theta_{x_2 x_1} & \cos\theta_{x_2 y_1} & \cos\theta_{x_2 z_1} \\ \cos\theta_{y_2 x_1} & \cos\theta_{y_2 y_1} & \cos\theta_{y_2 z_1} \\ \cos\theta_{z_2 x_1} & \cos\theta_{z_2 y_1} & \cos\theta_{z_2 z_1} \end{bmatrix} \quad (3.1)$$

Properties of the direction cosine matrix

The transformation matrix (often called the direction cosine matrix, regardless of how it is derived or represented) does not depend on the vector $\mathbf{v}$ for its existence. It is therefore the same no matter what we choose for $\mathbf{v}$. If we take

$$\{v\}_1 = \begin{Bmatrix} 1 \\ 0 \\ 0 \end{Bmatrix}$$

Then $\{v\}_2$ is just the first column of the direction cosine matrix:

$$\{v\}_2 = \begin{Bmatrix} \cos\theta_{x_2 x_1} \\ \cos\theta_{y_2 x_1} \\ \cos\theta_{z_2 x_1} \end{Bmatrix}$$

Since the length of $\mathbf{v}$ is 1, we have shown the well-known property of direction cosines, that

$$\cos^2\theta_{x_2 x_1} + \cos^2\theta_{y_2 x_1} + \cos^2\theta_{z_2 x_1} = 1$$

The same obviously holds true for any column of the direction cosine matrix.

If we need to go the other way, $\{v\}_1 = T_{1,2}\{v\}_2$, it is clear that $T_{1,2} = T_{2,1}^{-1}$, and we might be tempted to invert the direction cosine matrix. On the other hand, had we begun by assuming $\{v\}_2$ was known, and figuring out how to get $\{v\}_1 = T_{1,2}\{v\}_2$ using similar arguments to the above, we would have arrived at

$$\begin{Bmatrix} v_{x_1} \\ v_{y_1} \\ v_{z_1} \end{Bmatrix} = \begin{bmatrix} \cos\theta_{x_1 x_2} & \cos\theta_{x_1 y_2} & \cos\theta_{x_1 z_2} \\ \cos\theta_{y_1 x_2} & \cos\theta_{y_1 y_2} & \cos\theta_{y_1 z_2} \\ \cos\theta_{z_1 x_2} & \cos\theta_{z_1 y_2} & \cos\theta_{z_1 z_2} \end{bmatrix} \begin{Bmatrix} v_{x_2} \\ v_{y_2} \\ v_{z_2} \end{Bmatrix}$$

Here we observe that $\cos\theta_{x_1 x_2}$ is the same as $\cos\theta_{x_2 x_1}$, $\cos\theta_{x_1 y_2}$ is the same as $\cos\theta_{y_2 x_1}$, etc., since (even if we had defined positive rotations of these angles) the cosine is an even function. So the same transformation is

$$\begin{Bmatrix} v_{x_1} \\ v_{y_1} \\ v_{z_1} \end{Bmatrix} = \begin{bmatrix} \cos\theta_{x_2 x_1} & \cos\theta_{y_2 x_1} & \cos\theta_{z_2 x_1} \\ \cos\theta_{x_2 y_1} & \cos\theta_{y_2 y_1} & \cos\theta_{z_2 y_1} \\ \cos\theta_{x_2 z_1} & \cos\theta_{y_2 z_1} & \cos\theta_{z_2 z_1} \end{bmatrix} \begin{Bmatrix} v_{x_2} \\ v_{y_2} \\ v_{z_2} \end{Bmatrix}$$

Obviously the columns of $T_{1,2}$ are the rows of $T_{2,1}$ (and have unit length as well). This leads us to another nice property of these transformation matrices, that the inverse of the direction cosine matrix is equal to its transpose,

$$T_{1,2} = T_{2,1}^{-1} = T_{2,1}^T$$

Coordinate System Transformations

Since $T_{2,1}T_{2,1}^{-1} = T_{2,1}T_{2,1}^T = I_3$, the 3 × 3 identity matrix, it must be true that the scalar (dot) product of any row of $T_{2,1}$ with any other row must be zero. It is easy to show that $T_{1,2}T_{1,2}^{-1} = T_{1,2}T_{1,2}^T = I_3$, so the scalar (dot) product of any column of $T_{2,1}$ with any other column must be zero since the columns of $T_{2,1}$ are the rows of $T_{1,2}$. When the rows (or columns) of the direction cosine matrix are viewed as vectors, this means the rows (or the columns) form orthogonal bases for three-dimensional space.

Also, with $T_{2,1}T_{2,1}^T = I_3$, if we take the determinant of each side and note that $|T_{2,1}^T| = |T_{2,1}|$, we have

$$|T_{2,1}T_{2,1}^T| = |T_{2,1}||T_{2,1}^T|$$
$$= |T_{2,1}||T_{2,1}|$$
$$= |T_{2,1}|^2 = 1$$

The only conclusion we can reach at this point is that $|T_{2,1}| = \pm 1$. Because the identity matrix is a transformation matrix with determinant $+1$, and since all other transformations may be reached through continuous rotations from the identity transformation, it seems unreasonable to think that the sign of the determinant would be different for some rotations and not others. We will show in our discussion of Euler angles that the right answer is indeed $|T_{2,1}| = +1$.

While we have nine variables in $T_{2,1} = \{t_{ij}\}, i, j = 1 \ldots 3$, there are six nonlinear constraining equations based on the orthogonality of the rows (or columns) of the matrix. For the rows these are:

$$t_{11}^2 + t_{12}^2 + t_{13}^2 = 1$$
$$t_{21}^2 + t_{22}^2 + t_{23}^2 = 1$$
$$t_{31}^2 + t_{32}^2 + t_{33}^2 = 1$$
$$t_{11}t_{21} + t_{12}t_{22} + t_{13}t_{23} = 0$$
$$t_{11}t_{31} + t_{12}t_{32} + t_{13}t_{33} = 0$$
$$t_{21}t_{31} + t_{22}t_{32} + t_{23}t_{33} = 0$$

The result of these constraints is that there are only three independent variables. In principle all nine may be derived given any three that do not violate a constraint.

3.2.3 Euler angles

If there are only three independent variables in the direction cosine matrix, then we should be able to express each of the t_{ij} in terms of some set of three (not necesarily unique) independent variables. One means of determining these variables is by use of a famous theorem due to the Swiss mathematician Leonhard Euler (1707–1783). Briefly, this theorem holds that any arbitrarily oriented reference frame may be placed in alignment with (made to have axes parallel to) any other reference frame by three successive rotations about the axes of the reference frame. The order of selection of axes in these rotations is arbitrary, but the same axis may not be used twice in succesion. The rotation sequences are usually denoted by three numbers, 1 for x, 2 for y, and 3 for z. The 12 valid sequences are 123, 121, 131, 132, 213, 212, 231, 232, 312, 313, 321, and 323. The angles through which these rotations are performed (defined as positive according to the right-hand rule for right-handed coordinate systems) are called generically *Euler angles*.

The rotation sequence most often used in flight dynamics is the 321, or $z - y - x$. Considering a rotation from F_1 to F_2 the first rotation (Figure 3.4) is about z_1 through an angle θ_z which is positive according to the right-hand rule about the z_1-axis. With two rotations to go, the resulting alignment in general is oriented with neither F_1 nor F_2, but some intermediate reference frame (the first of two) denoted F'. Since the rotation was about z_1, z' is parallel to it but neither of the other two primed axes are.

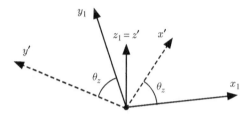

Figure 3.4 Rotation through θ_z.

The next rotation (Figure 3.5) is through an angle θ_y about the axis $y\prime$ of the first intermediate reference frame to the second intermediate reference frame, F''. Note that $y'' = y'$, and neither y'' or z'' are necessarily axes of either F_1 or F_2.

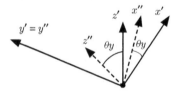

Figure 3.5 Rotation through θ_y.

The final rotation (Figure 3.6) is about x'' through angle θ_x and the final alignment is parallel to the axes of F_2.

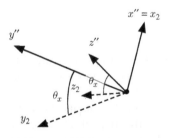

Figure 3.6 Rotation through θ_x.

Now assuming we know the angles θ_x, θ_y, and θ_z we need to relate them to the elements of the direction cosine matrix $T_{2,1}$. We will do this by seeing how the arbitrary vector

Coordinate System Transformations

v is represented in each of the intermediate and final reference frames in terms of its representation in the prior reference frame.

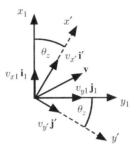

Figure 3.7 Rotation from F_1 to F'.

Consider first the rotation about z_1 (Figure 3.7). In terms of the direction cosines previously defined, the angles between the axes are as follows: between z_1 and z' it is zero; between either z and any x or y it is 90 deg; between x_1 and x' or y_1 and y' it is θ_z; between x_1 and y' it is 90 deg $+ \theta_z$; and between y_1 and x' it is 90 deg $- \theta_z$. We therefore may write

$$\begin{Bmatrix} v_{x'} \\ v_{y'} \\ v_{z'} \end{Bmatrix} = \begin{bmatrix} \cos\theta_{x'x_1} & \cos\theta_{x'y_1} & \cos\theta_{x'z_1} \\ \cos\theta_{y'x_1} & \cos\theta_{y'y_1} & \cos\theta_{y'z_1} \\ \cos\theta_{z'x_1} & \cos\theta_{z'y_1} & \cos\theta_{z'z_1} \end{bmatrix} \begin{Bmatrix} v_{x_1} \\ v_{y_1} \\ v_{z_1} \end{Bmatrix}$$

$$= \begin{bmatrix} \cos\theta_z & \cos(90\deg - \theta_z) & \cos 90\deg \\ \cos(90\deg + \theta_z) & \cos\theta_z & \cos 90\deg \\ \cos 90\deg & \cos 90\deg & \cos 0 \end{bmatrix} \begin{Bmatrix} v_{x_1} \\ v_{y_1} \\ v_{z_1} \end{Bmatrix}$$

$$= \begin{bmatrix} \cos\theta_z & \sin\theta_z & 0 \\ -\sin\theta_z & \cos\theta_z & 0 \\ 0 & 0 & 1 \end{bmatrix} \begin{Bmatrix} v_{x_1} \\ v_{y_1} \\ v_{z_1} \end{Bmatrix}$$

In short, $\{v\}' = T_{F',1}\{v\}_1$ in which

$$T_{F',1} = \begin{bmatrix} \cos\theta_z & \sin\theta_z & 0 \\ -\sin\theta_z & \cos\theta_z & 0 \\ 0 & 0 & 1 \end{bmatrix} \quad (3.2)$$

From the rotation about y' we get $\{v\}'' = T_{F'',F'}\{v\}'$ in which

$$T_{F'',F'} = \begin{bmatrix} \cos\theta_y & 0 & -\sin\theta_y \\ 0 & 1 & 0 \\ \sin\theta_y & 0 & \cos\theta_y \end{bmatrix} \quad (3.3)$$

From the rotation about x'', finally, $\{v\}_2 = T_{2,F''}\{v\}''$ with

$$T_{2,F''} = \begin{bmatrix} 1 & 0 & 0 \\ 0 & \cos\theta_x & \sin\theta_x \\ 0 & -\sin\theta_x & \cos\theta_x \end{bmatrix} \quad (3.4)$$

We now cascade the relationships: $\{v\}_2 = T_{2,F''}\{v\}''$, $\{v\}'' = T_{F'',F'}\{v\}'$ to arrive first at $\{v\}_2 = T_{2,F''}T_{F'',F'}\{v\}'$ and finally have the result

$$\{v\}_2 = T_{2,F''}T_{F'',F'}T_{F',1}\{v\}_1$$

The transformation matrix we seek is the product of the three sequential transformations (in the correct order!), or $T_{2,1} = T_{2,F''}T_{F'',F'}T_{F',1}$. The details are slightly tedious, but the result is

$$T_{2,1} = \begin{bmatrix} \cos\theta_y\cos\theta_z & \cos\theta_y\sin\theta_z & -\sin\theta_y \\ \begin{pmatrix}\sin\theta_x\sin\theta_y\cos\theta_z \\ -\cos\theta_x\sin\theta_z\end{pmatrix} & \begin{pmatrix}\sin\theta_x\sin\theta_y\sin\theta_z \\ +\cos\theta_x\cos\theta_z\end{pmatrix} & \sin\theta_x\cos\theta_y \\ \begin{pmatrix}\cos\theta_x\sin\theta_y\cos\theta_z \\ +\sin\theta_x\sin\theta_z\end{pmatrix} & \begin{pmatrix}\cos\theta_x\sin\theta_y\sin\theta_z \\ -\sin\theta_x\cos\theta_z\end{pmatrix} & \cos\theta_x\cos\theta_y \end{bmatrix} \quad (3.5)$$

Since this is just a different way of representing the direction cosine matrix, it must be true that $\cos\theta_y\cos\theta_z = \cos\theta_{x_2 x_1}$, $\cos\theta_y\sin\theta_z = \cos\theta_{x_2 y_1}$, etc. At this point we may observe that since $T_{2,1} = T_{2,F''}T_{F'',F'}T_{F',1}$, we must have

$$|T_{2,1}| = |T_{2,F''}T_{F'',F'}T_{F',1}| = |T_{2,F''}||T_{F'',F'}||T_{F',1}|$$

The determinant of each of the three intermediate transformations is easily verified to be $+1$, so that we have the expected result,

$$|T_{2,1}| = +1$$

It is very important to note that the Euler angles θ_x, θ_y, and θ_z have been defined for a rotation *from* F_1 *to* F_2 using a 321 rotation sequence. Thus, a 321 rotation *from* F_2 *to* F_1 ($T_{1,2}$) with suitably defined angles (say, ϕ_x, ϕ_y, and ϕ_z) would have the same form as given for $T_{2,1}$, but the angles involved would be physically different from those in $T_{2,1}$. So, for the rotation from F_1 to F_2 using a 321 rotation sequence:

$$T_{1,2} = \begin{bmatrix} \cos\phi_y\cos\phi_z & \cos\phi_y\sin\phi_z & -\sin\phi_y \\ \begin{pmatrix}\sin\phi_x\sin\phi_y\cos\phi_z \\ -\cos\phi_x\sin\phi_z\end{pmatrix} & \begin{pmatrix}\sin\phi_x\sin\phi_y\sin\phi_z \\ +\cos\phi_x\cos\phi_z\end{pmatrix} & \sin\phi_x\cos\phi_y \\ \begin{pmatrix}\cos\phi_x\sin\phi_y\cos\phi_z \\ +\sin\phi_x\sin\phi_z\end{pmatrix} & \begin{pmatrix}\cos\phi_x\sin\phi_y\sin\phi_z \\ -\sin\phi_x\cos\phi_z\end{pmatrix} & \cos\phi_x\cos\phi_y \end{bmatrix}$$

This is just Equation 3.5 with ϕ replacing θ. Alternatively we may note that $T_{1,2} = T_{2,1}^{-1} = T_{2,1}^T$ and may write the matrix

$$T_{1,2} = \begin{bmatrix} \cos\theta_y\cos\theta_z & \begin{pmatrix}\sin\theta_x\sin\theta_y\cos\theta_z \\ -\cos\theta_x\sin\theta_z\end{pmatrix} & \begin{pmatrix}\cos\theta_x\sin\theta_y\cos\theta_z \\ +\sin\theta_x\sin\theta_z\end{pmatrix} \\ \cos\theta_y\sin\theta_z & \begin{pmatrix}\sin\theta_x\sin\theta_y\sin\theta_z \\ +\cos\theta_x\cos\theta_z\end{pmatrix} & \begin{pmatrix}\cos\theta_x\sin\theta_y\sin\theta_z \\ -\sin\theta_x\cos\theta_z\end{pmatrix} \\ -\sin\theta_y & \sin\theta_x\cos\theta_y & \cos\theta_x\cos\theta_y \end{bmatrix}$$

The two matrices are the same, only the definitions of the angles are different. Clearly the relationships among the two sets of angles are non-trivial.

In short, the definitions of Euler angles are unique to the rotation sequence used (321, 213, etc.) and the decision as to which frame one goes from and which one goes to in that sequence (F1 to F2 or F2 to F1). We will normally define our Euler angles going in only one direction using a 321 rotation sequence, and rely on relationships like $T_{1,2} = T_{2,1}^{-1} = T_{2,1}^{T}$ to obtain the other.

3.2.4 Euler parameters

The derivation of Euler parameters is given in Appendix C. Euler parameters are based on the observation that any two coordinate systems are instantaneously related by a single rotation about some axis that has the same representation in each system. The axis, called the eigenaxis, has direction cosines ξ, ζ, and χ; the angle of rotation is η. Then, with the definition of the Euler parameters:

$$\begin{aligned} q_0 &\doteq \cos(\eta/2) \\ q_1 &\doteq \xi \sin(\eta/2) \\ q_2 &\doteq \zeta \sin(\eta/2) \\ q_3 &\doteq \chi \sin(\eta/2) \end{aligned} \quad (3.6)$$

the transformation matrix becomes:

$$T_{2,1} = \begin{bmatrix} (q_0^2 + q_1^2 - q_2^2 - q_3^2) & 2(q_1 q_2 + q_0 q_3) & 2(q_1 q_3 - q_0 q_2) \\ 2(q_1 q_2 - q_0 q_3) & (q_0^2 - q_1^2 + q_2^2 - q_3^2) & 2(q_2 q_3 + q_0 q_1) \\ 2(q_1 q_3 + q_0 q_2) & 2(q_2 q_3 - q_0 q_1) & (q_0^2 - q_1^2 - q_2^2 + q_3^2) \end{bmatrix} \quad (3.7)$$

Euler parameters have one great disadvantage relative to Euler angles: Euler angles may in most cases be easily visualized. If one is given values of θ_x, θ_y, and θ_z, then if they are not too large it is not hard to visualize the relative orientation of two coordinate systems. A given set of Euler parameters, however, conveys almost no information about how the systems are related. Thus even in applications in which Euler parameters are preferred (such as in flight simulation), the results are very often converted into Euler angles for ease of interpretation and visualization.

The direct approach to convert a set of Euler parameters to the corresponding set of Euler angles is to equate corresponding elements of the two representations of the transformation matrix. We may combine the (2,3) and (3,3) entries to obtain

$$\frac{\sin \theta_x \cos \theta_y}{\cos \theta_x \cos \theta_y} = \tan \theta_x = \frac{2(q_2 q_3 + q_0 q_1)}{q_0^2 - q_1^2 - q_2^2 + q_3^2}$$

$$\theta_x = \tan^{-1}\left(\frac{t_{23}}{t_{33}}\right) = \tan^{-1}\left[\frac{2(q_2 q_3 + q_0 q_1)}{q_0^2 - q_1^2 - q_2^2 + q_3^2}\right], \quad -\pi \leq \theta_x < \pi$$

From the (1,3) entry we have

$$-\sin \theta_y = 2(q_1 q_3 - q_0 q_2)$$

$$\theta_y = -\sin^{-1}(t_{13}) = -\sin^{-1}(2q_1 q_3 - 2q_0 q_2), \quad -\pi/2 \leq \theta_y \leq \pi/2$$

Finally we use the (1,1) and (1,2) entries to yield

$$\frac{\cos\theta_y \sin\theta_z}{\cos\theta_y \cos\theta_z} = \tan\theta_z = \frac{2(q_1 q_2 + q_0 q_3)}{q_0^2 + q_1^2 - q_2^2 - q_3^2}$$

$$\theta_z = \tan^{-1}\left(\frac{t_{12}}{t_{11}}\right) = \tan^{-1}\left[\frac{2(q_1 q_2 + q_0 q_3)}{q_0^2 + q_1^2 - q_2^2 - q_3^2}\right], 0 \leq \theta_z \leq 2\pi$$

With the arctangent functions one has to be careful to not divide by zero when evaluating the argument. Most software libraries have the two-argument arctangent function which avoids this problem and helps keep track of the quadrant. In summary,

$$\theta_x = \tan^{-1}\left[\frac{2(q_2 q_3 + q_0 q_1)}{q_0^2 - q_1^2 - q_2^2 + q_3^2}\right], -\pi \leq \theta_x < \pi$$

$$\theta_y = -\sin^{-1}(2q_1 q_3 - 2q_0 q_2), -\pi/2 \leq \theta_y \leq \pi/2 \qquad (3.8)$$

$$\theta_z = \tan^{-1}\left[\frac{2(q_1 q_2 + q_0 q_3)}{q_0^2 + q_1^2 - q_2^2 - q_3^2}\right], 0 \leq \theta_z \leq 2\pi$$

Going the other way, from Euler angles to Euler parameters, has been studied extensively. It may be done in a somewhat similar manner to equating elements of the transformation matrix. Another more elegant method due to Junkins and Turner (1978) yields a very nice result, here presented without proof:

$$q_0 = \cos(\theta_z/2)\cos(\theta_y/2)\cos(\theta_x/2) + \sin(\theta_z/2)\sin(\theta_y/2)\sin(\theta_x/2)$$
$$q_1 = \cos(\theta_z/2)\cos(\theta_y/2)\sin(\theta_x/2) - \sin(\theta_z/2)\sin(\theta_y/2)\cos(\theta_x/2)$$
$$q_2 = \cos(\theta_z/2)\sin(\theta_y/2)\cos(\theta_x/2) + \sin(\theta_z/2)\cos(\theta_y/2)\sin(\theta_x/2) \qquad (3.9)$$
$$q_3 = \sin(\theta_z/2)\cos(\theta_y/2)\cos(\theta_x/2) - \cos(\theta_z/2)\sin(\theta_y/2)\sin(\theta_x/2)$$

Just one final note on Euler parameters. Frequently in the literature Euler parameters are referred to as quaternions. However, quaternions are actually defined as half-scalar, half-vector entities in some coordinate system as $\tilde{Q} = q_0 + q_1 \mathbf{i} + q_2 \mathbf{j} + q_3 \mathbf{k}$. The algebra of quaternions is useful for proving theorems regarding Euler parameters, but will not be used in this book.

3.3 Transformations of Systems of Equations

Suppose we have a linear system of equations in some reference frame F_1:

$$\{\mathbf{y}\}_1 = A_1 \{\mathbf{x}\}_1$$

We know the transformation to F_2 ($T_{2,1}$) and wish to represent the same equations in that reference frame. Each side of $\{\mathbf{y}\}_1 = A_1\{\mathbf{x}\}_1$ is a vector in F_1, so we transform both sides as $T_{2,1}\{\mathbf{y}\}_1 = \{\mathbf{y}\}_2 = T_{2,1} A_1 \{\mathbf{x}\}_1$. Then we may insert the identity matrix in the form of $T_{2,1}^T T_{2,1}$ in the right-hand side to yield $\{\mathbf{y}\}_2 = T_{2,1} A_1 T_{2,1}^T T_{2,1} \{\mathbf{x}\}_1$. The reason for doing this is to transform the vector $\{x\}_1$ to $\{x\}_2 = T_{2,1}\{x\}_1$. Grouping terms we have the result,

$$\{y\}_2 = T_{2,1} A_1 T_{2,1}^T \{x\}_2 = A_2\{x\}_2$$

With the conclusion that the transformation of a matrix A_1 from F_1 to F_2 (or the equivalent operation performed by A_1 in F_2) is given by

$$A_2 = T_{2,1} A_1 T_{2,1}^T \tag{3.10}$$

This is sometimes spoken of as the transformation of a matrix. Such transformations may be very useful. For example, if we could find F_2 and $T_{2,1}$ such that A_2 is diagonal, then the system of equations $\{y\}_2 = A_2\{x\}_2$ is very easy to solve. The original variables can then be recovered by $\{y\}_1 = T_{2,1}^T\{y\}_2$, $\{x\}_1 = T_{2,1}^T\{x\}_2$.

3.4 Customs and Conventions

3.4.1 Names of Euler angles

Some transformation matrices occur frequently, and the 321 Euler angles associated with them are given special symbols. These are summarized as follows:

T_{F_2,F_1}	θ_x	θ_y	θ_z
$T_{B,H}$	ϕ	θ	ψ
$T_{W,H}$	μ	γ	χ
$T_{B,W}$	0	α	$-\beta$

(3.11)

3.4.2 Principal values of Euler angles

The principal range of values of the Euler angles is fixed largely by convention and may vary according to the application. In this book the convention is

$$\begin{aligned} -\pi &\leq \theta_x < \pi \\ -\pi/2 &\leq \theta_y \leq \pi/2 \\ 0 &\leq \theta_z < 2\pi \end{aligned} \tag{3.12}$$

These ranges are most frequently used in flight dynamics, although occasionally one sees $-\pi < \theta_z \leq \pi$ and $0 < \theta_x \leq 2\pi$.

Problems

1. One often hears in casual conversation that γ is the difference between θ and α.
   (a) Use the relationship $T_{W,H} = T_{W,B} T_{B,H}$ to find and expression for γ in terms of the angles on the right-hand side of the expression.
   (b) Using that result, under what conditions is it true that $\gamma = \theta - \alpha$?
2. Consider any two body-fixed coordinate systems, F_{B_1} and F_{B_2}, where F_{B_2} is related to F_{B_1} by a positive rotation about the y_{B_1} axis through an angle Θ.

(a) Determine the nine angles between the axes of F_{B_1} and those of F_{B_2}. Use this information to write down T_{B_2,B_1} using the direction cosine matrix.

(b) For a 321 Euler angle sequence from F_{B_1} to F_{B_2}, determine the three angles θ_x, θ_y, and θ_z. Use this information to write down T_{B_2,B_1} using the Euler angle representation of the direction cosine matrix.

(c) Determine the eigenaxis and angle that relates F_{B_2} to F_{B_1}. Use this information to write down T_{B_2,B_1} using the Euler parameter representation of the direction cosine matrix.

3. Consider the previously defined coordinate systems F_{EC} and F_E. Denote latitude by the symbol λ and longitude by the symbol μ. Measure latitude positive north and negative south of the equator, $-90 \deg \leq \lambda \leq +90 \deg$. Measure longitude positive east and negative west of zero longitude, $-180 \deg < \mu \leq +180 \deg$.

(a) Given the latitude λ_E and longitude μ_E of F_E, find a transformation matrix from F_{EC} to F_E ($T_{E,EC}$) that is a function only of λ_E and μ_E.

(b) Using your answer to part 3a, transform the Earth's rotation vector in F_{EC} to its representation in Earth-fixed coordinates at Blacksburg, Virginia, USA ($F_E = F_{BBurg}$). That is, transform $\{\omega_E\}_{EC}$ to $\{\omega_E\}_{BB}$. Use $\lambda_{BBurg} = 37°12.442''N$, $\mu_{BBurg} = 80°24.446''W$.

4. Consider the Euler angles involved in a 321 rotation from F_W to F_B.

(a) For the initial (F_W) and each intermediate coordinate system involved in this transformation, determine which, if any, of the axes lie in one of the three planes formed by pairs of axes of F_B. Describe and sketch each case identified and explain why it is so.

(b) What is the value of θ_x in the transformation? Explain.

(c) Compare these rotations with the definitions of the aerodynamic angles α and β, and verify that the names of the Euler angles given for $T_{B,W}$ are correct.

(d) Verify that the transformation matrix from wind to body axes is

$$T_{B,W} = \begin{bmatrix} \cos\alpha\cos\beta & -\cos\alpha\sin\beta & -\sin\alpha \\ \sin\beta & \cos\beta & 0 \\ \sin\alpha\cos\beta & -\sin\alpha\sin\beta & \cos\alpha \end{bmatrix} \quad (3.13)$$

5. Find a direction cosine matrix T such that

$$T\mathbf{v} = \mathbf{e}_2$$

where

$$\mathbf{v} = \begin{Bmatrix} 0.6 \\ 0 \\ 0.8 \end{Bmatrix} \quad \mathbf{e}_2 = \begin{Bmatrix} 0 \\ 1 \\ 0 \end{Bmatrix}$$

The direction cosine matrix must have all the properties described in the text.

6. Using suitably defined angles, find the transformation matrix $T_{2,1}$ for a 212 sequence of Euler angle rotations. Show all work.

7. Consider two *left-handed* coordinate systems.

(a) Using the relationships between the respective axes of the two coordinate systems, derive the direction cosine matrix for left-handed systems.

Coordinate System Transformations

(b) How does this result differ from the direction cosine matrix for right-handed systems?

8. A transformation matrix is given by:

$$T_{2,1} = \begin{bmatrix} 2/3 & 1/3 & 2/3 \\ -2/3 & 2/3 & 1/3 \\ -1/3 & -2/3 & 2/3 \end{bmatrix}$$

Find solutions for the eigenaxis e_η and the rotation angle η that relate the two reference frames.

9. Refer to Figure 3.8. A model of an aircraft is placed in a wind tunnel with coordinate system (subscript WT) as shown, with x_{WT} pointed opposite the direction of the airflow V. Beginning from a position where the body-fixed axes are aligned with the wind-tunnel axes, two rotations of the model are performed in this sequence: (1) a rotation θ_y about y_{WT}, and (2) a rotation θ_x about x_B. Describe how you would evaluate the aerodynamic angles α and β given the values of θ_x and θ_y.

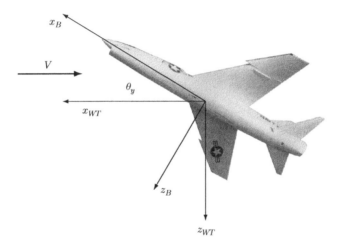

Figure 3.8 Problem 9. Note axes x_B, x_{WT}, and z_{WT} are in the plane of the page, and z_B is not (aircraft has been rotated about x_B through angle θ_x).

Reference

Junkins, J.L. and Turner, J.D. (1978) Optimal continuous torque attitude maneuvers, AIAA-AAS Astrodynamics Conference, Palo Alto, California, August 1978.

4

Rotating Coordinate Systems

4.1 General

Refer to Figure 4.1 throughout the following discussion. At any instant we have $\{v\}_2 = T_{2,1}\{v\}_1$ and $\{v\}_1 = T_{2,1}^T\{v\}_2$. If F_2 is rotating with respect to F_1 with angular rate ω_2^1 then the relative orientation of the axes of the two reference frames must be changing, and $T_{2,1}$ is time-varying. To find the rate of change of $T_{2,1}$ we will consider particular vectors and see how the relative rotation of the reference frames affects their representation. By then considering arbitrary vectors we may infer $\dot{T}_{2,1}$.

We use the dot notation and associated subscript to indicate the time-derivative of a quantity as seen from a particular reference frame. Thus the notation $\dot{\mathbf{v}}_2$ indicates the rate of change of the vector $\mathbf{v}$ relative to F_2. The entity is itself another vector and may be represented in any coordinate system, so that $\{\dot{\mathbf{v}}_2\}_1$ means the vector defined as the rate of change of $\mathbf{v}$ relative to F_2; once that vector is defined it is represented by its components in F_1. The simplest case is that of a vector $\mathbf{v}$ whose components are given in a particular reference frame, whose derivative is taken with respect to that reference frame, and then represented in that reference frame. In this case the result is found by taking the derivative of each of the components:

$$\{\mathbf{v}\}_2 = v_{x_2}\mathbf{i}_2 + v_{y_2}\mathbf{j}_2 + v_{z_2}\mathbf{k}_2$$

$$\{\dot{\mathbf{v}}_2\}_2 = \dot{v}_{x_2}\mathbf{i}_2 + \dot{v}_{y_2}\mathbf{j}_2 + \dot{v}_{z_2}\mathbf{k}_2$$

Having found this vector and its representation in F_2 we could then calculate (at the instant the derivative is valid)

$$\{\dot{\mathbf{v}}_2\}_1 = T_{1,2}\{\dot{\mathbf{v}}_2\}_2$$

Note that in general this is *not* the rate of change of $\mathbf{v}$ relative to F_1.

With the notation established, consider $\{\mathbf{v}\}_2 = T_{2,1}\{\mathbf{v}\}_1$. At first we might just write down $\dot{\mathbf{v}}_2 = \dot{T}_{2,1}\mathbf{v}_1 + T_{2,1}\dot{\mathbf{v}}_1$ (?), but it is by no means clear who is taking the derivatives of the vectors and where they are represented. What we want is $\{\dot{\mathbf{v}}_2\}_2$, so we will use the expression $\{\mathbf{v}\}_2 = v_{x_2}\mathbf{i}_2 + v_{y_2}\mathbf{j}_2 + v_{z_2}\mathbf{k}_2$ and see what the derivative of each term is. We have

$$\begin{Bmatrix} v_{x_2} \\ v_{y_2} \\ v_{z_2} \end{Bmatrix} = \begin{bmatrix} t_{11} & t_{12} & t_{13} \\ t_{21} & t_{22} & t_{23} \\ t_{31} & t_{32} & t_{33} \end{bmatrix}_{2,1} \begin{Bmatrix} v_{x_1} \\ v_{y_1} \\ v_{z_1} \end{Bmatrix}$$

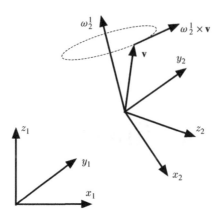

Figure 4.1 Vector **v** fixed in F_2. The dashed line is the path of the tip of **v** as F_2 rotates with angular rate ω with respect to F_1.

The components are each similar to $v_{x_2} = t_{11} v_{x_1} + t_{12} v_{y_1} + t_{13} v_{z_1}$. The derivative of this component is

$$\dot{v}_{x_2} = \dot{t}_{11} v_{x_1} + t_{11} \dot{v}_{x_1} + \dot{t}_{12} v_{y_1} + t_{12} \dot{v}_{y_1} + \dot{t}_{13} v_{z_1} + t_{13} \dot{v}_{z_1}$$

The terms $\dot{t}_{ij}$ are clearly the elements of $\dot{T}_{2,1}$, and the terms $\dot{v}_{x_1}, \dot{v}_{y_1}, \dot{v}_{z_1}$ are those from $\{\dot{\mathbf{v}}_1\}_1 = \dot{v}_{x_1} \mathbf{i}_1 + \dot{v}_{y_1} \mathbf{j}_1 + \dot{v}_{z_1} \mathbf{k}_1$. Reassembling we have

$$\{\dot{\mathbf{v}}_2\}_2 = \dot{T}_{2,1} \{\mathbf{v}\}_1 + T_{2,1} \{\dot{\mathbf{v}}_1\}_1$$

In words, then, the vector defined as the rate of change of **v** relative to F_2 and represented by its components in F_2 equals the rate of change of the transformation times **v** represented in F_1, plus the transformation of the vector that is the rate of change of **v** relative to F_1 and represented by its components in F_1.

Working from $\{\mathbf{v}\}_1 = T_{2,1}^T \{\mathbf{v}\}_2$, it is easy to show that

$$\{\dot{\mathbf{v}}\}_1 = \dot{T}_{2,1}^T \{\mathbf{v}\}_2 + T_{2,1}^T \{\dot{\mathbf{v}}_2\}_2 \tag{4.1}$$

Now recalling that the direction cosine matrix depends only on the relative orientation of two reference frames, and not on any particular vector in either one, it follows that the above equations involving $T_{2,1}$ and $\dot{T}_{2,1}$ must hold for any choice of **v**. So we may pick **v** fixed with respect to F_1 so that $\{\dot{\mathbf{v}}_1\}_1 = 0$, or fixed in F_2 so that $\{\dot{\mathbf{v}}_2\}_2 = 0$. Let $\{\dot{\mathbf{v}}_2\}_2 = 0$, and consider what $\{\dot{\mathbf{v}}_1\}_1$ looks like:

$$\{\dot{\mathbf{v}}_1\}_1 = \dot{T}_{2,1}^T \{\mathbf{v}\}_2 + T_{2,1}^T \{\dot{\mathbf{v}}_2\}_2 = \dot{T}_{2,1}^T \{\mathbf{v}\}_2$$

$$\dot{T}_{2,1}^T \{\mathbf{v}\}_2 = \{\omega_2^1\}_1 \times \{\mathbf{v}\}_1 \tag{4.2}$$

To more easily manipulate this equation we will replace the cross product operation with the product of a matrix and a vector. Consider $\mathbf{u} \times \mathbf{v}$, with

$$\mathbf{u} = \begin{Bmatrix} u_x \\ u_y \\ u_z \end{Bmatrix}, \quad \mathbf{v} = \begin{Bmatrix} v_x \\ v_y \\ v_z \end{Bmatrix}$$

Rotating Coordinate Systems

Define the matrix U as

$$U = \begin{bmatrix} 0 & -u_z & u_y \\ u_z & 0 & -u_x \\ -u_y & u_x & 0 \end{bmatrix}$$

It is then easy to verify that $\mathbf{u} \times \mathbf{v} = U\mathbf{v}$. This matrix is *skew symmetric*, meaning that $U^T = -U$. We may therefore rewrite $\dot{T}_{2,1}^T \{v\}_2 = \{\omega_2^1\}_1 \times \{v\}_1$ as

$$\dot{T}_{2,1}^T \{v\}_2 = \{\Omega_2^1\}_1 \{v\}_1$$

We then replace $\{v\}_1$ with $\{v\}_1 = T_{1,2}\{v\}_2 = T_{2,1}^T \{v\}_2$,

$$\dot{T}_{2,1}^T \{v\}_2 = \{\Omega_2^1\}_1 T_{2,1}^T \{v\}_2$$

This must hold for all $\{v\}_2$, so

$$\dot{T}_{2,1}^T = \{\Omega_2^1\}_1 T_{2,1}^T$$

Transposing each side,

$$\dot{T}_{2,1} = T_{2,1} \{\Omega_2^1\}_1^T = -T_{2,1}\{\Omega_2^1\}_1$$

Now we go the other direction. If F_2 is rotating with respect to F_1 with angular rate ω_2^1 then F_1 is rotating with respect to F_2 with angular rate $\omega_1^2 = -\omega_2^1$. We start with $\{\dot{v}_2\}_2 = \dot{T}_{2,1}\{v\}_1 + T_{2,1}\{\dot{v}_1\}_1$. We pick v fixed in F_1 so that $\{\dot{v}_1\}_1 = 0$. Then $\{\dot{v}_2\}_2 = \dot{T}_{2,1}\{v\}_1 = \{\omega_1^2\}_2 \times \{v\}_2 = \{\Omega_1^2\}_2 \{v\}_2$, and finally

$$\dot{T}_{2,1} = \{\Omega_1^2\}_2 T_{2,1} = -\{\Omega_2^1\}_2 T_{2,1}$$

This gives us four ways of evaluating the derivative of the direction cosine matrix in terms of the relative rotation of two reference frames:

$$\begin{aligned}\dot{T}_{2,1} &= T_{2,1}\{\Omega_1^2\}_1 \\ &= -T_{2,1}\{\Omega_2^1\}_1 \\ &= \{\Omega_1^2\}_2 T_{2,1} \\ &= -\{\Omega_2^1\}_2 T_{2,1}\end{aligned} \quad (4.3)$$

Which expression is to be used depends on how ω is most naturally expressed. For reasons that will become clear later, we will most often assume we know $\{\omega_2^1\}_2$ or equivalently $\{\Omega_2^1\}_2$. In the following application of the equations developed we will therefore use:

$$\dot{T}_{2,1} = -\{\Omega_2^1\}_2 T_{2,1}$$

With this result we can easily write an expression for the transformation of the derivative of a vector. Using $\{\dot{v}_2\}_2 = \dot{T}_{2,1}\{v\}_1 + T_{2,1}\{\dot{v}_1\}_1$ we have

$$\begin{aligned}\{\dot{v}_2\}_2 &= T_{2,1}\{\dot{v}_1\}_1 - \{\Omega_2^1\}_2 T_{2,1}\{v\}_1 \\ &= T_{2,1}\{\dot{v}_1\}_1 - \{\Omega_2^1\}_2\{v\}_2\end{aligned} \quad (4.4)$$

To avoid overworking the subscripts we will define:

$$\{\omega_2^1\}_2 = \begin{Bmatrix} \omega_x \\ \omega_y \\ \omega_z \end{Bmatrix}$$

$$\{\Omega_2^1\}_2 = \begin{bmatrix} 0 & -\omega_z & \omega_y \\ \omega_z & 0 & -\omega_x \\ -\omega_y & \omega_x & 0 \end{bmatrix}$$

4.2 Direction Cosines

Here we have

$$T_{2,1} = \begin{bmatrix} \cos\theta_{x2x1} & \cos\theta_{x2y1} & \cos\theta_{x2z1} \\ \cos\theta_{y2x1} & \cos\theta_{y2y1} & \cos\theta_{y2z1} \\ \cos\theta_{z2x1} & \cos\theta_{z2y1} & \cos\theta_{z2z1} \end{bmatrix}$$

Term-by-term differentiation yields $\dot{T}_{2,1} = \dot{t}_{ij}$, $i, j = 1 \ldots 3$ where, for example, $\dot{t}_{11} = -\dot\theta_{x2x1} \sin\theta_{x2x1}$.

The right-hand side of the equation is

$$\dot{T}_{2,1} = -\begin{bmatrix} 0 & -\omega_z & \omega_y \\ \omega_z & 0 & -\omega_x \\ -\omega_y & \omega_x & 0 \end{bmatrix} \begin{bmatrix} \cos\theta_{x2x1} & \cos\theta_{x2y1} & \cos\theta_{x2z1} \\ \cos\theta_{y2x1} & \cos\theta_{y2y1} & \cos\theta_{y2z1} \\ \cos\theta_{z2x1} & \cos\theta_{z2y1} & \cos\theta_{z2z1} \end{bmatrix}$$

We could multiply these matrices to evaluate each $\dot{t}_{ij}$, for example

$$\dot{t}_{11} = -\dot\theta_{x2x1} \sin\theta_{x2x1} = \omega_z \cos\theta_{y2x1} - \omega_y \cos\theta_{z2x1}$$

This would result in nine nonlinear ordinary differential equations. The six nonlinear constraining equations based on the orthogonality of the rows (or columns) of the matrix could probably be used to simplify the relationships, but not without some trouble. The use of direction cosines in evaluating $\dot{T}_{2,1}$ is rare and has nothing to offer over the use of Euler angles or Euler parameters, and will not be used in this book.

4.3 Euler Angles

The Euler angle representation of $T_{2,1}$ (for a 321 rotation from F_1 to F_2) is

$$T_{2,1} = \begin{bmatrix} (\cos\theta_y \cos\theta_z) & (\cos\theta_y \sin\theta_z) & (-\sin\theta_y) \\ \begin{pmatrix} \sin\theta_x \sin\theta_y \cos\theta_z \\ -\cos\theta_x \sin\theta_z \end{pmatrix} & \begin{pmatrix} \sin\theta_x \sin\theta_y \sin\theta_z \\ +\cos\theta_x \cos\theta_z \end{pmatrix} & (\sin\theta_x \cos\theta_y) \\ \begin{pmatrix} \cos\theta_x \sin\theta_y \cos\theta_z \\ +\sin\theta_x \sin\theta_z \end{pmatrix} & \begin{pmatrix} \cos\theta_x \sin\theta_y \sin\theta_z \\ -\sin\theta_x \cos\theta_z \end{pmatrix} & (\cos\theta_x \cos\theta_y) \end{bmatrix}$$

Rotating Coordinate Systems

Clearly we will have some trouble performing the element-by-element differentiation of this matrix for the left-hand side of $\dot{T}_{2,1} = -\{\Omega_2^1\}_2 T_{2,1}$. The right-hand side is not much better, and equating the various elements to find expressions for $\dot{\theta}_x$, $\dot{\theta}_y$, and $\dot{\theta}_z$ would be tedious. It is easier to proceed in a different way. First we note that the axes z_1, y', and x'' are not orthogonal, but can be used to form a basis for the space in which the rotations occur. That is, we may represent ω_2^1 in terms of the unit vectors in the directions z_1, y', and x''. The vector ω_2^1 is made up of a rate $\dot{\theta}_z$ which is about z_1, plus a rate $\dot{\theta}_y$ which is about y', and $\dot{\theta}_x$ which is about x'', or

$$\omega_2^1 = \dot{\theta}_z \mathbf{k}_1 + \dot{\theta}_y \mathbf{j}' + \dot{\theta}_x \mathbf{i}''$$

There are some savings in noting that $\mathbf{k}_1 = \mathbf{k}'$, $\mathbf{j}' = \mathbf{j}''$, and $\mathbf{i}'' = \mathbf{i}_2$, and so

$$\omega_2^1 = \dot{\theta}_z \mathbf{k}' + \dot{\theta}_y \mathbf{j}'' + \dot{\theta}_x \mathbf{i}_2$$

We require $\{\omega_2^1\}_2$, so we transform the unit vectors $\mathbf{k}'$ and $\mathbf{j}''$ ($\mathbf{i}_2$ is good the way it is) into F_2 using the intermediate rotations previously developed (Equations 3.3 and 3.4). That is,

$$\{\mathbf{k}'\}_2 = T_{2,F'}\{\mathbf{k}'\}_{F'}$$

$$\{\mathbf{j}''\}_2 = T_{2,F''}\{\mathbf{j}''\}_{F''}$$

This is straightforward, since,

$$\{\mathbf{k}'\}_{F'} = \begin{Bmatrix} 0 \\ 0 \\ 1 \end{Bmatrix}, \quad \{\mathbf{j}''\}_{F''} = \begin{Bmatrix} 0 \\ 1 \\ 0 \end{Bmatrix}$$

Also we have

$$T_{2,F''} = \begin{bmatrix} 1 & 0 & 0 \\ 0 & \cos\theta_x & \sin\theta_x \\ 0 & -\sin\theta_x & \cos\theta_x \end{bmatrix}$$

$$T_{2,F'} = T_{2,F''} T_{F'',F'}$$

$$= \begin{bmatrix} 1 & 0 & 0 \\ 0 & \cos\theta_x & \sin\theta_x \\ 0 & -\sin\theta_x & \cos\theta_x \end{bmatrix} \begin{bmatrix} \cos\theta_y & 0 & -\sin\theta_y \\ 0 & 1 & 0 \\ \sin\theta_y & 0 & \cos\theta_y \end{bmatrix}$$

$$= \begin{bmatrix} \cos\theta_y & 0 & -\sin\theta_y \\ \sin\theta_x \sin\theta_y & \cos\theta_x & \sin\theta_x \cos\theta_y \\ \cos\theta_x \sin\theta_y & -\sin\theta_x & \cos\theta_x \cos\theta_y \end{bmatrix}$$

From this we evaluate

$$\{\omega_2^1\}_2 = \dot{\theta}_z \{\mathbf{k}'\}_2 + \dot{\theta}_y \{\mathbf{j}''\}_2 + \dot{\theta}_x \mathbf{i}_2$$

$$= \dot{\theta}_z T_{2,F'}\{\mathbf{k}'\}_{F'} + \dot{\theta}_y T_{2,F''}\{\mathbf{j}''\}_{F''} + \dot{\theta}_x \mathbf{i}_2$$

$$= \dot{\theta}_z \begin{Bmatrix} -\sin\theta_y \\ \sin\theta_x \cos\theta_y \\ \cos\theta_x \cos\theta_y \end{Bmatrix} + \dot{\theta}_y \begin{Bmatrix} 0 \\ \cos\theta_x \\ -\sin\theta_x \end{Bmatrix} + \dot{\theta}_x \begin{Bmatrix} 1 \\ 0 \\ 0 \end{Bmatrix}$$

With $\{\omega_2^1\}_2^T = \{\omega_x, \omega_y, \omega_z\}$ we evaluate each term as

$$\{\omega_2^1\}_2 = \begin{Bmatrix} \omega_x \\ \omega_y \\ \omega_z \end{Bmatrix} = \begin{Bmatrix} -\sin\theta_y \dot\theta_z + \dot\theta_x \\ \sin\theta_x \cos\theta_y \dot\theta_z + \cos\theta_x \dot\theta_y \\ \cos\theta_x \cos\theta_y \dot\theta_z - \sin\theta_x \dot\theta_y \end{Bmatrix}$$

We may rewrite this as a matrix times a vector as

$$\{\omega_2^1\}_2 = \begin{Bmatrix} \omega_x \\ \omega_y \\ \omega_z \end{Bmatrix} = \begin{bmatrix} 1 & 0 & -\sin\theta_y \\ 0 & \cos\theta_x & \sin\theta_x \cos\theta_y \\ 0 & -\sin\theta_x & \cos\theta_x \cos\theta_y \end{bmatrix} \begin{Bmatrix} \dot\theta_x \\ \dot\theta_y \\ \dot\theta_z \end{Bmatrix} \quad (4.5)$$

We wish to solve this equation for $\dot\theta_x$, $\dot\theta_y$, and $\dot\theta_z$ in terms of the Euler angles and rotational components. The determinant of the matrix on the right-hand side is easily verified to be $\cos\theta_y$, which means the inverse does not exist if $\theta_y = \pm 90$ deg. This condition results in $\dot\theta_x$ and $\dot\theta_z$ being undefined. If $\theta_y \neq \pm 90$ deg, we have

$$\begin{Bmatrix} \dot\theta_x \\ \dot\theta_y \\ \dot\theta_z \end{Bmatrix} = \begin{bmatrix} 1 & \sin\theta_x \tan\theta_y & \cos\theta_x \tan\theta_y \\ 0 & \cos\theta_x & -\sin\theta_x \\ 0 & \sin\theta_x \sec\theta_y & \cos\theta_x \sec\theta_y \end{bmatrix} \{\omega_2^1\}_2 \quad (4.6)$$

This completely general result applies to any two coordinate systems with a defined 321 transformation. All that is needed to apply it to a specific case are (1) the right names for the Euler angles (e.g., θ, ϕ, and ψ for local horizontal to body) and (2) the right relationships for the relative angular rotation rates $\{\omega_2^1\}_2$ (e.g., p, q, and r for the inertial to body transformation).

The singularity involving θ_y is the fatal flaw of which we spoke earlier. It is also true that no matter what sequence is taken for the Euler angle rotations, the angle of the second rotation displays a similar singularity at either zero or ± 90 deg. To avoid this, one could use the nine direction cosines themselves, or use Euler parameters.

4.4 Euler Parameters

In evaluating the left- and right-hand sides of $\dot T_{2,1} = -\{\Omega_2^1\}_2 T_{2,1}$, we have

$T_{2,1} =$

$$\begin{bmatrix} (q_0^2 + q_1^2 - q_2^2 - q_3^2) & 2(q_1 q_2 + q_0 q_3) & 2(q_1 q_3 - q_0 q_2) \\ 2(q_1 q_2 - q_0 q_3) & (q_0^2 - q_1^2 + q_2^2 - q_3^2) & 2(q_2 q_3 + q_0 q_1) \\ 2(q_1 q_3 + q_0 q_2) & 2(q_2 q_3 - q_0 q_1) & (q_0^2 - q_1^2 - q_2^2 + q_3^2) \end{bmatrix}$$

Clearly it should not be necessary to take the derivative of each of the elements in this matrix. We will use the three diagonal terms, combined with $q_0^2 + q_1^2 + q_2^2 + q_3^2 = 1$ to eliminate q_0. For the (1,1) entry we have

$$t_{11} = q_0^2 + q_1^2 - q_2^2 - q_3^2$$
$$= (1 - q_1^2 - q_2^2 - q_3^2) + q_1^2 - q_2^2 - q_3^2$$
$$= 1 - 2q_2^2 - 2q_3^2$$

Rotating Coordinate Systems

Similarly for the (2,2) and (3,3) entries,

$$t_{22} = q_0^2 - q_1^2 + q_2^2 - q_3^2 = 1 - 2q_1^2 - 2q_3^2$$
$$t_{33} = q_0^2 - q_1^2 - q_2^2 + q_3^2 = 1 - 2q_1^2 - 2q_2^2$$

The derivatives of these three terms are

$$dt_{11}/dt = -4(q_2\dot{q}_2 + q_3\dot{q}_3)$$
$$dt_{22}/dt = -4(q_1\dot{q}_1 + q_3\dot{q}_3)$$
$$dt_{33}/dt = -4(q_1\dot{q}_1 + q_2\dot{q}_2)$$

Now the right-hand side of $\dot{T}_{2,1} = -\{\Omega_2^1\}_2 T_{2,1}$ is

$$\{\Omega_2^1\}_2 T_{2,1} =$$

$$\begin{bmatrix} 0 & -\omega_z & \omega_y \\ \omega_z & 0 & -\omega_x \\ -\omega_y & \omega_x & 0 \end{bmatrix} \begin{bmatrix} \begin{pmatrix} q_0^2 + q_1^2 \\ -q_2^2 - q_3^2 \end{pmatrix} & 2(q_1q_2 + q_0q_3) & 2(q_1q_3 - q_0q_2) \\ 2(q_1q_2 - q_0q_3) & \begin{pmatrix} q_0^2 - q_1^2 \\ +q_2^2 - q_3^2 \end{pmatrix} & 2(q_2q_3 + q_0q_1) \\ 2(q_1q_3 + q_0q_2) & 2(q_2q_3 - q_0q_1) & \begin{pmatrix} q_0^2 - q_1^2 \\ -q_2^2 + q_3^2 \end{pmatrix} \end{bmatrix}$$

Evaluating just the diagonal terms,

$$\{\Omega_2^1\}_2 T_{2,1} =$$

$$\begin{bmatrix} \begin{pmatrix} -2\omega_z(q_1q_2 - q_0q_3) \\ +2\omega_y(q_1q_3 + q_0q_2) \end{pmatrix} & ? & ? \\ ? & \begin{pmatrix} 2\omega_z(q_1q_2 + q_0q_3) \\ -2\omega_x(q_2q_3 - q_0q_1) \end{pmatrix} & ? \\ ? & ? & \begin{pmatrix} -2\omega_y(q_1q_3 - q_0q_2) \\ +2\omega_x(q_2q_3 + q_0q_1) \end{pmatrix} \end{bmatrix}$$

Equating the diagonal terms on the left- and right-hand sides,

$$4(q_2\dot{q}_2 + q_3\dot{q}_3) = -2\omega_z(q_1q_2 - q_0q_3) + 2\omega_y(q_1q_3 + q_0q_2)$$
$$4(q_1\dot{q}_1 + q_3\dot{q}_3) = 2\omega_z(q_1q_2 + q_0q_3) - 2\omega_x(q_2q_3 - q_0q_1)$$
$$4(q_1\dot{q}_1 + q_2\dot{q}_2) = -2\omega_y(q_1q_3 - q_0q_2) + 2\omega_x(q_2q_3 + q_0q_1)$$

Now placing these expressions in matrix form, we have

$$\begin{bmatrix} 0 & q_2 & q_3 \\ q_1 & 0 & q_3 \\ q_1 & q_2 & 0 \end{bmatrix} \begin{Bmatrix} \dot{q}_1 \\ \dot{q}_2 \\ \dot{q}_3 \end{Bmatrix} =$$

$$\frac{1}{2} \begin{bmatrix} 0 & (q_1q_3 + q_0q_2) & (q_0q_3 - q_1q_2) \\ (q_0q_1 - q_2q_3) & 0 & (q_1q_2 + q_0q_3) \\ (q_2q_3 + q_0q_1) & (q_0q_2 - q_1q_3) & 0 \end{bmatrix} \begin{Bmatrix} \omega_x \\ \omega_y \\ \omega_z \end{Bmatrix}$$

Solving this set of equations is not as messy as it first appears, and it simplifies very nicely. The result is

$$\begin{Bmatrix} \dot{q}_1 \\ \dot{q}_2 \\ \dot{q}_3 \end{Bmatrix} = \frac{1}{2} \begin{bmatrix} q_0 & -q_3 & q_2 \\ q_3 & q_0 & -q_1 \\ -q_2 & q_1 & q_0 \end{bmatrix} \begin{Bmatrix} \omega_x \\ \omega_y \\ \omega_z \end{Bmatrix}$$

We may recover $\dot{q}_0$ by using $q_0^2 = 1 - q_1^2 - q_2^2 - q_3^2$ whence

$$q_0 \dot{q}_0 = -q_1 \dot{q}_1 - q_2 \dot{q}_2 - q_3 \dot{q}_3$$

Substitutions and a little manipulation yield

$$\dot{q}_0 = \frac{1}{2}(-q_1 \omega_x - q_2 \omega_y - q_3 \omega_z)$$

$$\begin{Bmatrix} \dot{q}_0 \\ \dot{q}_1 \\ \dot{q}_2 \\ \dot{q}_3 \end{Bmatrix} = \frac{1}{2} \begin{bmatrix} -q_1 & -q_2 & -q_3 \\ q_0 & -q_3 & q_2 \\ q_3 & q_0 & -q_1 \\ -q_2 & q_1 & q_0 \end{bmatrix} \begin{Bmatrix} \omega_x \\ \omega_y \\ \omega_z \end{Bmatrix}$$

This expression can also be written as

$$\begin{Bmatrix} \dot{q}_0 \\ \dot{q}_1 \\ \dot{q}_2 \\ \dot{q}_3 \end{Bmatrix} = \frac{1}{2} \begin{bmatrix} -q_0 & -q_1 & -q_2 & -q_3 \\ -q_1 & q_0 & -q_3 & q_2 \\ -q_2 & q_3 & q_0 & -q_1 \\ -q_3 & -q_2 & q_1 & q_0 \end{bmatrix} \begin{Bmatrix} 0 \\ \omega_x \\ \omega_y \\ \omega_z \end{Bmatrix} \quad (4.7)$$

It is easy to verify that there are no singularities in this expression.

4.5 Customs and Conventions

4.5.1 Angular velocity components

Inertial angle rates in body and wind axes occur frequently, and the components associated with them are given special symbols. These are as follows:

$$\{\omega_B\}_B = \begin{Bmatrix} p \\ q \\ r \end{Bmatrix}$$

$$\{\omega_W\}_W = \begin{Bmatrix} p_W \\ q_W \\ r_W \end{Bmatrix}$$

Problems

1. In this problem we wish to determine the components of inertial rotation of a local horizontal reference frame F_H in terms of the Earth's rotation and the movement of the reference frame over the surface of the Earth. It may be shown that the substitutions

for the Euler angles $\theta_z = \mu_H$, $\theta_y = -(\lambda_H + 90 \deg)$, and $\theta_x = 0$ will get the correct transformation:

$$T_{H,EC} = \begin{bmatrix} -\sin\lambda_H \cos\mu_H & -\sin\lambda_H \sin\mu_H & \cos\lambda_H \\ -\sin\mu_H & \cos\mu_H & 0 \\ -\cos\lambda_H \cos\mu_H & -\cos\lambda_H \sin\mu_H & -\sin\lambda_H \end{bmatrix}$$

(a) Use a generic expression for $\{\omega_H^{EC}\}_H$,

$$\{\omega_H^{EC}\}_H = \begin{Bmatrix} \omega_x \\ \omega_y \\ \omega_z \end{Bmatrix}$$

Determine $\{\omega_H^{EC}\}_H$ in terms of present latitude and longitude, and rate of change of latitude and longitude (λ_H, μ_H, $\dot\lambda_H$, and $\dot\mu_H$).

(b) Find $\{\omega_{EC}\}_H$ for arbitrary λ_H and μ_H. That is, determine the inertial rotation of F_{EC} as represented in F_H, in terms of present latitude and longitude, and ω_E. (ω_E is the magnitude of the Earth's inertial rotation rate.)

(c) Use the two previous results to determine the inertial rotation of F_H in F_H, $\{\omega_H\}_H$.

2. Consider a 212 sequence of Euler angle rotations from F_1 to F_2 with angles θ_{y_1}, θ_x, θ_{y_2}. Use $\{\omega_2^1\}_2^T = \{\omega_x, \omega_y, \omega_z\}$. Find $\dot\theta_{y_1}$, $\dot\theta_x$, and $\dot\theta_{y_2}$ in terms of θ_{y_1}, θ_x, θ_{y_2}, and the components of $\{\omega_2^1\}_2$.

3. Define the Euler angles in a 321 rotation from F_1 to F_2 as θ_z, θ_y, and θ_x. We showed that

$$\{\omega_2^1\}_2 = \begin{Bmatrix} \omega_x \\ \omega_y \\ \omega_z \end{Bmatrix} = \begin{bmatrix} 1 & 0 & -\sin\theta_y \\ 0 & \cos\theta_x & \sin\theta_x \cos\theta_y \\ 0 & -\sin\theta_x & \cos\theta_x \cos\theta_y \end{bmatrix} \begin{Bmatrix} \dot\theta_x \\ \dot\theta_y \\ \dot\theta_z \end{Bmatrix}$$

(a) Apply this relationship to evaluate $\{\omega_B^W\}_B$ (B refers to an arbitrary F_B and W to F_W). Suggestion: apply the Euler angles that correspond to the rotation from F_W to F_B.

(b) Note that $\{\omega_W\}_W = \{\omega_B\}_W - \{\omega_B^W\}_W = T_{W,B}[\{\omega_B\}_B - \{\omega_B^W\}_B]$. Use $\{\omega_B^W\}_B$ from part a and evaluate the right-hand side of the expression

$$\{\omega_W\}_W = \begin{Bmatrix} p_W \\ q_W \\ r_W \end{Bmatrix} = T_{W,B}[\{\omega_B\}_B - \{\omega_B^W\}_B]$$

(c) Solve the equations from part b to find expressions for $\dot\alpha$ and $\dot\beta$. Suggestion: of the three scalar equations that result the first should be

$$p_W = p\cos\alpha \cos\beta + \sin\beta(q - \dot\alpha) + r\sin\alpha \cos\beta$$

The other two equations should yield expressions that begin $\dot\alpha = q - \cdots$ and $\dot\beta = r_W + \cdots$.

4. Consider the matrix:
$$Q = \begin{bmatrix} -q_0 & -q_1 & -q_2 & -q_3 \\ -q_1 & q_0 & -q_3 & q_2 \\ -q_2 & q_3 & q_0 & -q_1 \\ -q_3 & -q_2 & q_1 & q_0 \end{bmatrix}$$

Considering the rows and columns as vectors, determine the length of each row and each column. Evaluate (numerically) the scalar (dot) product of any two rows, and of any two columns. Using these results, find Q^{-1} and then solve the Euler parameter rate equations for ω_x, ω_y, and ω_z. Express them as a matrix equation, and also as a scalar equations, such as
$$\omega_x = 2(q_0 \dot{q}_1 + \cdots)$$

5. An airplane is in a spin with constant $\phi = -50°$ and constant $\theta = -50°$. See Figures 4.2 and 4.3. The rotation rate is constant as well, $\dot{\psi} = 30\,\text{deg/s}$. At time $t = 0$

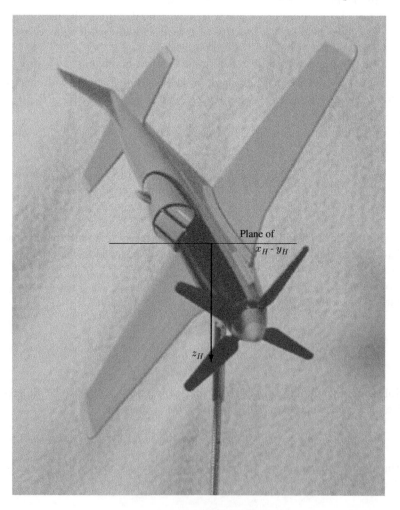

Figure 4.2 Problem 5 at $t = 0$.

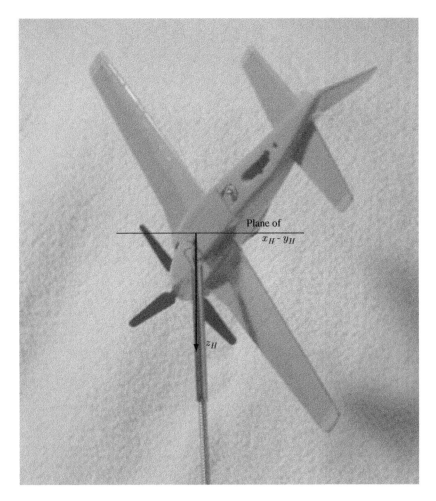

Figure 4.3 Problem 5 at a later time.

the aircraft passes through due north, $\psi = 0$. Define the orientation of the body axes relative to F_H at this instant as F_1. The aircraft continues to rotate about an axis pointed toward the ground. Six seconds later the body axes are oriented differently; define this orientation as F_2. Find the Euler parameters $q_0 \ldots q_3$ that define the transformation matrix between these two coordinate systems, T_{21}.

5

Inertial Accelerations

5.1 General

We need expressions for the inertial accelerations of our aircraft in order to apply Newton's Second Law. Our point of view will be from some local coordinate system, usually attached to the aircraft. The problem is that this local coordinate system rotates with respect to Earth, which rotates with respect to inertial space. The first problem we will tackle is to relate inertial acceleration to our locally observed acceleration. This will re-introduce the familiar concepts of tangential, coriolis, and centripetal accelerations. We will then address the way in which inertial acceleration is related to the entire mass of the aircraft, its distribution, and our choice of coordinate systems.

5.2 Inertial Acceleration of a Point

5.2.1 Arbitrary moving reference frame

We begin by hypothesizing an inertial reference F_I and some moving reference frame F_M. The origin of F_M may be accelerated ($\mathbf{a}_M$) and the reference frame itself may be rotating ($\boldsymbol{\omega}_M$) with respect to inertial space, both assumed known. It is also assumed that we know how some point P moves around with respect to F_M. We are seeking an expression for the inertial acceleration of the point P expressed in F_M ($\{\mathbf{a}_P\}_M$). Later we will use this expression twice: first with the Earth-centered frame as F_M and a point on the Earth's surface as P, then with an Earth-fixed frame as F_M and the aircraft CG as P. These two results will enable us to write the inertial acceleration of the CG in a local coordinate system. The same results will be used to evaluate certain rotational accelerations and decide whether they are small enough to be neglected.

Figure 5.1 shows a generic picture of F_M and P in inertial space. It is clear from this figure that we may express the inertial position of P as $\mathbf{r}_P = \mathbf{r}_M + \mathbf{r}_P^M$. The inertial velocity of P is the first derivative of $\mathbf{r}_P$ with respect to time as seen from the inertial frame, and its inertial acceleration is the second derivative, $\mathbf{v}_P = \dot{\mathbf{r}}_P$ and $\mathbf{a}_P = \dot{\mathbf{v}}_P$. What we are

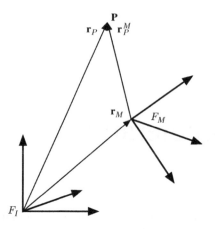

Figure 5.1 A point in inertial space.

after is $\{\mathbf{a}_P\}_M$, but first we formulate $\{\mathbf{v}_P\}_M$.

$$\begin{aligned}\mathbf{v}_P &= \frac{d}{dt}\mathbf{r}_P \\ &= \frac{d}{dt}\left(\mathbf{r}_M + \mathbf{r}_P^M\right) \\ &= \frac{d}{dt}\mathbf{r}_M + \frac{d}{dt}\mathbf{r}_P^M\end{aligned}$$

The first term on the right is the inertial velocity of the moving reference frame, or $\mathbf{v}_M$, and the second is the derivative of the vector $\mathbf{r}_P^M$ that is in a rotating reference frame. With the dot notation for time derivatives we have $\mathbf{v}_P = \mathbf{v}_M + \dot{\mathbf{r}}_P^M$. Assume these vectors are expressed in F_I ($\{\mathbf{v}_P\}_I$, $\{\mathbf{v}_M\}_I$, and $\{\dot{\mathbf{r}}_P^M\}_I$ with the I subscript just to be clear) then we may transform them to F_M using $T_{M,I}$:

$$\begin{aligned}\{\mathbf{v}_P\}_M &= T_{M,I}\{\mathbf{v}_P\}_I \\ &= T_{M,I}\{\mathbf{v}_M\}_I + T_{M,I}\left\{\frac{d\mathbf{r}_P^M}{dt}\bigg|_I\right\}_I \\ &= \{\mathbf{v}_M\}_M + T_{M,I}\left\{\frac{d\mathbf{r}_P^M}{dt}\bigg|_I\right\}_I\end{aligned}$$

The expression $T_{M,I}\{\mathbf{v}_M\}_I$ is a straightforward transformation, $T_{M,I}\{\mathbf{v}_M\}_I = \{\mathbf{v}_M\}_M$. The other term on the right is the transformation of the inertial derivative of a vector. We can express the derivative in the moving reference frame using the previously derived result (note we use $\mathbf{u}$ instead of $\mathbf{v}$ for our arbitrary vector, since $\mathbf{v}$ is velocity in this chapter), $\{\dot{\mathbf{u}}_2\}_2 = T_{2,1}\{\dot{\mathbf{u}}_1\}_1 - \{\Omega_2^1\}_2\{\mathbf{u}\}_2$, or $T_{2,1}\{\dot{\mathbf{u}}_1\}_1 = \{\dot{\mathbf{u}}_2\}_2 + \{\Omega_2^1\}_2\{\mathbf{u}\}_2$. Here we let $F_1 = F_I$, $F_2 = F_M$, $\mathbf{u}_1 = \{\mathbf{r}_P^M\}_I$, and $\mathbf{u}_2 = \{\mathbf{r}_P^M\}_M$. This results in

$$T_{M,I}\left\{\frac{d\mathbf{r}_P^M}{dt}\bigg|_I\right\}_I = \left\{\frac{d\mathbf{r}_P^M}{dt}\bigg|_M\right\}_M + \{\Omega_M\}_M\{\mathbf{r}_P^M\}_M$$

Inertial Accelerations

The first term on the right seems to have overworked the superscripts and subscripts a bit. In words, it means the derivative of the vector $\mathbf{r}_P^M$ as seen from F_M, with the answer expressed in F_M coordinates. This is just the velocity of P relative to F_M expressed in F_M coordinates, $\{\mathbf{v}_P^M\}_M$. We therefore have

$$\{\mathbf{v}_P\}_M = \{\mathbf{v}_M\}_M + \{\mathbf{v}_P^M\}_M + \{\Omega_M\}_M \{\mathbf{r}_P^M\}_M \tag{5.1}$$

This is probably more familiar with the cross product,

$$\{\Omega_M\}_M \{\mathbf{r}_P^M\}_M = \{\boldsymbol{\omega}_M\}_M \times \{\mathbf{r}_P^M\}_M$$

To express the inertial acceleration of P in F_M we need the transformation of the inertial derivative of $\mathbf{v}_P$. Using our previous result for the needed transformation we have

$$\{\mathbf{a}_P\}_M = T_{M,I} \left\{ \frac{d\mathbf{v}_P}{dt}_I \right\}_I$$

$$= \left\{ \frac{d\mathbf{v}_P}{dt}_M \right\}_M + \{\Omega_M\}_M \{\mathbf{v}_P\}_M$$

Again the overworked superscripts and subscripts. This one means the derivative of the vector $\mathbf{v}_P$ as seen from F_M, with the answer expressed in F_M coordinates. We already have $\mathbf{v}_P$ in F_M coordinates, $\{\mathbf{v}_P\}_M = \{\mathbf{v}_M\}_M + \{\mathbf{v}_P^M\}_M + \{\Omega_M\}_M\{\mathbf{r}_P^M\}_M$, so:

$$\left\{ \frac{d\mathbf{v}_P}{dt}_M \right\}_M = \left\{ \frac{d\mathbf{v}_M}{dt}_M \right\}_M + \{\mathbf{a}_P^M\}_M + \left[\frac{d(\{\Omega_M\}_M\{\mathbf{r}_P^M\}_M)}{dt}_M \right]_M$$

$$= \left\{ \begin{matrix} [T_{M,I}\frac{d\mathbf{v}_M}{dt} - \{\Omega_M\}_M\{\mathbf{v}_M\}_M] + \{\mathbf{a}_P^M\}_M \\ + [\{\dot{\Omega}_M\}_M\{\mathbf{r}_P^M\}_M + \{\Omega_M\}_M\{\mathbf{v}_P^M\}_M] \end{matrix} \right\}$$

$$= \left\{ \begin{matrix} [\{\mathbf{a}_M\}_M - \{\Omega_M\}_M\{\mathbf{v}_M\}_M] + \{\mathbf{a}_P^M\}_M \\ + [\{\dot{\Omega}_M\}_M\{\mathbf{r}_P^M\}_M + \{\Omega_M\}_M\{\mathbf{v}_P^M\}_M] \end{matrix} \right\}$$

The other term is $\{\Omega_M\}_M\{\mathbf{v}_P\}_M$, whence

$$\{\Omega_M\}_M\{\mathbf{v}_P\}_M = \{\Omega_M\}_M [\{\mathbf{v}_M\}_M + \{\mathbf{v}_P^M\}_M + \{\Omega_M\}_M\{\mathbf{r}_P^M\}_M]$$

$$= \{\Omega_M\}_M\{\mathbf{v}_M\}_M + \{\Omega_M\}_M\{\mathbf{v}_P^M\}_M + \{\Omega_M\}_M^2\{\mathbf{r}_P^M\}_M$$

Assembling and canceling the two $\{\Omega_M\}_M\{\mathbf{v}_M\}_M$ terms that appear,

$$\{\mathbf{a}_P\}_M = \{\mathbf{a}_M\}_M + \{\mathbf{a}_P^M\}_M$$
$$+ \{\dot{\Omega}_M\}_M\{\mathbf{r}_P^M\}_M$$
$$+ 2\{\Omega_M\}_M\{\mathbf{v}_P^M\}_M$$
$$+ \{\Omega_M\}_M^2\{\mathbf{r}_P^M\}_M \tag{5.2}$$

In words Equation 5.2 says that the inertial acceleration of a point P is the inertial acceleration of the (origin of the) moving reference frame, plus the acceleration of the point relative to the moving reference frame, plus the tangential, coriolis, and centripetal accelerations at the point. All quantities are expressed in the coordinates of the moving reference frame.

Equation 5.2 has wide application in the study of dyanmics, including aircraft and spacecraft dynamics. It is also useful for analyzing all sorts of curious kinematic problems designed to confound undergraduate engineering students, often involving ants crawling across turntables or rotating cones. It is sometimes referred to as the Basic Kinematic Equation and given its own acronym, the 'BKE'.

5.2.2 Earth-centered moving reference frame

As promised we now will use the expression for $\{a_P\}_M$ twice: first with the Earth-centered frame as F_M and a point on the Earth's surface as P, then with an Earth-fixed frame as F_M and the aircraft CG as P. In the first case we have $F_M = F_{EC}$ and $P = O_E$, or, to save subscripts, $P = E$, understood to be the origin of some F_E. Substituting we have

$$\{a_E\}_{EC} = \{a_{EC}\}_{EC} + \{a_E^{EC}\}_{EC}$$
$$+ \{\dot{\Omega}_{EC}\}_{EC} \{r_E^{EC}\}_{EC}$$
$$+ 2\{\Omega_{EC}\}_{EC} \{v_E^{EC}\}_{EC}$$
$$+ \{\Omega_{EC}\}_{EC}^2 \{r_E^{EC}\}_{EC}$$

On the right-hand side, a_{EC} is the inertial accleration of the Earth's center. If we neglect the annual rotation of the Earth about the sun, and any inertial acceleration the sun may have, this term is zero. The term r_E^{EC} is the position vector of the Earth-fixed origin from the center of the Earth and (barring earthquakes) is constant. Therefore v_E^{EC} and a_E^{EC} are zero. If the Earth's rotation is constant (it changes only on a geological scale) then $\dot{\Omega}_{EC}$ is zero as well. This leaves just the centripetal acceleration,

$$\{a_E\}_{EC} = \{\Omega_{EC}\}_{EC}^2 \{r_E^{EC}\}_{EC}$$

The rotation of the Earth is about 360 deg per 24 hours, or 7.27×10^{-5} rad/s. If we take the diameter of the Earth to be 6875.5 nautical miles then r_E^{EC} has magnitude of roughly 2.09×10^7 ft. For F_E on the equator the magnitude of the centripetal acceleration is about 0.11 ft/s^2, and it is zero at the poles.

5.2.3 Earth-fixed moving reference frame

Now with an Earth-fixed frame as F_M and the aircraft CG as P, we substitute $F_M = F_E$ and $P = C$ and write

$$\{a_C\}_E = \{a_E\}_E + \{a_C^E\}_E$$
$$+ \{\dot{\Omega}_E\}_E \{r_C^E\}_E$$
$$+ 2\{\Omega_E\}_E \{v_C^E\}_E$$
$$+ \{\Omega_E\}_E^2 \{r_C^E\}_E$$

Inertial Accelerations

The coordinate system F_E is fixed relative to F_{EC} and thus rotates with it, so $\dot{\Omega}_E$ is zero. This leaves

$$\{\mathbf{a}_C\}_E = \{\mathbf{a}_E\}_E + \{\mathbf{a}_C^E\}_E$$
$$+ 2\{\Omega_E\}_E \{\mathbf{v}_C^E\}_E$$
$$+ \{\Omega_E\}_E^2 \{\mathbf{r}_C^E\}_E$$

The question is, do we really need to keep track of all of the terms? We need the inertial acceleration of our aircraft's CG for Newton's Second Law, and it would be convenient to ignore the rotational accelerations and just use the local acceleration $\mathbf{a}_C^E$ for our aircraft's inertial acceleration. For instance, the centripetal acceleration $\mathbf{a}_E$ was just derived in F_{EC}, so $\{\mathbf{a}_E\}_E = T_{E,EC}\{\mathbf{a}_E\}_{EC}$, and it has maximum magnitude of 0.11 ft/s². Whether it and other terms should be retained depends on what kind of problem is being addressed.

The terms $\mathbf{r}_C^E$, $\mathbf{v}_C^E$, and $\mathbf{a}_C^E$ are, respectively, the position, velocity, and acceleration of our aircraft's CG relative to a point on the Earth, and are what would be measured by a ground tracking station at F_E's origin. The position $\mathbf{r}_C^E$ may be on the order of hundreds of nautical miles; if it is very large then the problem should probably be referred to an Earth-centered reference frame. With 7.27×10^{-5} rad/s for the magnitude of Ω_E and 1.5×10^6 ft for the magnitude of $\mathbf{r}_C^E$, the maximum centripetal component of acceleration is on the order of 8×10^{-3} ft/s². The velocity could easily be greater than 1000 kts, and may be much greater if, for example, a re-entry vehicle is being tracked. For rough estimates, the maximum coriolis acceleration at a speed of 2000 ft/s is around 0.3 ft/s².

In problems that span a large time or distance or both, and in which great accuracy is required, each term may be significant. In calculations for artillery projectiles they are all retained. If the objective is to understand the short-term dynamics of an aircraft in response to atmospheric disturbances or control inputs, it is usually assumed that $\{\mathbf{a}_C\}_E = \{\mathbf{a}_C^E\}_E$. Because this expression ignores all terms resulting from the rotation of the Earth and does not depend on the Earth's curvature, it is commonly called the flat-Earth approximation.

$$\{\mathbf{a}_C\}_E = \{\mathbf{a}_C^E\}_E \quad \text{(Flat, non-rotating Earth)} \tag{5.3}$$

5.3 Inertial Acceleration of a Mass

The previous equations dealt only with a point moving with respect to a rotating reference frame in inertial space. Newton's Second Law applies to an infinitesimal mass dm at such a point. Any physical body, such as an aircraft, has finite mass that is the summation of an infinite number of such infinitesimal masses. We begin by examining how Newton's Second Law applies to one such infinitesimal mass, and sum these relationships over the whole mass. Figure 5.2 is the same as Figure 5.1, except now the point P is replaced by dm located at that point. The mass dm is acted on by infinitesimal force $d\mathbf{f}$ and has inertial velocity $\mathbf{v}_{dm}$. We replace F_M by a coordinate system with its origin at the CG of the aircraft for reasons that will become clear shortly. Here we have used F_B, but any suitably defined coordinate system would do as well.

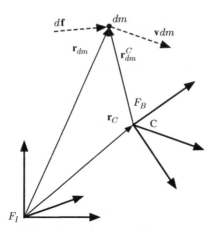

Figure 5.2 A particle in inertial space.

5.3.1 Linear acceleration

Newton tells us that the infinitesimal force $d\mathbf{f}$ is proportional to the (inertial) time rate of change of the momentum of dm, or

$$d\mathbf{f} = \dot{\mathbf{v}}_{dm} dm$$

We wish to integrate both sides of this equation over the entire mass of the aircraft, and relate the result to the acceleration of the CG. First note that the CG is defined such that

$$\int_m \mathbf{r}_{dm}^C dm = 0$$

Now, $\dot{\mathbf{v}}_{dm} = \ddot{\mathbf{r}}_{dm}$ so we relate $\mathbf{r}_{dm}$ to $\mathbf{r}_{dm}^C$ and $\mathbf{r}_C$ to obtain $\mathbf{r}_{dm} = \mathbf{r}_{dm}^C + \mathbf{r}_C$. We will form the integrals and then differentiate with respect to time as we go. First,

$$\int_m \mathbf{r}_{dm} dm = \int_m \mathbf{r}_{dm}^C dm + \int_m \mathbf{r}_C dm$$

On the right-hand side, the first integral is zero because we have chosen our coordinate system's origin to be at the CG. In the second integral, we observe that, for a given mass distribution, the inertial position of the CG is fixed and so $\mathbf{r}_C$ can be taken out of the integral. The result is

$$\int_m \mathbf{r}_{dm} dm = \mathbf{r}_C \int_m dm = m\mathbf{r}_C$$

Differentiating both sides with respect to time, and noting that the integrals do not depend on time,

$$m\dot{\mathbf{r}}_C = m\mathbf{v}_C = \int_m \dot{\mathbf{r}}_{dm} dm = \int_m \mathbf{v}_{dm} dm$$

Differentiating again,

$$m\dot{\mathbf{v}}_C = m\mathbf{a}_C = \int_m \dot{\mathbf{v}}_{dm} dm$$

Inertial Accelerations

But we have $d\mathbf{f} = \dot{\mathbf{v}}_{dm}dm$, so

$$m\mathbf{a}_C = \int_m d\mathbf{f}$$

The integral on the right accounts for all the forces acting on particles of mass in the body of the aircraft. The internal forces are taken to cancel one another out, so the net result is the vector sum of all the externally applied forces, $\mathbf{F}$. In other words,

$$\mathbf{F} = m\mathbf{a}_C \tag{5.4}$$

The freshman form of Newton's Second Law is valid for a mass only if applied to the mass center of gravity. We drop the subscript C and write the inertial equation as

$$\mathbf{F} = m\mathbf{a} = m\dot{\mathbf{v}}$$

In the body axes we have $\{\mathbf{F}\}_B = T_{B,I}\mathbf{F}$, or

$$\{\mathbf{F}\}_B = mT_{B,I}\dot{\mathbf{v}}$$
$$= m\{\dot{\mathbf{v}}_B\}_B + m\{\Omega_B\}_B\{\mathbf{v}\}_B \tag{5.5}$$

(Referenced to CG.)

5.3.2 Rotational acceleration

We also will need the rotational equivalent of Newton's Second Law. The moment of $d\mathbf{f}$ about O_I is $d\mathbf{M}$, and

$$d\mathbf{M} = \mathbf{r}_{dm} \times d\mathbf{f} = \mathbf{r}_{dm} \times \dot{\mathbf{v}}_{dm}dm$$

The inertial angular momentum of dm is denoted $d\mathbf{h}$ and is

$$d\mathbf{h} = \mathbf{r}_{dm} \times \mathbf{v}_{dm}dm$$

Relating the two through the Second Law,

$$d\mathbf{M} = \mathbf{r}_{dm} \times d\mathbf{f} = \frac{d}{dt}d\mathbf{h} = \frac{d}{dt}\left(\mathbf{r}_{dm} \times \mathbf{v}_{dm}\right)dm$$

To get the moment and time rate of change of angular momentum with respect to the CG, we again integrate and use $\mathbf{r}_{dm} = \mathbf{r}_{dm}^C + \mathbf{r}_C$. We have

$$\int_m \mathbf{r}_{dm} \times d\mathbf{f} = \frac{d}{dt}\int_m \left(\mathbf{r}_{dm} \times \mathbf{v}_{dm}\right)dm$$

Working first on the left-hand side,

$$\int_m \mathbf{r}_{dm} \times d\mathbf{f} = \int_m \left(\mathbf{r}_{dm}^C + \mathbf{r}_C\right) \times d\mathbf{f}$$
$$= \int_m \left(\mathbf{r}_{dm}^C \times d\mathbf{f}\right) + \int_m \left(\mathbf{r}_C \times d\mathbf{f}\right)$$
$$= \int_m \left(\mathbf{r}_{dm}^C \times d\mathbf{f}\right) + \mathbf{r}_C \times \int_m d\mathbf{f}$$
$$= \mathbf{M}_C + \left(\mathbf{r}_C \times \mathbf{F}\right)$$

Here we have defined $\mathbf{M}_C$ to be the moment about the CG.

On the right-hand side,

$$\frac{d}{dt}\int_m (\mathbf{r}_{dm} \times \mathbf{v}_{dm})\,dm = \frac{d}{dt}\int_m (\mathbf{r}^C_{dm} \times \mathbf{v}_{dm})\,dm + \frac{d}{dt}\int_m (\mathbf{r}_C \times \mathbf{v}_{dm})\,dm$$

$$= \frac{d}{dt}\mathbf{h}_C + \frac{d}{dt}\left[\mathbf{r}_C \times \int_m \mathbf{v}_{dm}\,dm\right]$$

We have defined $\mathbf{h}_C$ as the angular momentum with respect to the CG. For the remaining term we have

$$\frac{d}{dt}\left[\mathbf{r}_C \times \int_m \mathbf{v}_{dm}\,dm\right] = \mathbf{v}_C \times \int_m \mathbf{v}_{dm}\,dm + \mathbf{r}_C \times \int_m \dot{\mathbf{v}}_{dm}\,dm$$

We previously showed that

$$m\mathbf{v}_C = \int_m \mathbf{v}_{dm}\,dm$$

and

$$m\mathbf{a}_C = \int_m \dot{\mathbf{v}}_{dm}\,dm$$

so

$$\frac{d}{dt}\left[\mathbf{r}_C \times \int_m \mathbf{v}_{dm}\right] = (\mathbf{v}_C \times m\mathbf{v}_C) + (\mathbf{r}_C \times m\mathbf{a}_C)$$

$$= \mathbf{r}_C \times m\mathbf{a}_C$$

$$= \mathbf{r}_C \times \mathbf{F}$$

On the left,

$$\int_m (\mathbf{r}_{dm} \times d\mathbf{f}) = \mathbf{M}_C + (\mathbf{r}_C \times \mathbf{F})$$

and on the right,

$$\frac{d}{dt}\int_m (\mathbf{r}_{dm} \times \mathbf{v}_{dm})\,dm = \frac{d}{dt}\mathbf{h}_C + (\mathbf{r}_C \times \mathbf{F})$$

Upon equating the left and right sides,

$$\mathbf{M}_C = \frac{d}{dt}\mathbf{h}_C \qquad (5.6)$$

That is, the externally applied moments about the CG are equal to the inertial time-rate of change of angular momentum about the CG. We drop the subscript C and use the dot notation so that $\mathbf{M} = \dot{\mathbf{h}}$. In the body-fixed coordinate system this is:

$$\{\mathbf{M}\}_B = T_{B,I}\mathbf{M} = T_{B,I}\dot{\mathbf{h}}$$

$$= \{\dot{\mathbf{h}}_B\}_B + \{\Omega_B\}_B \mathbf{h}_B$$

To evaluate the angular momentum we have

$$\mathbf{h} = \int_m (\mathbf{r}^C_{dm} \times \mathbf{v}_{dm})\,dm = \int_m (\mathbf{R}_{dm}\mathbf{v}_{dm})\,dm$$

Inertial Accelerations

In this expression $\mathbf{R}_{dm}$ is the matrix equivalent of the operation $[\mathbf{r}_{dm}^C \times]$ (the superscript C will be dropped in all subsequent discussion, as everything will be referenced to the CG). The vector $\mathbf{v}_{dm}$ is the inertial velocity of dm, or $\mathbf{v}_{dm} = \dot{\mathbf{r}}dm$. To relate the angular momentum to the moving reference frame we use $\{\dot{\mathbf{r}}_{dmB}\}_B = T_{B,I}\{\dot{\mathbf{r}}_{dm}\} - \{\Omega_B\}_B\{\mathbf{r}_{dm}\}_B$. The angular momentum in inertial coordinates is then

$$\mathbf{h} = \int_m (R_{dm}\mathbf{v}_{dm})\,dm$$

$$= \int_m (R_{dm}\dot{\mathbf{r}}_{dm})\,dm$$

$$= \int_m \{R_{dm}T_{I,B}[\{\dot{\mathbf{r}}_{dmB}\}_B + \{\Omega_B\}_B\{\mathbf{r}_{dm}\}_B]\}\,dm$$

$$= \int_m R_{dm}T_{I,B}\{\dot{\mathbf{r}}_{dmB}\}_B\,dm + \int_m R_{dm}T_{I,B}\{\Omega_B\}_B\{\mathbf{r}_{dm}\}_B\,dm$$

The same angular momentum in body coordinates is just $\{\mathbf{h}\}_B = T_{B,I}\mathbf{h}$:

$$\{\mathbf{h}\}_B = T_{B,I}\mathbf{h}$$

$$= \int_m T_{B,I}R_{dm}T_{I,B}\{\dot{\mathbf{r}}_{dmB}\}_B\,dm + \int_m T_{B,I}R_{dm}T_{I,B}\{\Omega_B\}_B\{\mathbf{r}_{dm}\}_B\,dm$$

$$= \int_m \{R_{dm}\}_B\{\dot{\mathbf{r}}_{dmB}\}_B\,dm + \int_m \{R_{dm}\}_B\{\Omega_B\}_B\{\mathbf{r}_{dm}\}_B\,dm$$

We note that

$$\{\Omega_B\}_B\{\mathbf{r}_{dmB}\}_B = \{\omega_B\}_B \times \{\mathbf{r}_{dm}\}_B$$
$$= -\{\mathbf{r}_{dm}\}_B \times \{\omega_B\}_B$$
$$= -\{R_{dm}\}_B\{\omega_B\}_B$$

and

$$\{R_{dm}\}_B\{R_{dm}\}_B = \{R_{dm}\}_B^2$$

So,

$$\{\mathbf{h}\}_B = \int_m \{\mathbf{r}_{dm}\}_B \times \{\dot{\mathbf{r}}_{dmB}\}_B\,dm - \int_m \{R_{dm}\}_B^2\{\omega_B\}_B\,dm$$

Now, since $\{\omega_B\}_B$ does not affect the integration, we move it outside the integral,

$$\{\mathbf{h}\}_B = \int_m \{\mathbf{r}_{dm}\}_B \times \{\dot{\mathbf{r}}_{dmB}\}_B\,dm + \left[-\int_m \{R_{dm}\}_B^2\,dm\right]\{\omega_B\}_B$$

The first integral depends on $\{\dot{\mathbf{r}}_{dmB}\}_B$ which is the rate of change of $\mathbf{r}_{dm}$ as seen from F_B with components in F_B. If we denote

$$\{\mathbf{r}_{dm}\}_B \equiv \begin{Bmatrix} x_{dm} \\ y_{dm} \\ z_{dm} \end{Bmatrix}$$

then
$$\{\dot{\mathbf{r}}_{dm_B}\}_B = \begin{Bmatrix} \dot{x}_{dm} \\ \dot{y}_{dm} \\ \dot{y}_{dm} \end{Bmatrix}$$

$$\mathbf{h}_B^* \equiv \int_m \{\mathbf{r}_{dm}\}_B \times \{\dot{\mathbf{r}}_{dmB}\}_B dm$$

The other integral may be evaluated term-by-term. First,

$$-\{R_{dm}\}_B^2 = -\begin{bmatrix} 0 & -z_{dm} & y_{dm} \\ z_{dm} & 0 & -x_{dm} \\ -y_{dm} & x_{dm} & 0 \end{bmatrix}^2$$

$$= \begin{bmatrix} (y_{dm}^2 + z_{dm}^2) & -x_{dm}y_{dm} & -x_{dm}z_{dm} \\ -x_{dm}y_{dm} & (x_{dm}^2 + z_{dm}^2) & -y_{dm}z_{dm} \\ -x_{dm}z_{dm} & -y_{dm}z_{dm} & (x_{dm}^2 + y_{dm}^2) \end{bmatrix}$$

The integral of this matrix is the matrix of the integrals of its terms, so

$$-\int_m \{R_{dm}\}_B^2 dm =$$

$$\begin{bmatrix} \int_m (y_{dm}^2 + z_{dm}^2) dm & -\int_m x_{dm}y_{dm} dm & -\int_m x_{dm}z_{dm} dm \\ -\int_m x_{dm}y_{dm} dm & \int_m (x_{dm}^2 + z_{dm}^2) dm & -\int_m y_{dm}z_{dm} dm \\ -\int_m x_{dm}z_{dm} dm & -\int_m y_{dm}z_{dm} dm & \int_m (x_{dm}^2 + y_{dm}^2) dm \end{bmatrix}$$

We recognize the integrals on the right as the moments of inertia in the particular body axis system we have chosen,

$$-\int_m \{R_{dm}\}_B^2 dm = \begin{bmatrix} I_{xx} & -I_{xy} & -I_{xz} \\ -I_{xy} & I_{yy} & -I_{yz} \\ -I_{xz} & -I_{yz} & I_{zz} \end{bmatrix} \equiv I_B$$

Finally, for the angular momentum about the CG in body axes,

$$\mathbf{h}_B = \mathbf{h}_B^* + I_B \{\omega_B\}_B$$

This is the expression we needed for $\{\mathbf{M}\}_B = \{\dot{\mathbf{h}}_B\}_B + \{\Omega_B\}_B \{\mathbf{h}\}_B$. Everything is in body coordinates, so we have

$$\{\mathbf{M}\}_B = [\dot{\mathbf{h}}_B^* + \dot{I}_B \{\omega_B\}_B + I_B \{\dot{\omega}_B\}_B] + [\{\Omega_B\}_B \mathbf{h}_B^* + \{\Omega_B\}_B I_B \{\omega_B\}_B]$$

If we have a rigid body then $\mathbf{h}_B^*$, $\dot{\mathbf{h}}_B^*$, and $\dot{I}_B$ vanish leaving

$$\{\mathbf{M}\}_B = I_B \{\dot{\omega}_B\}_B + \{\Omega_B\}_B I_B \{\omega_B\}_B \qquad (5.7)$$

5.4 States

The position and velocity variables for which we have derived differential equations are collectively referred to as the *states* of the aircraft. The states are a minimum set of variables that completely describe the aircraft position, orientation, and velocity. The basic states derived so far are the inertial linear velocity, the inertial angular velocity, and a set of three transformation variables that uniquely describe the orientation of the aircraft with respect to inertial space. For the latter we usually take Euler angles, since they already constitute the minimum set of orientation variables. The only thing missing is the inertial position of the aircraft which is simply the integral of the inertial velocity. The 12 basic states of an aircraft may therefore be written as the scalar components of $\mathbf{v}_C$, $\boldsymbol{\omega}_B$, and $\mathbf{r}_C$, plus the three Euler angles that define $T_{B,I}$.

5.5 Customs and Conventions

5.5.1 Linear velocity components

Inertial velocities in body and wind axes occur frequently, and the components associated with them are given special symbols. These are shown in Figures 5.3 and 5.4, and are defined as follows:

$$\{\mathbf{v}_C\}_B = \begin{Bmatrix} u \\ v \\ w \end{Bmatrix}, \quad \{\mathbf{v}_C\}_W = \begin{Bmatrix} V_c \\ 0 \\ 0 \end{Bmatrix}$$

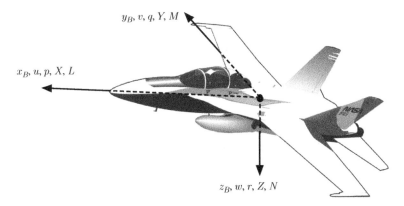

Figure 5.3 Linear velocity (u, v, w), angular velocity (p, q, r), aerodynamic force (X, Y, Z), and aerodynamic moment (L, M, N) components in body axes. Source: NASA.

The position and velocity of the aircraft *CG* with respect to some Earth-fixed coordinate system F_E is obviously of importance in navigation. The position $\{\mathbf{r}_C^E\}_E$ is denoted by x_E, y_E, and z_E, and the velocity

$$\{\mathbf{v}_C^E\}_E = \left\{\frac{d\mathbf{r}_C^E}{dt}_E\right\}_E = \begin{Bmatrix} \dot{x}_E \\ \dot{y}_E \\ \dot{z}_E \end{Bmatrix}$$

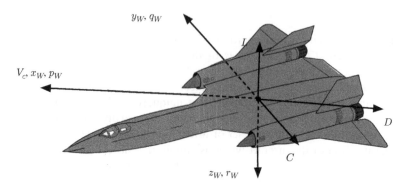

Figure 5.4 Linear velocity (V_c), angular velocity (p_W, q_W, r_W), and aerodynamic force (D, C, L) components in wind axes. Source: NASA.

With the flat-Earth assumption the aircraft's inertial velocity is the same as its velocity with respect to F_E, and since F_E is always parallel to F_H under this assumption,

$$\begin{Bmatrix} \dot{x}_E \\ \dot{y}_E \\ \dot{z}_E \end{Bmatrix} = T_{B,H}^T \begin{Bmatrix} u \\ v \\ w \end{Bmatrix} = T_{W,H}^T \begin{Bmatrix} V_c \\ 0 \\ 0 \end{Bmatrix} \quad \text{(Flat Earth)}$$

5.5.2 Angular velocity components

Angular velocity components were previously defined and are:

$$\{\omega_B\}_B = \begin{Bmatrix} p \\ q \\ r \end{Bmatrix}, \quad \{\omega_W\}_W = \begin{Bmatrix} p_W \\ q_W \\ r_W \end{Bmatrix}$$

5.5.3 Forces

The external forces that accelerate an aircraft are made up of its weight, thrust of the propulsive system, and aerodynamic forces.

The *weight* is naturally represented in the local horizontal reference frame, and is converted by the appropriate transformation to the desired system:

$$\{\mathbf{W}\}_H = \begin{Bmatrix} 0 \\ 0 \\ mg \end{Bmatrix}$$

$$\{\mathbf{W}\}_B = \begin{Bmatrix} -mg \sin \theta \\ mg \sin \phi \cos \theta \\ mg \cos \phi \cos \theta \end{Bmatrix}, \quad \{\mathbf{W}\}_W = \begin{Bmatrix} -mg \sin \gamma \\ mg \sin \mu \cos \gamma \\ mg \cos \mu \cos \gamma \end{Bmatrix}$$

The *thrust* is usually naturally represented in some body axis system. The direction of thrust may vary with respect to that system (as in the case of vectored thrust, such

Inertial Accelerations

as the Harrier or F-35C). If the thrust vector **T** varies in its relationship to the aircraft then a separate coordinate system analogous to the wind axes should be defined. For the analysis of a particular flight condition the thrust vector **T** is normally considered fixed with respect to the body. The direction of thrust usually lies in the plane of symmetry but may not align with the x-axis. The symbol ϵ_T will be used to denote the difference between **T** and x_B, positive as shown in Figure 5.5.

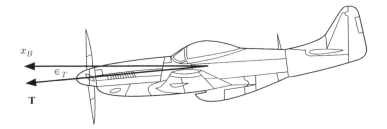

Figure 5.5 Thrust vector (orientation contrived for illustration). Source: NASA.

In this special case we have

$$\{\mathbf{T}\}_B = \begin{Bmatrix} T \cos \epsilon_T \\ 0 \\ T \sin \epsilon_T \end{Bmatrix}$$

$$\{\mathbf{T}\}_W = T_{W,B} \begin{Bmatrix} T \cos \epsilon_T \\ 0 \\ T \sin \epsilon_T \end{Bmatrix} = \begin{Bmatrix} T \cos \beta \cos (\epsilon_T - \alpha) \\ -T \sin \beta \cos (\epsilon_T - \alpha) \\ T \sin (\epsilon_T - \alpha) \end{Bmatrix}$$

We will represent the *Aerodynamic Forces* by the vector $\mathbf{F}_A$. From the point of view of the aerodynamicist the aerodynamic forces are naturally represented in the wind axes as the familiar components lift, drag, and side force. The customary definitions of lift (L) and drag (D) have positive lift in the negative z_W direction and positive drag in the negative x_W direction. We will adopt the symbol C for the aerodynamic side force and for consistency define positive side force in the negative y_W direction, as shown in Figure 5.4. Thus,

$$\{\mathbf{F}_A\}_W = \begin{Bmatrix} -D \\ -C \\ -L \end{Bmatrix}$$

The relationship (angle-of-attack and sideslip) of the wind axes to some body-axis system greatly influences the magnitude of the aerodynamic forces, so the wind axes alone are insufficient to characterize these forces. Representation of the aerodynamic forces in the body axes is at least as 'natural' as their representation in wind axes, and these forces have their own symbols in body axes, X, Y, and Z. The relationships are shown in Figure 5.3 and defined as follows:

$$\{\mathbf{F}_A\}_B = \begin{Bmatrix} X \\ Y \\ Z \end{Bmatrix}$$

The two representations are obviously related by $\{\mathbf{F}_A\}_B = T_{B,W}\{\mathbf{F}_A\}_W$, so that data given in either system may be represented in the other. In fact, aerodynamic data are often presented in a hybrid system, in which lift and drag are the familiar wind-axis quantities, but the side force is in the body-axis system. The reason for using this non-orthogonal system is related to the way the forces are actually measured in wind tunnel experiments. Clearly this representation will suffer at very high angles of sideslip β.

5.5.4 Moments

Since the gravity force acts through the *CG* it does not generate any moments about it. Aerodynamic moments about the *CG* are normally represented in a body-axis system for two reasons: first, this is how they are normally measured in wind-tunnel experiments, and second, the moments of inertia of the aircraft are reasonably constant in body axes and accelerations are easier to formulate than in wind axes. The components of aerodynamic moments $\mathbf{M}_A$ about the *CG* in body axes are given the names L, M, and N. The relationships are shown in Figure 5.3 and defined as follows:

$$\{\mathbf{M}_A\}_B = \begin{Bmatrix} L \\ M \\ N \end{Bmatrix}$$

Thrust-generated moments about the *CG* will arise if the net thrust vector **T** does not pass directly through the *CG*. In aircraft with variable thrust vectoring this is intentional, and serves as an additional control. In multi-engine aircraft the loss of one or more engines can create quite large moments. Even if the net thrust vector is fixed with respect to the body, the moment it generates will be proportional to the magnitude of the thrust, which can vary greatly. Thrust generated moments will be denoted $\mathbf{M}_T$, and will normally be available in the body axes, $\{\mathbf{M}_T\}_B$,

$$\{\mathbf{M}_T\}_B = \begin{Bmatrix} L_T \\ M_T \\ N_T \end{Bmatrix}$$

5.5.5 Groupings

Longitudinal

The assumption that the aircraft has a plane of symmetry leads to a grouping of variables that are associated with motion in that plane. These variables are the X and Z forces, the pitching moment M, the velocities u and w, and the pitch rate q. Collectively these variables are called *longitudinal* variables.

Lateral–directional

Variables associated with motion about a body-fixed x-axis, loosely those thought of as rolling the aircraft, are called *lateral* variables. These are the rolling moment L and the roll rate p. The remaining variables are the side force Y, the yawing moment N, and

Inertial Accelerations

the yaw rate r, collectively called *directional* variables. For reasons we will see later there are strong relationships between lateral and directional variables, and they are often grouped together as *lateral/directional* variables, or in the jargon, *lat–dir* variables.

Symmetric flight

If an aircraft experiences no sideslip β it is said to be in *symmetric* flight. The longitudinal motion of an aircraft is often analyzed under the assumption of no sideslip and, for purposes of such analysis, the lift L and drag D are spoken of as longitudinal variables as well.

Problems

1. Assume a flat-Earth, stationary atmosphere, and symmetric aircraft. A fighter aircraft has an accelerometer mounted at body-fixed coordinates r_x, r_y, and r_z. At a particular instant while maneuvering the accelerometer measures inertial accelerations in the body-axis system with components g_x, g_y, and g_z. Gyroscopes mounted at the center of gravity measure p, q, r, $\dot{p}$, $\dot{q}$, and $\dot{r}$. Write an expression for the inertial acceleration of the aircraft at its center of gravity, expressed in F_B.

2. A centrifuge is used to simulate aircraft accelerations. The pilot occupies a gondola that rotates with velocity ω_G^E in the horizontal plane at a distance r from the center of rotation (see Figure 5.6). Clearly define any coordinate systems used to answer the following questions.
   (a) Assuming ω_G^E is constant, determine the acceleration at the gondola.
   (b) Assuming ω_G^E is time-varying, determine the acceleration at the gondola.
   (c) The pilot in the gondola leans forward to retrieve a dropped pencil. In so doing his head moves with velocity $\mathbf{v}_h^G$ relative to the gondola. The motion is directed toward the center of rotation. Determine any additional accelerations the pilot will experience.

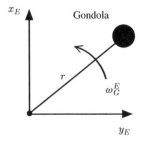

Figure 5.6 Centrifuge (problem 2).

3. Using the customary notation for body-axis forces and velocities, evaluate the force equations
$$\{\mathbf{F}\}_B = m\{\dot{\mathbf{v}}_B\}_B + m\{\Omega_B\}_B \{\mathbf{v}\}_B$$

The vector $\{v\}$ refers to the inertial velocity of the aircraft at the CG. Write out the answer as three scalar equations of the form $\dot{u} = \ldots$, $\dot{v} = \ldots$, and $\dot{w} = \ldots$

4. Assume an aircraft is symmetric about the plane defined by $x_B - z_B$. The coordinates x_{dm}, y_{dm}, and z_{dm} are measured in some body-fixed coordinate system. Explain why I_{xy} and I_{yz} vanish.

5. Using the customary notation for body-axis moments and velocities, evaluate the moment equations

$$\{M\}_B = I_B\{\dot{\omega}_B\}_B + \{\Omega_B\}_B I_B \{\omega_B\}_B$$

Write out the answer as three scalar equations of the form $\dot{p} = \ldots$, etc. Assume the aircraft is symmetric about the plane of $x_B - z_B$.

6. Even though the wind-axis system is body-carried and not body-fixed, the force equations may still be applied. Using the customary notation for wind-axis forces and velocities, evaluate the force equations

$$\{F\}_W = m\{\dot{v}_W\}_W + m\{\Omega_W\}_W \{v\}_W$$

7. Assume a rigid aircraft with plane of symmetry $x_B - z_B$. The sideslip and angle-of-attack with respect to a given body-axis system are free to vary. Consider the wind-axis moment equation,

$$\{M\}_W = \dot{h}^* + \dot{I}_W\{\omega_W\}_W + I_W\{\dot{\omega}_W\}_W + \{\Omega_W\}_W h^* + \{\Omega_W\}_W I_W\{\omega_W\}_W$$

Identify each term that vanishes in body axes but not in wind axes, and briefly explain why they do not vanish.

8. Assume a rigid aircraft with plane of symmetry $x_B - z_B$. Consider two body-fixed coordinate systems, F_{B_1} and F_{B_2}, related by a positive rotation about y_B through angle ϵ, positive from F_{B_1} to F_{B_2}. Further assume the axes of F_{B_1} are F_P, the principal axes of the aircraft, and that the values of the principal moments of inertia I_{xp}, I_{yp}, and I_{zp} of the inertia matrix I_P ($= I_{B_1}$) are known. Transform the inertia matrix I_P into F_{B_2}, and determine expressions for each of the moments of inertia of the aircraft in F_{B_2} in terms of the principal moments of inertia.

9. Assume $F_B \equiv F_P$. Find an expression for $\dot{I}_{xy}$ in F_W, in terms of α, β, $\dot{\alpha}$, $\dot{\beta}$, and the principal moments of inertia.

10. Given the names of the linear velocity components in wind and body axes, and

$$T_{B,W} = \begin{bmatrix} \cos\alpha\cos\beta & -\cos\alpha\sin\beta & -\sin\alpha \\ \sin\beta & \cos\beta & 0 \\ \sin\alpha\cos\beta & -\sin\alpha\sin\beta & \cos\alpha \end{bmatrix}$$

Verify that these definitions and relationships are consistent with the common definitions of α and β, that is,

$$\alpha = \tan^{-1}\left(\frac{w}{u}\right), \quad \beta = \sin^{-1}\left(\frac{v}{\sqrt{u^2 + v^2 + w^2}}\right)$$

6

Forces and Moments

6.1 General

In the previous section we derived $\mathbf{F} = m\dot{\mathbf{v}}$ and $\mathbf{M} = \dot{\mathbf{h}}$, valid if we are referring the problem to the *CG* of the aircraft. The aim is to solve these sets of equations for the derivative terms, and integrate them to yield velocity and position. On the left-hand side of these sets of equations we have $\mathbf{F}$ and $\mathbf{M}$ with major contributors from aerodynamic terms $\mathbf{F}_A$ and $\mathbf{M}_A$, and the thrust $\mathbf{T}$. These forces and moments in turn depend to a large extent on the very quantities we are trying to determine. For example, if everything else is held constant the aerodynamic forces and moments are functions of the velocity V. If we have any hope of solving the differential equations we have developed then we need some way to calculate the forces and moments.

The dependencies of forces and moments on other variables is usually very complex, and not amenable to setting forth analytical expressions that describe them. It is difficult enough to determine accurately the dependencies in steady cases, yet we need to know how forces and moments vary during dynamic changes in the flight condition that probably involve unsteady flow phenomena. Clearly some approximations will have to be made.

No attempt will be made here to present a detailed discussion of the sources of dependencies of the forces and moments. Instead sweeping generalizations and hand-waving will be offered to provide a high-level rationale for the method we will adopt in characterizing those forces and moments. We will then examine how such data are likely to be made available and different ways in which the data can be used. Thereafter, when necessary, we will assume that such data are available.

6.1.1 Assumptions

Precise calculations of the external forces and moments an aircraft experiences will not be available. The magnitude of the errors we will introduce will vary from problem to problem, but in most cases will dwarf the error introduced by ignoring centripetal and coriolis accelerations, and the variation of gravity's magnitude with altitude. It is unnecessary and over-complicating to continue to relate the aircraft motion to the inertial reference frame. We therefore adopt the flat-Earth approximation for the remainder of our study.

Aerodynamic forces and moments depend on the interaction of the aircraft with the atmosphere, which may be in motion relative to the Earth. The large-scale motion of the atmosphere is of importance in navigation problems, but in flight dynamics and control it is the local motion of the atmosphere that is of concern. For the time being we will assume the atmosphere is stationary with respect to the Earth. Later we will introduce local variations to examine their influence on the aircraft's responses.

The upshot of these assumptions is that all the equations previously derived for the inertial motion of an aircraft may be used to describe its relationship to the atmosphere.

6.1.2 State variables

Among the variables we will consider are the states of the aircraft. Generally these are all the velocity and position variables for which we have derived differential equations, namely V, α, β, p, q, r, θ, ϕ, ψ, x_E, y_E, and h. Instead of V, α, and β we could use u, v, and w, but V, α, and β are preferred. (Force and moment dependencies on linear velocities are almost always formulated in terms of the wind, axis representation, V, α, and β. This is because it is easier in experiments to hold two of these quantities constant while varying the third, than it is with u, v, and w.)

From the list of variables we may remove from consideration the Euler angles θ, ϕ, and ψ as they are only used to keep track of the orientation of the gravity vector relative to the body. We may also remove x_E and y_E since (barring Bermuda triangle phenomena) the location of an aircraft with respect to the Earth's surface should not affect the forces or moments acting on it. The altitude h must be retained, since it influences the properties of the air mass, which will effect the aerodynamic forces and moments.

6.1.3 State rates

Simple dependencies on states are often inadequate to describe the variation of forces and moments when angle-of-attack and sideslip are varying with time. Generally speaking this is because the flow field about the aircraft will not adjust immediately to such variations. This is a complicated phenomenon for which an approximation is required. The usual approximation is to add to the list of independent variables the time rate of change of angle-of-attack and sideslip, $\dot{\alpha}$ and $\dot{\beta}$. The former, $\dot{\alpha}$, appears quite frequently in aerodynamic force and moment data, while $\dot{\beta}$ is more rarely used.

6.1.4 Flight controls

In addition to the state variables, we will also consider as separate influences the aircraft *flight controls*. If terminology becomes a problem we will refer to them as the *control effectors* to distinguish them from the pilot's control stick and rudder pedals, which are *control inceptors*.

A flight control is any device or mechanism that may be used to directly change the external forces or moments. If we ignore any dynamics associated with the acceleration of the aircraft engine then the throttle is a flight control. The typical set of aerodynamic flight controls are the elevator, ailerons, and rudder. These will be referred to as the primary flight controls. Most aircraft also have secondary controls such as flaps and speed brakes (Figure 6.1) that are used only in particular flight regimes.

Positive deflections of the primary flight controls will be determined by a right-hand rule in which the thumb points along the axis of the primary moment being generated, and

Figure 6.1 Secondary flight controls, flaps, and speedbrake. Source: NASA.

the curled fingers point in the direction of positive deflection. Thus δ_a is positive when the right aileron is trailing-edge down (TED) and the left is up (TEU), δ_e is positive when the surface is TED, and δ_r is positive when the surface is trailing-edge right (TER).

Use caution when using data from different sources, as these sign conventions are *not* universal.

Modern aircraft may have more and different primary flight controls than the 'classical' three controls. Possibilities include maneuvering trailing or leading edge flaps, canards, flaperons, thrust vectoring, and side-force generators. See, for example, Figure 6.2. These additional controllers are more or less redundant, meaning there are more controls than

Figure 6.2 F-15 ACTIVE (Advanced Control Technology for Integrated Vehicles). Each control surface could be deflected independently, as could the thrust-vectoring vanes. Source: NASA.

moments to be generated. We will therefore combine any redundant controls and refer to them according to the moments they are intended to produce. The nomenclature to be used will be:

δ_ℓ Roll control
δ_m Pitch control
δ_n Yaw control

Additionally we will denote δ_T as the thrust control and generally ignore the dynamics of the engine. When discussing a particular aircraft we will replace the generic aerodynamic flight controls with more descriptive terms, such as δ_a (ailerons) for δ_ℓ, and δ_e (elevator) or δ_{HT} (horizontal tail) for δ_m.

6.1.5 Independent variables

We therefore assume the external forces and moments of the aircraft are functions of the independent variables $V, \alpha, \beta, p, q, r, h, \dot{\alpha}, \dot{\beta}, \delta_T, \delta_\ell, \delta_m$, and δ_n. We will discuss the dependency on each variable as it affects the force or moment while holding all other variables constant. Care should be taken not to mix wind and body axis representations of velocity. In the wind axes, velocity relative to the body is completely described by V, α, and β, and in the body axes by u, v, and w. So if we are speaking of the dependency of lift on α, we mean we are holding V and β constant. On the other hand, if we speak of the dependency of lift on the velocity component w, then we are holding u and v constant.

6.2 Non-Dimensionalization

In basic aerodynamics we learned that forces are made non-dimensional by dividing them by the dynamic pressure $\bar{q}$ and a characteristic area S, and moments by $\bar{q}$, S, and a characteristic length. The length is the wing mean geometric chord $\bar{c}$ for the pitching moment and the span b for rolling and yawing moments. The dynamic pressure depends on the local density ρ of the atmosphere and the velocity V of the aircraft relative to the atmosphere,

$$\bar{q} = \frac{1}{2}\rho V^2$$

The non-dimensional quantity is denoted as a coefficient C plus a subscript identifying the force or moment in question. Thus,

$$C_D = \frac{D}{\bar{q}S} \quad C_L = \frac{L}{\bar{q}S} \quad C_C = \frac{C}{\bar{q}S}$$

$$C_X = \frac{X}{\bar{q}S} \quad C_Y = \frac{Y}{\bar{q}S} \quad C_Z = \frac{Z}{\bar{q}S}$$

$$C_T = \frac{T}{\bar{q}S} \quad\quad\quad C_W = \frac{W}{\bar{q}S} = \frac{mg}{\bar{q}S}$$

$$C_\ell = \frac{L}{\bar{q}Sb} \quad C_m = \frac{M}{\bar{q}S\bar{c}} \quad C_n = \frac{N}{\bar{q}Sb}$$

Theory then tells us that the non-dimensional coefficients are dependent themselves on other non-dimensional quantities such as Mach number or angle-of-attack. We will write the dependencies of forces and moments as the dependency of the associated non-dimensional coefficient on non-dimensional quantities, plus the dependence on the dynamic pressure. For example, in $D = C_D \bar{q} S$, C_D will be a function of several non-dimensional quantities that are derived from the states, state-rates, and controls. All of the dimensional forces and moments will depend on h (density as a function of altitude) and V through the dynamic pressure $\bar{q}$.

The non-dimensional quantities that are derived from V, p, q, r, $\dot{\alpha}$, and $\dot{\beta}$ (all angular measurements in radians) are defined as follows:

$M = V/a$ (a is the local speed of sound in the atmosphere), and

$$\hat{p} = \frac{pb}{2V} \quad \hat{q} = \frac{q\bar{c}}{2V} \quad \hat{r} = \frac{rb}{2V}$$

$$\hat{\dot{\alpha}} = \frac{\dot{\alpha}\bar{c}}{2V} \quad \hat{\dot{\beta}} = \frac{\dot{\beta}b}{2V}$$

Because dependency on V appears explicitly in the dynamic pressure term, separate from the Mach dependency of the coefficients, we will later require one more non-dimensional velocity term. This term is valid only when the current velocity V is compared with some reference velocity V_{Ref}, and is defined as

$$\hat{V} = \frac{V}{V_{Ref}}$$

6.3 Non-Dimensional Coefficient Dependencies

6.3.1 General

It is generally true that anything that changes the pressure distribution about an aircraft will cause changes in all the forces and moments. Thus a case could be made that each of the non-dimensional coefficients are dependent on all the listed states and controls. Clearly some of these effects are more important than others, and this is largely determined by common sense and experiment. The dependencies listed below are just those usually considered to have dominant effects for 'conventional' aircraft. A specific aircraft in a particular flight condition may exhibit different dependencies, and other data may have to be acquired to adequately describe the effects.

The force coefficients described below are the wind-axis representations of lift and drag (C_L and C_D) and the body-axis representation of side force (C_Y) since these are the forces most frequently measured in experiments. The body-axis force coefficients are related to the wind-axis representations by transformation $T_{B,W}$ (since all dimensional force variables are non-dimensionalized by the same quantity). However, the mixed system of C_D, C_Y, and C_L does not so obviously transform to C_X, C_Y, and C_Z. However, it may be shown that

$$\begin{Bmatrix} C_X \\ C_Y \\ C_Z \end{Bmatrix} = \begin{bmatrix} \cos\alpha \sec\beta & -\cos\alpha \tan\beta & -\sin\alpha \\ 0 & 1 & 0 \\ \sin\alpha \sec\beta & -\sin\alpha \tan\beta & \cos\alpha \end{bmatrix} \begin{Bmatrix} -C_D \\ C_Y \\ -C_L \end{Bmatrix} \quad (6.1)$$

In the general case C_X and C_Z each must be a function of α and β, irregardless of the dependencies of C_D, C_Y, and C_L. The dependency of C_X and C_Z on β introduces an unnecessary complication in later analysis. To continue to focus on the essentials, we neglect that dependency and take the relationships to be

$$\begin{Bmatrix} C_X \\ C_Y \\ C_Z \end{Bmatrix} = \begin{bmatrix} \cos\alpha & 0 & -\sin\alpha \\ 0 & 1 & 0 \\ \sin\alpha & 0 & \cos\alpha \end{bmatrix} \begin{Bmatrix} -C_D \\ C_Y \\ -C_L \end{Bmatrix} \quad (6.2)$$

6.3.2 Altitude dependencies

The dimensional forces and moments are dependent on altitude through the dynamic pressure term, as previously discussed. There is another phenomenon known as *ground effect* that introduces an additional dependency of lift and pitching moment coefficients on altitude. Ground effect, loosely speaking, is a 'cushioning' that occurs when an aircraft is at an altitude roughly equal to half its wing span. If needed we will non-dimensionalize altitude by dividing by the chord length, $\hat{h} = h/\bar{c}$.

6.3.3 Velocity dependencies

Dependence of the non-dimensional coefficients on velocity is due primarily to Mach effects. However, the case of the thrust coefficient is somewhat different. Depending on the powerplant being considered, we may require that the thrust coefficient be a function of velocity (or the non-dimensional velocity, $\hat{V}$) directly. Consider for example a rocket powered aircraft: it is generally valid to consider the thrust T constant with velocity. Since $T = C_T \rho V^2 S/2$, C_T must decrease as V increases. For the case of the thrust coefficient C_T, we therefore assume a dependency on $\hat{V}$.

Mach dependency of the other coefficients is associated with compressibility phenomena, which can create large variations in the pressure distribution over the aircraft. This will affect all of the aerodynamic coefficients. Most of our reference conditions will be in symmetric flight, meaning the airflow over the aircraft is the same on the left and right sides. For a change in Mach from this condition, no additional sideforce, or rolling or yawing moments are generated. Even in asymmetric flight the effects of changes in Mach are relatively small and may be neglected. We therefore limit Mach dependency to longitudinal forces and moment, and do not consider it for the lateral/directional moments or force. Except for certain cases, at low subsonic speeds Mach dependency is usually ignored.

6.3.4 Angle-of-attack dependencies

The angle-of-attack of an aircraft greatly influences the entire flow field and is likely to affect all force and moment coefficients. In symmetric flight, that is, no sideslip, the effect of changes of angle-of-attack is generally the same on the left and right sides of the aircraft. In that case the rolling moment, or lateral coefficient C_ℓ, and the yawing moment and sideforce, or directional coefficients C_n and C_Y, will be unaffected. However, the coefficients C_L, C_D, and C_m have strong dependencies on angle-of-attack.

C_L dependency on α

The variation of lift coefficient of an airfoil with angle-of-attack should be familiar. Typically, the relationship is similar to that shown in Figure 6.3.

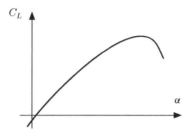

Figure 6.3 Typical lift vs. angle-of-attack curve.

All the surfaces of an aircraft – wing, horizontal tail, the fuselage itself – behave to some extent as airfoils, and each will have its own lift versus angle-of-attack relationship. The net effect is not the sum of all the contributions because of mutual interference within the flow field, but overall will have a similar shape to that of a single airfoil.

C_D dependency on α

The drag coefficient is usually thought of as being a function of the lift coefficent according to $C_D = C_d + kC_L^2$. To the extent that C_L is linear in α, this means C_D is a function of α^2, or roughly as shown in Figure 6.4.

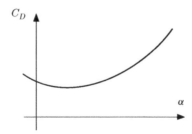

Figure 6.4 Typical drag vs. angle-of-attack curve.

C_m dependency on α

The pitching moment coefficient dependence on angle-of-attack arises because changes in the net aerodynamic forces in the plane of symmetry due to changes in angle-of-attack do not generally act directly through the aircraft CG. In particular, a change in angle-of-attack on the horizontal tail creates a change in lift on the tail, which is normally far removed from the CG, resulting in a change in pitching moment. In fact the horizontal tail is normally intended to provide moments that tend to counteract changes in angle-of-attack to provide static stability in pitch. This is seen in a typical C_m versus α relationship in which the slope is negative, indicating that positive changes in α result in negative (nose down) changes in C_m, as shown in Figure 6.5.

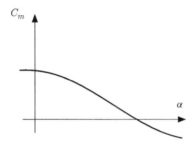

Figure 6.5 Typical pitching moment vs. angle-of-attack curve.

6.3.5 Sideslip dependencies

Sideslip affects an aircraft in a manner similar to angle-of-attack (but with a point-of-view from above the aircraft). The big differences are the absence of a large airfoil corresponding to the wing, the lack of symmetry about the plane of $x_B - y_B$, and the associated fact that the vertical tail is not symmetric about that plane. The fuselage acts roughly like a large, inefficient airfoil itself, but the primary influence of the fuselage is to partially alter the flow of air to surfaces on its lee (downwind) side. The vertical tail of course is an airfoil, and it generates aerodynamic forces in the presence of sideslip much like the horizontal tail.

C_Y dependency on β

The fuselage and vertical tail combine to produce sideforce in the presence of sideslip, acting as airfoils as just described. The picture for C_Y versus β is very much like that for C_L versus α, except the slope is normally less and the stall occurs earlier due to the inefficiency of the fuselage as an airfoil (Figure 6.6).

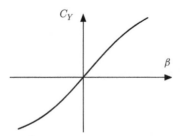

Figure 6.6 Typical yawing moment vs. sideslip curve.

C_n dependency on β

The generation of yawing moment in the presence of sideslip is analogous to the generation of pitching moment with angle-of-attack. For static directional stability, however, the slope of C_n versus β should be positive. This is different from the negative slope required of C_m versus α for static longitudinal stability because of the way positive sideslip is defined.

C_ℓ dependency on β

There are several consequences of sideslip that create rolling moments on aircraft. The most obvious is the fact that the vertical tail (fin) has more area above the x_B-axis than below it. With positive sideslip the fin, as an airfoil, generates an aerodynamic force F_f in the negative y_B direction, hence a negative rolling moment. Thus the fin normally generates a negative contribution to C_ℓ with positive β as shown in Figure 6.7.

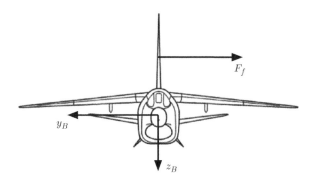

Figure 6.7 Vertical fin contribution to rolling moment. Source: NASA.

A more subtle contributor to rolling moment due to sideslip is wing *dihedral*. Dihedral is the angle at which the left and right wings are inclined upward (positive) or downward (negative). Negative dihedral is sometimes called *anhedral* and (rarely) *cathedral*. In the presence of positive sideslip the plane of the right wing sees an addition vertical component of air flow that is upward, thus increasing the angle of attack (and lift) on that wing. The opposite is true on the left wing. Like the fin, dihedral normally generates a negative contribution to C_ℓ with positive β. This is easily seen by resolving the relative airflow due to sideslip into spanwise and normal components, as shown in Figure 6.8

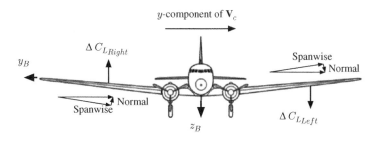

Figure 6.8 Effect of dihedral. The y-component of $\mathbf{V}_c$ is resolved into a spanwise flow and a normal component on each wing. Source: NASA.

Other major contributions to rolling moment due to sideslip result from wing sweep and from wing position relative to the fuselage (high wing or low). Of these a low wing position is the only one which tends to create positive rolling moments with positive sideslip. The actual slope of the C_ℓ versus β curve, however, is usually negative, as shown in Figure 6.9.

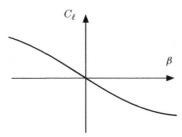

Figure 6.9 Typical rolling moment vs. sideslip curve

6.3.6 Angular velocity dependencies

The dependencies on $\dot{\alpha}$ and $\dot{\beta}$ were discussed in Section 6.1.3. As mentioned there, $\dot{\beta}$ dependencies are rarely seen in the analysis of 'conventional' aircraft. As for $\dot{\alpha}$ only lift and pitching moment are commonly assumed to have dependency (this in spite of the fact that drag is a function of lift). In the following we will therefore assume that C_L and C_m are functions of $\dot{\alpha}$, and that no coefficients are dependent on $\dot{\beta}$.

With respect to p, q, and r, all dependencies arise from similar phenomena: superimposed on the linear flow of the air mass over the aircraft are components created by the rotation within the air mass.

C_m dependency on $\hat{q}$

This effect is illustrated by considering the longitudinal case, with pitch rate q and the horizontal tail some distance ℓ_t behind the CG. The vertical velocity at the tail due to q is $V_t = \ell_t q$, as shown in Figure 6.10.

When the vertical velocity V_t is added to the free stream velocity, the effect is to increase the local angle-of-attack of the tail by some amount $\Delta \alpha_t$, shown in Figure 6.11.

The increase in angle-of-attack $\Delta \alpha_t$ at the tail will generally increase the lift, which will create a nose-down (negative) pitching moment.

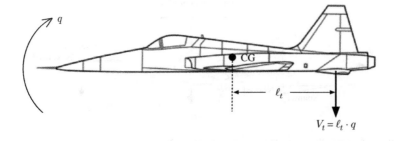

Figure 6.10 Vertical velocity at empenage induced by pitch rate. Source: NASA.

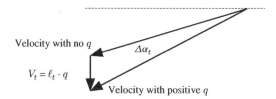

Figure 6.11 Change in angle-of-attack due to pitch rate.

C_ℓ and C_n dependencies on $\hat{p}$ and $\hat{r}$

Analogous to the effect of pitch rate upon the horizontal tail, a similar situation exists for the vertical tail in response to yaw rate r, and a negative yawing moment will be induced by positive yaw rate. In the case of the wings positive roll rate, p increases the local angle-of-attack along the right (down-going) wing and decreases it along the left (up-going) wing. The increased lift on the right wing plus the decreased lift on the left wing generates a negative rolling moment.

Changes in lift on the wing are accompanied by changes in drag as well. In the case of an aircraft with a positive roll rate, increased lift on the right wing increases its drag, while the decreased lift on the left wing decreases its drag. The differential in drag causes a yawing moment, in this case positive (nose right).

As in our discussion regarding rolling moments due to sideslip, the fact that the fin is not symmetrical about the $x_B - z_B$ plane means that any force generated by the fin will induce a rolling moment. Since a positive yaw rate generates a force in the positive y_B direction at some distance above the x_B-axis, a positive rolling moment will result.

C_L and C_Y dependencies on $\hat{p}$, $\hat{q}$, and $\hat{r}$

Since each of the rotary motions creates a change in aerodynamic forces (which generated the moments described), we include dependency of C_L on $\hat{q}$ and of C_Y on $\hat{p}$ and $\hat{r}$.

6.3.7 Control dependencies

C_T and C_m dependencies on δ_T

The throttle δ_T is assumed to directly control the thrust of the engine and, therefore, the coefficient of thrust. We further assume the thrust line lies in the plane of symmetry $x_B - z_B$ so that the direction of thrust lies in the plane of symmetry as previously discussed, with ϵ_T defined as the angle between **T** and x_B. In terms of the thrust coefficient C_T and the throttle δ_T,

$$\{\mathbf{T}\}_B = \bar{q}S \begin{Bmatrix} T = C_T\left(\hat{V}, \delta_T\right)\cos\epsilon_T \\ 0 \\ T = C_T\left(\hat{V}, \delta_T\right)\sin\epsilon_T \end{Bmatrix}$$

It is usually the case that the thrust-line does not act exactly through the center of gravity of the aircraft. Thus there is a pitching moment due to thrust, and a change in C_m due to change in δ_T.

C_ℓ, C_m, and C_n dependencies on δ_ℓ, δ_m, and δ_n

Primary effects
We have postulated generic aerodynamic controls that are intended as moment generators: δ_ℓ, δ_m, and δ_n. At a minimum therefore C_ℓ, C_m, and C_n will be functions of δ_ℓ, δ_m, and δ_n, respectively.

Coupled effects
Most pitching moment control effectors (elevators, horizontal tails, canards, etc.) when operated symmetrically (left–right) produce pure pitching moment, no rolling or yawing moments. There is, however, a good amount of out-of-axis effects from typical rolling and yawing moment control effectors such as ailerons, spoilers, and rudders. It is therefore common to find that there is a dependency of C_ℓ on δ_n and of C_n on δ_ℓ.

C_D, C_Y, and C_L dependencies on δ_ℓ, δ_m, and δ_n

Roll control
All aerodynamic controls generate forces to produce the desired moments. In the case of roll control δ_ℓ, for implementations such as ailerons the forces generated are almost a pure couple, leaving no resultant effect on net lift. Other implementations, such as spoilers, change the lift on one wing but not the other, and there will be a net change in lift. In the conventional case of ailerons assumed subsequently, there will be no dependency of any force coefficients on δ_ℓ.

Pitch control
Every practical implementation of aircraft pitch control δ_m generates an unbalanced lift force. We therefore assume C_L is dependent upon δ_m. Since a change in lift creates a change in drag, we further assume C_D is dependent upon δ_m.

Yaw control
Similar to pitch control, most implementations of yaw control create unbalanced side forces on the aircraft. We therefore take C_Y to be dependent upon δ_n.

6.3.8 Summary of dependencies

The force and moment coefficient dependencies assumed here and subsequently are representative of a 'conventional' aircraft. Any given aircraft may require the assumption of other dependencies to adequately represent its aerodynamic and thrust forces and moments. For our purposes, however, we assume the following:

$$\begin{aligned}
C_D &= C_D(M, \alpha, \delta_m) \\
C_Y &= C_Y(\beta, \hat{p}, \hat{r}, \delta_n) \\
C_L &= C_L(M, \alpha, \hat{\dot{\alpha}}, \hat{q}, \hat{h}, \delta_m) \\
C_\ell &= C_\ell(\beta, \hat{p}, \hat{r}, \delta_\ell, \delta_n) \\
C_m &= C_m(M, \alpha, \hat{\dot{\alpha}}, \hat{q}, \hat{h}, \delta_m, \delta_T) \\
C_n &= C_n(\beta, \hat{p}, \hat{r}, \delta_\ell, \delta_n) \\
C_T &= C_T(\hat{V}, \delta_T)
\end{aligned} \quad (6.3)$$

6.4 The Linear Assumption

Even though the force and moment dependencies listed above can be quite nonlinear, we can restrict our analysis to a range of variables in which the nonlinearity is only slight, and consider the dependencies to be linear. This is completely equivalent to writing each coefficient as a Taylor series expansion about some reference condition, and discarding all but the constant and first-order terms. Thus, for a coefficient C that is a function of several variables $v_1, v_2, \ldots, v_n$, this results in the linear approximation given by:

$$C(v_1, v_2, \ldots v_n) = C(v_1, v_2, \ldots v_n)_{Ref}$$
$$+ \left(\frac{\partial C}{\partial v_1}\right)_{Ref} \Delta v_1 + \left(\frac{\partial C}{\partial v_2}\right)_{Ref} \Delta v_2 \cdots + \left(\frac{\partial C}{\partial v_n}\right)_{Ref} \Delta v_n$$

In this expression, $\Delta v_i = v_i - v_{i_{Ref}}$. If $v_{i_{Ref}} = 0$, then $\Delta v_i = v_i$. The customary notation in flight dynamics is to abbreviate the partial derivative of a non-dimensional coefficient with respect to a non-dimensional variable using successive subscripts, for example,

$$\left(\frac{\partial C_L}{\partial \alpha}\right)_{Ref} \equiv C_{L_\alpha}$$

Using this notation, we may write:

$$C_D(M, \alpha, \delta_m) = C_D(M, \alpha, \delta_m)_{Ref} + C_{D_M}\Delta M + C_{D_\alpha}\Delta \alpha + C_{D_{\delta_m}}\Delta \delta_m$$

The other coefficients are represented in similar fashion.

With respect to the lateral–directional coefficients, the reference conditions are usually taken such that $C_{Y_{Ref}} = C_{\ell_{Ref}} = C_{n_{Ref}} = 0$. Due to the symmetry of the airplane, this condition generally requires $\beta_{Ref} = \hat{p}_{Ref} = \hat{r}_{Ref} = 0$. Suitably defined control deflection conventions also yield $\delta_{\ell_{Ref}} = \delta_{n_{Ref}} = 0$.

We have neglected the influence of Mach number on the lateral–directional coefficients; it is reasonable, however, to posit that the dependency of the coefficients on other states and controls will be influenced by changes in Mach number. This is accounted for by making some or all of the derivatives with respect to these states and controls functionally dependent on Mach. Thus one may see expressions such as

$$C_\ell = C_{\ell_\beta}(M)\Delta\beta + C_{\ell_p}(M)\Delta\hat{p} + \cdots$$

Here, $C_{\ell_\beta}(M)$ means C_{ℓ_β} is a function of Mach number, and similarly for the other derivatives.

6.5 Tabular Data

The force and moment coefficients are normally available in tabular form. Thus, for example, at a particular Mach number and control setting the drag coefficient may be measured at discrete values of angle-of-attack and the results tabulated. To obtain drag coefficients at angles-of-attack between the discrete values the data are interpolated. The interpolation is usually linear.

Within an interval of the table denote α_{Lower} the angle-of-attack at the lower end of the interval, α_{Upper} that at the upper end, and likewise for the coefficients $C_D(\alpha_{Lower})$ and $C_D(\alpha_{Upper})$. With linear interpolation for $C_D(\alpha)$ at arbitrary α in an interval we have:

$$\frac{C_D(\alpha) - C_D(\alpha_{Lower})}{C_D(\alpha_{Upper}) - C_D(\alpha_{Lower})} = \frac{\alpha - \alpha_{Lower}}{\alpha_{Upper} - \alpha_{Lower}}$$

This expression may be solved for the desired $C_D(\alpha)$,

$$C_D(\alpha) = C_D(\alpha_{Lower}) + \frac{\alpha - \alpha_{Lower}}{\alpha_{Upper} - \alpha_{Lower}}[C_D(\alpha_{Upper}) - C_D(\alpha_{Lower})]$$

This last expression is of the form

$$C_D = C_1(\alpha_{Lower}, \alpha_{Upper}) + C_2(\alpha_{Lower}, \alpha_{Upper})\alpha$$

The notation indicates that the values of C_1 and C_2 on the right-hand side are valid only in the range of some particular α_{Lower} and α_{Upper}. Instead of recording the tabulated data, one may pre-calculate and record the values of C_1 and C_2 for each interval.

Because each coefficient is a function of several variables, the tabulation must be repeated for various discrete values of each of the independent variables. Continuing the drag coefficient example, the coefficient is (assumed to be) a function of Mach, angle-of-attack, and control setting, $C_D(M, \alpha, \delta_m)$. Therefore the tabulation will consist of C_D at several discrete values (called *breakpoints*) of α, repeated for several discrete settings of δ_m, all at some specific Mach number. The two-level tabulation must then be repeated at other discrete values of Mach number. For arbitrary combinations of M, α, and δ_m the tables are interpolated first with respect to one of the variables, then with respect to a second, and finally with respect to the third.

Interpolations of functions of more than three variables become computationally intensive and require fairly large amounts of data storage for adequate representation of an aircraft's forces and moments. It is customary to provide data for the more highly non-linear dependencies, and to adjust the results as affine corrections for the remaining variables.

6.6 Customs and Conventions

The non-dimensional derivatives introduced in Section 6.4 are used quite frequently in flight dynamics. Collectively those derivatives taken with respect to states (or state rates) are called *stability derivatives,* and those with respect to controls are called *control derivatives.* The sign of certain of the stability derivatives is an indication of the static stability of an aircraft with respect to disturbances in some variable, such as the requirement that C_{M_α} be negative for static stability in pitch, or that C_{n_β} should be positive for directional static stability. The contribution of the other stability derivatives is to the dynamic stability of the aircraft, and will be addressed later. Among the stability derivatives those that are with respect to rotary motion are called *rotary derivatives,* or *damping derivatives.*

Problems

1. The A-4D *Skyhawk* has the following properties: $S = 260$ ft^2, $b = 27.5$ ft, and $\bar{c} = 10.8$ ft. If the aircraft is flying at $M = 0.4$ ft/s at sea-level ($a = 1116.4$ ft/s, $\rho = 0.0023769$ lb·s^2/ft^4), and $C_{m_q} = -3.6$ (dimensionless), determine the change in pitching moment (units ft · lb) that results from a change in pitch rate of $+10$ deg/s.
2. An aircraft has the following properties: $S = 500$ ft^2, $b = 50$ ft, and $\bar{c} = 10$ ft. If the aircraft is flying at $V = 222.5$ ft/s at sea-level ($\bar{q} = 57$ lb/ft^2), and $C_{\ell_p} = -0.40$, determine the change in rolling moment if the roll rate changes by 30 deg/s.
3. Consider the following data for drag coefficient versus angle-of-attack:

α deg	−5	0	5	10	15	20	25
C_D	0.044	0.032	0.035	0.045	0.059	0.077	0.101

   (a) Use linear interpolation to determine the drag coefficient at $\alpha = 12$ deg.
   (b) Find the least-squares solution for a quadratic curve $C_D = K_1 + K_2\alpha + K_3\alpha^2$ fit to the data and determine the drag coefficient at $\alpha = 12$ deg.
   (c) Determine the coefficients C_1 and C_2 in the expression $C_{D_{\alpha_1,\alpha_2}} = C_1 + C_2\alpha$ for each interval of α; that is, for $\alpha_1 = -5$ deg and $\alpha_2 = 0$ deg, then for $\alpha_1 = 0$ deg and $\alpha_2 = 5$ deg, etc. Use the results to determine the drag coefficient at $\alpha = 12$ deg.
4. Consider the following data for drag coefficient versus angle-of-attack at two different Mach numbers:

α, deg	−5	0	5	10	15	20	25
C_D, $M = 0.7$	0.044	0.032	0.035	0.045	0.059	0.077	0.101
C_D, $M = 0.8$	0.045	0.034	0.038	0.049	0.064	0.083	0.108

   (a) Using linear interpolation, determine the drag coefficient at $\alpha = 8$ deg and $M = 0.72$.
   (b) Evaluate the constants in the expression $C_{D_{\alpha_1,\alpha_2,M_1,M_2}} = C_1 + C_2\alpha + C_3M + C_4\alpha M$ for data in the range $\alpha_1 = 5$ deg, $\alpha_2 = 10$ deg, $M_1 = 0.7$, and $M_2 = 0.8$.
   (c) Using the results from part b., determine the drag coefficient at $\alpha = 8$ deg and $M = 0.72$.
5. Derive Equation 6.1.
6. The table below shows another way of modeling aerodynamic data. The columns correspond to sideslip β from -30 deg to $+30$ deg, and the rows represent angle-of-attack α from -10 deg to $+25$ deg. The table entries are the values of $C_{\ell_{\delta_a}}(\alpha, \beta)$ in the expression

$$C_\ell(\alpha, \beta, \hat{p}, \hat{r}, \delta_a, \delta_r) = C_\ell(\alpha, \beta, \hat{p}, \hat{r}, \delta_r) + C_{\ell_{\delta_a}}(\alpha, \beta)\delta_a$$

$\alpha \downarrow \beta \rightarrow$	-30	-20	-10	0	10	20	30
-10	-0.041	-0.041	-0.042	-0.040	-0.043	-0.044	-0.043
-5	-0.052	-0.053	-0.053	-0.052	-0.049	-0.048	-0.049
0	-0.053	-0.053	-0.052	-0.051	-0.048	-0.048	-0.047
5	-0.056	-0.053	-0.051	-0.052	-0.049	-0.047	-0.045
10	-0.050	-0.050	-0.049	-0.048	-0.043	-0.042	-0.042
15	-0.056	-0.051	-0.049	-0.048	-0.042	-0.041	-0.037
20	-0.082	-0.066	-0.043	-0.042	-0.042	-0.020	-0.030
25	-0.059	-0.043	-0.035	-0.037	-0.036	-0.028	-0.013

(a) For fixed values of α, β, $\hat{p}$, $\hat{r}$, and δ_r, is C_ℓ linear or nonlinear with respect to aileron deflections δ_a? Explain.
(b) Based on the given data, what is the likely sign convention used to define positive aileron deflections?
(c) The ailerons are limited to $\pm 30 \deg$ of deflection. The aircraft has a wing reference area $S = 400$ ft^2 and span $b = 32$ ft. If the aircraft is flying at $V = 500$ ft/s at sea-level on a standard day, what is the maximum rolling moment (in ft · lb) the ailerons can generate?

7

Equations of Motion

7.1 General

The rigid body equations of motion are the differential equations that describe the evolution of the 12 basic states of an aircraft: the scalar components of $\mathbf{v}_B$, $\boldsymbol{\omega}_B$, and $\mathbf{r}_C$, plus the three Euler angles that define $T_{B,I}$. Most of the differential equations needed were derived either in Chapter 4, 'Rotating Coordinate Systems' or as problems following Chapter 5, 'Inertial Accelerations'.

The usual coordinate systems in which we represent the equations of motion are the wind- and body-axis systems. Mixed systems using both coordinate systems are common. Any such set of equations of motion must be *complete*. This means that in every expression $d(State)/dt = \ldots$, everything on the right-hand side must be given by either an algebraic equation (including *variable = constant*), by another differential equation, or by some external input to the system (e.g., the pilot's application of controls).

7.2 Body-Axis Equations

7.2.1 Body-axis force equations

The body-axis force equations were previously derived, and are

$$\{\mathbf{F}\}_B = m\{\dot{\mathbf{v}}_B\}_B + m\{\Omega_B\}_B\{\mathbf{v}\}_B$$

In terms of the rates-of-change of the inertial components of velocity, as seen in the body-axes:

$$\{\dot{\mathbf{v}}_B\}_B = \frac{1}{m}\{\mathbf{F}\}_B - \{\Omega_B\}_B\{\mathbf{v}\}_B$$

All of the terms in this equation have been defined, so all we have to do is make the necessary substitutions and expand the equations to solve for $\dot{u}$, $\dot{v}$, and $\dot{w}$. On the right-hand side, the net applied force is comprised of aerodynamic forces, the aircraft weight, and the force due to thrust:

$$\{\mathbf{F}\}_B = \{\mathbf{F}_A\}_B + \{\mathbf{W}\}_B + \{\mathbf{T}\}_B$$

$$= \begin{Bmatrix} X \\ Y \\ Z \end{Bmatrix} + \begin{Bmatrix} -mg\sin\theta \\ mg\sin\phi\cos\theta \\ mg\cos\phi\cos\theta \end{Bmatrix} + \begin{Bmatrix} T\cos\epsilon \\ 0 \\ T\sin\epsilon \end{Bmatrix} \quad (7.1)$$

Aircraft Flight Dynamics and Control, First Edition. Wayne Durham.
© 2013 John Wiley & Sons, Ltd. Published 2013 by John Wiley & Sons, Ltd.

For the motion variables we substitute

$$\{\mathbf{v}\}_B = \begin{Bmatrix} u \\ v \\ w \end{Bmatrix}, \quad \{\dot{\mathbf{v}}_B\}_B = \begin{Bmatrix} \dot{u} \\ \dot{v} \\ \dot{w} \end{Bmatrix}$$

$$\{\Omega_B\}_B = \begin{bmatrix} 0 & -r & q \\ r & 0 & -p \\ -q & p & 0 \end{bmatrix}$$

The cross-product term is easily evaluated, and is

$$\{\Omega_B\}_B\{\mathbf{v}\}_B = \begin{Bmatrix} qw - rv \\ ru - pw \\ pv - qu \end{Bmatrix}$$

As a result we may write the body-axis force equations,

$$\dot{u} = \frac{1}{m}(X + T\cos\epsilon_T) - g\sin\theta + rv - qw \tag{7.2a}$$

$$\dot{v} = \frac{1}{m}(Y) + g\sin\phi\cos\theta + pw - ru \tag{7.2b}$$

$$\dot{w} = \frac{1}{m}(Z + T\sin\epsilon_T) + g\cos\phi\cos\theta + qu - pv \tag{7.2c}$$

7.2.2 Body-axis moment equations

The moment equations are given by

$$\{\mathbf{M}\}_B = I_B\{\dot{\boldsymbol{\omega}}_B\}_B + \{\Omega_B\}_B I_B\{\boldsymbol{\omega}_B\}_B$$

In terms of the rates of change of the inertial components of angular rotation, as seen in the body axes,

$$\{\dot{\boldsymbol{\omega}}_B\}_B = I_B^{-1}[\{\mathbf{M}\}_B - \{\Omega_B\}_B I_B\{\boldsymbol{\omega}_B\}_B]$$

The externally applied moments are those due to aerodynamics and thrust. We will ignore rolling and yawing moments due to thrust as these are special cases (note, however, that yawing moment due to thrust is an important consideration in multi-engine aircraft, and may have to be added in for analysis of the consequences of engine failure). As a result the moments are

$$\{\mathbf{M}\}_B = \begin{Bmatrix} L \\ M + M_T \\ N \end{Bmatrix}$$

We assume a plane of symmetry so in the inertia matrix the cross-products involving y become zero,

$$I_B = \begin{bmatrix} I_{xx} & 0 & -I_{xz} \\ 0 & I_{yy} & 0 \\ -I_{xz} & 0 & I_{zz} \end{bmatrix}$$

Equations of Motion

The inverse is then given by

$$I_B^{-1} = \frac{1}{I_D}\begin{bmatrix} I_{zz} & 0 & I_{xz} \\ 0 & I_D/I_{yy} & 0 \\ I_{xz} & 0 & I_{xx} \end{bmatrix}$$

Here, $I_D = I_{xx}I_{zz} - I_{xz}^2$. The rest is just substitution and expansion:

$$\dot{p} = \frac{I_{zz}}{I_D}[L + I_{xz}pq - (I_{zz} - I_{yy})qr] + \frac{I_{xz}}{I_D}[N - I_{xz}qr - (I_{yy} - I_{xx})pq] \quad (7.3a)$$

$$\dot{q} = \frac{1}{I_{yy}}[M + M_T - (I_{xx} - I_{zz})pr - I_{xz}(p^2 - r^2)] \quad (7.3b)$$

$$\dot{r} = \frac{I_{xz}}{I_D}[L + I_{xz}pq - (I_{zz} - I_{yy})qr] + \frac{I_{xx}}{I_D}[N - I_{xz}qr - (I_{yy} - I_{xx})pq] \quad (7.3c)$$

If we are lucky enough to be working the problem in principal axes, then

$$\dot{p} = [L - (I_{zp} - I_{yp})qr]/I_{xp}$$
$$\dot{q} = [M + M_T - (I_{xp} - I_{zp})pr]/I_{yp}$$
$$\dot{r} = [N - (I_{yp} - I_{xp})pq]/I_{zp}$$

7.2.3 Body-axis orientation equations (kinematic equations)

The required equations were derived in Chapter 4, 'Rotating Coordinate Systems'. We will use the Euler angle relationships, with the Euler angles (flat-Earth assumption) from the transformation from the local horizontal to the body axes. The resulting matrix equation is

$$\begin{Bmatrix} \dot{\phi} \\ \dot{\theta} \\ \dot{\psi} \end{Bmatrix} = \begin{bmatrix} 1 & \sin\phi \tan\theta & \cos\phi \tan\theta \\ 0 & \cos\phi & -\sin\phi \\ 0 & \sin\phi \sec\theta & \cos\phi \sec\theta \end{bmatrix} \begin{Bmatrix} p \\ q \\ r \end{Bmatrix}$$

As three scalar equations we have

$$\dot{\phi} = p + (q\sin\phi + r\cos\phi)\tan\theta \quad (7.4a)$$

$$\dot{\theta} = q\cos\phi - r\sin\phi \quad (7.4b)$$

$$\dot{\psi} = (q\sin\phi + r\cos\phi)\sec\theta \quad (7.4c)$$

7.2.4 Body-axis navigation equations

The position of the aircraft relative to the Earth is found by integrating the aircraft velocity along its path, or by representing the velocity in Earth-fixed coordinates and integrating each component. The latter is easier, and is given by

$$\begin{Bmatrix} \dot{x}_E \\ \dot{y}_E \\ \dot{z}_E \end{Bmatrix} = T_{H,B} \begin{Bmatrix} u \\ v \\ w \end{Bmatrix}$$

The needed transformation is $T_{B,H}^T$. Expanding the equations,

$$\dot{x}_E = u(\cos\theta\cos\psi) + v(\sin\phi\sin\theta\cos\psi - \cos\phi\sin\psi)$$
$$+ w(\cos\phi\sin\theta\cos\psi + \sin\phi\sin\psi) \tag{7.5a}$$
$$\dot{y}_E = u(\cos\theta\sin\psi) + v(\sin\phi\sin\theta\sin\psi + \cos\phi\cos\psi)$$
$$+ w(\cos\phi\sin\theta\sin\psi - \sin\phi\cos\psi) \tag{7.5b}$$
$$\dot{h} = -\dot{z}_E = u\sin\theta - v\sin\phi\cos\theta - w\cos\phi\cos\theta \tag{7.5c}$$

7.3 Wind-Axis Equations

7.3.1 Wind-axis force equations

Development of the wind-axis force equations begins similarly to the body-axis equations, with

$$\{\dot{\mathbf{v}}_W\}_W = \frac{1}{m}\{\mathbf{F}\}_W - \{\mathbf{\Omega}_W\}_W\{\mathbf{v}\}_W$$

The external forces are

$$\{\mathbf{F}_A\}_W = \begin{Bmatrix} -D \\ -C \\ -L \end{Bmatrix}$$

$$\{\mathbf{T}\}_W = T_{W,B}\begin{Bmatrix} T\cos\epsilon_T \\ 0 \\ T\sin\epsilon_T \end{Bmatrix} = \begin{Bmatrix} T\cos\beta\cos(\epsilon_T - \alpha) \\ -T\sin\beta\cos(\epsilon_T - \alpha) \\ T\sin(\epsilon_T - \alpha) \end{Bmatrix}$$

$$\{\mathbf{W}_W\} = \begin{Bmatrix} -mg\sin\gamma \\ mg\sin\mu\cos\gamma \\ mg\cos\mu\cos\gamma \end{Bmatrix}$$

Again inserting the linear and angular velocity components,

$$\{\mathbf{v}\}_W = \begin{Bmatrix} V \\ 0 \\ 0 \end{Bmatrix}$$

$$\{\boldsymbol{\omega}_W\}_W = \begin{Bmatrix} p_W \\ q_W \\ r_W \end{Bmatrix}$$

$$\{\mathbf{\Omega}_W\}_W\{\mathbf{v}\}_W = \begin{Bmatrix} 0 \\ Vr_W \\ -Vq_W \end{Bmatrix}$$

The resulting scalar equations are

$$\dot{V} = \frac{1}{m}[-D - mg\sin\gamma + T\cos\beta\cos(\epsilon_T - \alpha)] \tag{7.6a}$$

Equations of Motion

$$r_W = \frac{1}{mV}[-C - mg \sin \mu \cos \gamma - T \sin \beta \cos(\epsilon_T - \alpha)] \quad (7.6b)$$

$$q_W = \frac{1}{mV}[L - mg \cos \mu \cos \gamma - T \sin(\epsilon_T - \alpha)] \quad (7.6c)$$

This is a little different from the body-axis results, in that we have one differential equation and two algebraic equations. In body axes we had differential equations for each of the three velocity components so that angle-of-attack and sideslip could be calculated at any instant for use in determining forces and moments. What is needed for the wind-axis force equations are separate relationships for α and β.

We know that the angle rates $\dot{\alpha}$ and $\dot{\beta}$ will be related to the relative rotation of the wind and body axes through the previously derived result

$$\{\omega_2^1\}_2 = \begin{bmatrix} 1 & 0 & -\sin \theta_y \\ 0 & \cos \theta_x & \sin \theta_x \cos \theta_y \\ 0 & -\sin \theta_x & \cos \theta_x \cos \theta_y \end{bmatrix} \begin{Bmatrix} \dot{\theta}_x \\ \dot{\theta}_y \\ \dot{\theta}_z \end{Bmatrix}$$

We substitute $(\theta_x, \theta_y, \theta_z) = (0, \alpha, -\beta)$ to arrive at

$$\{\omega_B^W\}_B = \begin{bmatrix} 1 & 0 & -\sin \alpha \\ 0 & 1 & 0 \\ 0 & 0 & \cos \alpha \end{bmatrix} \begin{Bmatrix} 0 \\ \dot{\alpha} \\ -\dot{\beta} \end{Bmatrix} = \begin{Bmatrix} -\dot{\beta} \sin \alpha \\ \dot{\alpha} \\ \dot{\beta} \cos \alpha \end{Bmatrix}$$

This is the rate at which the body axes rotate relative to the wind axes. From the body-axis moment equations we can determine $\{\omega_B\}_B$. We already have algebraic expressions for two components of $\{\omega_W\}_W$ (q_W and r_W, Equation 7.6) so with a little luck we should be able to assemble these results to get equations for $\dot{\alpha}$ and $\dot{\beta}$. We begin with

$$\{\omega_W\}_W = \{\omega_B\}_W - \{\omega_B^W\}_W$$
$$= T_{W,B}[\{\omega_B\}_B - \{\omega_B^W\}_B]$$

In terms of their components these equations are

$$\begin{Bmatrix} p_W \\ q_W \\ r_W \end{Bmatrix} = T_{W,B}\left[\begin{Bmatrix} p \\ q \\ r \end{Bmatrix} - \begin{Bmatrix} -\dot{\beta} \sin \alpha \\ \dot{\alpha} \\ \dot{\beta} \cos \alpha \end{Bmatrix}\right]$$

$$= \begin{Bmatrix} p \cos \alpha \cos \beta + \sin \beta (q - \dot{\alpha}) + r \sin \alpha \cos \beta \\ -p \cos \alpha \sin \beta + \cos \beta (q - \dot{\alpha}) - r \sin \alpha \sin \beta \\ -p \sin \alpha + r \cos \alpha + \dot{\beta} \end{Bmatrix}$$

These equations are easily solved for p_W and two differential equations for $\dot{\alpha}$ and $\dot{\beta}$:

$$p_W = p \cos \alpha \cos \beta + \sin \beta (q - \dot{\alpha}) + r \sin \alpha \cos \beta \quad (7.7a)$$

$$\dot{\alpha} = q - \sec \beta (q_W + p \cos \alpha \sin \beta + r \sin \alpha \sin \beta) \quad (7.7b)$$

$$\dot{\beta} = r_W + p \sin \alpha - r \cos \alpha \quad (7.7c)$$

The three differential equations for use in wind axes are therefore

$$\dot{V} = \frac{1}{m}[-D - mg \sin \gamma + T \cos \beta \cos(\epsilon_T - \alpha)] \qquad (7.8a)$$

$$\dot{\alpha} = q - \sec \beta (q_W + p \cos \alpha \sin \beta + r \sin \alpha \sin \beta) \qquad (7.8b)$$

$$\dot{\beta} = r_W + p \sin \alpha - r \cos \alpha \qquad (7.8c)$$

The wind axis angular rates to be used in these equations are given by the two algebraic equations from Equation 7.6,

$$q_W = \frac{1}{mV}[L - mg \cos \mu \cos \gamma - T \sin(\epsilon_T - \alpha)]$$

$$r_W = \frac{1}{mV}[-C + mg \sin \mu \cos \gamma - T \sin \beta \cos(\epsilon_T - \alpha)]$$

As a result *we may mix the wind-axis force equations with the body-axis moment equations, and never have to worry about the wind-axis moment equations.*

It should be noted that the differential equation for $\dot{\alpha}$ depends on q_W, which depends on C_L through the lift L. If C_L is dependent on $\dot{\alpha}$ then this dependency must be considered in solving for $\dot{\alpha}$. If C_L has some simple dependency on $\dot{\alpha}$, such as a linear relationship, then it may be easy to factor out $\dot{\alpha}$ and combine it and its factor with the explicit $\dot{\alpha}$ on the left-hand side of the equation. Moreover, if additionally C_m is dependent on $\dot{\alpha}$, then the body-axis pitching moment equation will contain not only $\dot{q}$ but $\dot{\alpha}$ as well, in which case the wind-axis $\dot{\alpha}$ equation must be solved simultaneously with the body-axis pitching moment equation to determine two equations for $\dot{\alpha}$ and $\dot{q}$.

If neither C_L nor C_m is functionally dependent on $\dot{\alpha}$ then a somewhat simpler approach is possible. Rather than use the mixed formulations in Equation 7.8 we may relate $\dot{V}$, $\dot{\alpha}$, and $\dot{\beta}$ directly to the body-axis force equations. With a few derivatives we proceed as follows:

$$\alpha \equiv \tan^{-1}\left(\frac{w}{u}\right) \Rightarrow \dot{\alpha} = \frac{u\dot{w} - w\dot{u}}{u^2 + w^2} \qquad (7.9a)$$

$$V \equiv \sqrt{u^2 + v^2 + w^2} \Rightarrow \dot{V} = \frac{u\dot{u} + v\dot{v} + w\dot{w}}{\sqrt{u^2 + v^2 + w^2}} = \frac{u\dot{u} + v\dot{v} + w\dot{w}}{V} \qquad (7.9b)$$

The $\dot{\beta}$ expression is best formulated in terms of V and $\dot{V}$

$$\beta \equiv \sin^{-1}\left(\frac{v}{V}\right) \Rightarrow \dot{\beta} = \frac{V\dot{v} - v\dot{V}}{V\sqrt{u^2 + w^2}} \qquad (7.9c)$$

Some care must be taken when mixing the two systems (u, v, w) and (V, α, β). It should be understood that one first evaluates $\dot{u}$, $\dot{v}$, and $\dot{w}$ from Equation 7.2, then applies those results to Equations 7.9a and 7.9b, and then V and $\dot{V}$ from Equation 7.9b are applied to Equation 7.9c.

7.3.2 Wind-axis orientation equations (kinematic equations)

With the appropriate substitutions,

$$\begin{Bmatrix} \dot{\mu} \\ \dot{\gamma} \\ \dot{\chi} \end{Bmatrix} = \begin{bmatrix} 1 & \sin \mu \tan \gamma & \cos \mu \tan \gamma \\ 0 & \cos \mu & -\sin \mu \\ 0 & \sin \mu \sec \gamma & \cos \mu \sec \gamma \end{bmatrix} \begin{Bmatrix} p_W \\ q_W \\ r_W \end{Bmatrix}$$

Equations of Motion

The previously derived results for p_W, q_W, and r_W (Equations 7.6 and 7.7) may be used. In scalar form,

$$\dot{\mu} = p_W + (q_W \sin\mu + r_W \cos\mu)\tan\gamma \tag{7.10a}$$

$$\dot{\gamma} = q_W \cos\mu - r_W \sin\mu \tag{7.10b}$$

$$\dot{\chi} = (q_W \sin\mu + r_W \cos\mu)\sec\gamma \tag{7.10c}$$

7.3.3 Wind-axis navigation equations

This proceeds exactly like the body-axis equations, only with simpler results (flat Earth):

$$\begin{Bmatrix} \dot{x}_E \\ \dot{y}_E \\ \dot{z}_E \end{Bmatrix} = T_{H,W} \begin{Bmatrix} V \\ 0 \\ 0 \end{Bmatrix}$$

$$\dot{x}_E = V \cos\gamma \cos\chi \tag{7.11a}$$

$$\dot{y}_E = V \cos\gamma \sin\chi \tag{7.11b}$$

$$\dot{z}_E = -V \sin\gamma \tag{7.11c}$$

7.4 Steady-State Solutions

7.4.1 General

The equations of motion developed above are all nonlinear first order ordinary differential equations. In addition they are highly coupled, that is, each differential equation depends upon variables which are described by other differential equations. Analytical solutions to such equations are not known. We may, however, gain some insight into the equations of motion by examining steady-state solutions, which then are algebraic equations. The term 'steady-state' as applied to the equations of motion is not quite accurate. If taken literally it means that the time rate of change of each state is zero. Considering just the navigation equations it is clear that this makes for an uninteresting case, since it implies the aircraft is parked somewhere. As the term is normally used 'steady-state' should be taken to mean the linear and angular motion variables are constant. That is,

$$\dot{u} = \dot{v} = \dot{w} = \dot{p} = \dot{q} = \dot{r} = 0 \tag{7.12}$$

and

$$\dot{V} = \dot{\alpha} = \dot{\beta} = p_W = q_W = r_W = 0 \tag{7.13}$$

In addition we will require that all the controls be constant, or

$$\delta_\ell, \delta_m, \delta_n, \delta_T : Constant \tag{7.14}$$

These conditions alone can tell us a lot about the force and moment equations. The fact that all 12 equations are coupled has implications for the other equations as well, which we will examine.

Forces and moments

Consider the aerodynamic and thrust forces and moments. With the exception of altitude (density) dependencies each is dependent only on independent variables that are constant in steady-state flight. If we restrict our analysis to flight at nearly constant altitude then we may conclude that all these forces and moments are constant as well:

$$D, C, L : Constant$$
$$X, Y, Z : Constant$$
$$L, M, N : Constant$$
$$T, M_T : Constant \tag{7.15}$$

Euler angles

In steady-state flight the body axis force equations become

$$X + T\cos\epsilon_T - mg\sin\theta + m(rv - qw) = 0 \tag{7.16a}$$
$$Y + mg\sin\phi\cos\theta + m(pw - ru) = 0 \tag{7.16b}$$
$$Z + T\sin\epsilon_T + mg\cos\phi\cos\theta + m(qu - pv) = 0 \tag{7.16c}$$

In these equations the only terms that have not been assumed or shown to be constant are those involving θ and ϕ. It is easy to show that they must be constant as well. Aside from the mathematics, the physical reason is that if θ or ϕ change then the orientation of the gravity vector with respect to the aircraft changes and the forces are no longer balanced. The Euler angle ψ does not appear in the equations, but with θ, ϕ, q, and r all constant, the kinematic equations (Equation 7.4) tell us that $\dot{\psi}$ must be constant as well. We therefore add to our list of conditions for steady-state flight,

$$\dot{\phi} = 0 \tag{7.17a}$$
$$\dot{\theta} = 0 \tag{7.17b}$$
$$\dot{\psi} = Constant \tag{7.17c}$$

By similar reasoning with the wind-axis force equations we arrive at

$$\dot{\mu} = 0 \tag{7.18a}$$
$$\dot{\gamma} = 0 \tag{7.18b}$$
$$\dot{\chi} = Constant \tag{7.18c}$$

With only one of the three Euler angles permitted to vary with time, the relationships between the body- and wind-axis rates is easily evaluated:

$$p = -\dot{\psi}\sin\theta = Constant \tag{7.19a}$$
$$q = \dot{\psi}\sin\phi\cos\theta = Constant \tag{7.19b}$$
$$r = \dot{\psi}\cos\phi\cos\theta = Constant \tag{7.19c}$$

Equations of Motion

$$p_W = -\dot{\chi}\sin\gamma = \text{Constant} \tag{7.20a}$$

$$q_W = \dot{\chi}\sin\mu\cos\gamma = \text{Constant} \tag{7.20b}$$

$$r_W = \dot{\chi}\cos\mu\cos\gamma = \text{Constant} \tag{7.20c}$$

7.4.2 Special cases

Straight flight

Straight flight means the aircraft is not turning relative to the Earth. This condition is satisfied only if $\dot{\psi} = 0$ and $\dot{\chi} = 0$. From our discussion above, this implies that $p = q = r = 0$ and $p_W = q_W = r_W = 0$.

The longitudinal equations (body-axis moment, wind-axis force) in steady, straight flight reduce to

$$M + M_T = 0$$

$$-D + T\cos\beta\cos(\epsilon_T - \alpha) = mg\sin\gamma$$

$$L - T\sin(\epsilon_T - \alpha) = mg\cos\mu\cos\gamma$$

Throughout the discussion in the text it is assumed there are no externally applied asymmetric moments. This would be the case, for example, if a multi-engine aircraft experienced an engine failure on one wing or the other. Such abnormal flight conditions require additional analysis.

In the lateral–directional moment equations (body axes) we have $I_{zz}L + I_{xz}N = 0$ and $I_{xz}L + I_{xx}N = 0$. From these relationships it is easy to show that $L = N = 0$. Also, the body-axis side force equation yields $Y = -mg\sin\phi\cos\theta$. Thus, the general requirements and conditions for straight flight are

$$C_m(M, \alpha, \delta_m) + C_{m_T}(\hat{V}, \delta_T) = 0 \tag{7.21a}$$

$$C_D(M, \alpha, \delta_m) - C_T(\hat{V}, \delta_T)\cos\beta\cos(\epsilon_T - \alpha) + mg\sin\gamma/\bar{q}S = 0 \tag{7.21b}$$

$$C_L(M, \alpha, \delta_m) - C_T(\hat{V}, \delta_T)\sin(\epsilon_T - \alpha) - mg\cos\mu\cos\gamma/\bar{q}S = 0 \tag{7.21c}$$

and

$$C_\ell(\beta, \delta_\ell, \delta_n) = 0 \tag{7.22a}$$

$$C_n(\beta, \delta_\ell, \delta_n) = 0 \tag{7.22b}$$

$$C_Y(\beta, \delta_n) + mg\sin\phi\cos\theta/\bar{q}S = 0 \tag{7.22c}$$

Equations 7.21 and 7.22 are six equations that may be satisfied for various combinations of the ten independent variables V, α, β, ϕ, θ, δ_ℓ, δ_m, δ_n, δ_T, and h. Note that V and h determine M and $\bar{q}$, and that when α, β, ϕ, and θ are specified then μ and γ are determined through $T_{H,W} = T_{H,B}T_{B,W}$.

Symmetric flight

Symmetric flight means the aircraft has no sideslip, or

$$\beta = v = 0 \qquad (7.23)$$

Balanced flight

Balanced flight is related to and often confused with symmetric flight. Almost every airplane built has a so-called 'slip indicator' (the slip indicator is normally combined as a single instrument with a gyroscopic turn indicator, all of which is referred to as the turn-and-slip indicator). It consists of a ball in a curved tube mounted on the pilot's intrument panel. The tube is filled with a fluid that dampens the motion of the ball, and prevents it from sticking so the ball is free to move within the tube, as depicted in Figure 7.1.

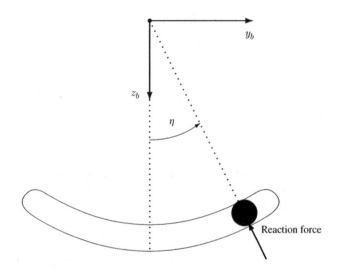

Figure 7.1 Aircraft slip indicator.

Balanced flight is normally taken to mean flight in which the ball in the slip indicator is centered. The confusion between symmetric and balanced flight comes from the fact that, despite its name, the slip indicator does not in general measure sideslip. Under normal conditions a centered ball means there is very little sideslip, however.

To analyze the conditions of steady, balanced flight, assume the tube is a circular arc, and measure the displacement of the ball from the center of the arc by the angle η, positive in the sense shown. The instrument is mounted in the y–z plane of the body-fixed coordinate system as shown; assume the ball is always at the center of mass of the airplane.

With small thrust angle ($\epsilon_T \approx 0$) the aircraft force equations in body axes (Equations 7.2) are

$$\dot{u} = \frac{1}{m}(X + T) - g \sin\theta + (rv - qw)$$

Equations of Motion

$$\dot{v} = \frac{Y}{m} + g\sin\phi\cos\theta + (pw - ru)$$

$$\dot{w} = \frac{Z}{m} + g\cos\phi\cos\theta + (qu - pv)$$

The ball is a rigid body in steady flight, and it follows the same trajectory as the CG of the aircraft. Moreover, if a coordinate system for the ball is taken as parallel to the body-axis coordinate system, then the quantities $\dot{u}$, $\dot{v}$, $\dot{w}$, u, v, w, p, q, and r are the same for the ball as for the aircraft, since these accelerations and velocities are those of the coordinate system itself. Likewise, the angles ϕ and θ describe the orientation of the coordinate system, so they are the same for ball and aircraft.

With subscript 'b' to denote the ball, the force equations of the ball in body axes are:

$$\dot{u}_b = \frac{X_b}{m_b} - g\sin\theta_b + (r_b v_b - q_b w_b)$$

$$\dot{v}_b = \frac{Y_b}{m_b} + g\sin\phi_b\cos\theta_b + (p_b w_b - r_b u_b)$$

$$\dot{w}_b = \frac{Z_b}{m_b} + g\cos\phi_b\cos\theta_b + (q_b u_b - p_b v_b)$$

Therefore,

$$\frac{X_b}{m_b} = \dot{u}_b + g\sin\theta_b - (r_b v_b - q_b w_b) = \dot{u} + g\sin\theta - (rv - qw) = \frac{(X+T)}{m}$$

$$\frac{Y_b}{m_b} = \dot{v}_b - g\sin\phi_b\cos\theta_b - (p_b w_b - r_b u_b) = \dot{v} - g\sin\phi\cos\theta - (pw - ru) = \frac{Y}{m}$$

$$\frac{Z_b}{m_b} = \dot{w}_b - g\cos\phi_b\cos\theta_b - (q_b u_b - p_b v_b) = \dot{w} - g\cos\phi\cos\theta - (qu - pv) = \frac{Z}{m}$$

$$\frac{(X+T)}{m} = \frac{X_b}{m_b}$$

$$\frac{Y}{m} = \frac{Y_b}{m_b}$$

$$\frac{Z}{m} = \frac{Z_b}{m_b}$$

Since motion of the ball is constrained to the y–z plane, the X force equations are always satisfied. From the Y and Z force equations we arrive at

$$\frac{Y_b}{m_b} = \frac{Y}{m}, \quad \frac{Z_b}{m_b} = \frac{Z}{m} \Rightarrow \frac{Y_b}{Z_b} = \frac{Y}{Z}$$

Hence,

$$\tan\eta = \frac{Y_b}{Z_b} = \frac{Y}{Z}$$

From this we conclude that the slip indicator is actually a body-axis side-force indicator. The side force Y consists of aerodynamic and thrust forces, but does not include the component of weight. If there is no component of thrust in the y-body direction then the ball deflection is proportional to aerodynamic side force only. For our purposes, then,

$$Y = 0 \quad \text{(Balanced flight)} \tag{7.28}$$

The aerodynamic side force Y is typically modeled as

$$Y = \bar{q} S C_y(\beta, \hat{p}, \hat{r}, \delta_r)$$

The effect of β is usually the much larger than that of the other independent variables, so the ball actually does respond more to sideslip than to other effects. Moreover, in straight flight $p = r = 0$, and often the effect of δ_r may be neglected. Under these conditions and assumptions the slip indicator is appropriately named.

Straight, symmetric flight

We combine the requirements for straight and symmetric flight (with symmetric thrust), for which Equations 7.22 become

$$C_\ell(\delta_\ell, \delta_n) = 0 \tag{7.29a}$$

$$C_n(\delta_\ell, \delta_n) = 0 \tag{7.29b}$$

$$C_Y(\delta_n) + mg \sin\phi \cos\theta / \bar{q} S = 0 \tag{7.29c}$$

From these equations the implication is that given some δ_n which satisfies the side-force requirement, there is some rolling moment controller δ_ℓ setting that will simultaneously satisfy both the rolling and yawing moment equations. Even if such a combination of controls existed (and that is unlikely) it is clear that we are talking about a very particular case. In general, then, we will require $\delta_\ell = 0$, $\delta_n = 0$, and $Y = 0$.

Since we now require $Y = 0$, we have $mg \sin\phi \cos\theta = 0$ which, for $\theta \neq \pm 90$ deg, means $\phi = 0$. Applying $\beta = 0$ and $\phi = 0$ to $T_{H,W} = T_{H,B} T_{B,W}$, and equating the (3,2) entries on the left and right sides yields $\sin\mu \cos\gamma = 0$, which, for $\gamma \neq \pm 90$ deg, means $\mu = 0$.

Summarizing the conditions and requirements for straight, symmetric flight,

$$C_\ell = C_n = C_Y = 0$$

$$\beta = v = \phi = \mu = \delta_\ell = \delta_n = 0$$

Turning, balanced flight

With $\dot{v} = 0$ and $Y = 0$, the body-axis side-force equation becomes

$$g \sin\phi \cos\theta = ru - pw$$

Equations of Motion

Using the kinematic requirements for steady flight, Equations 7.19a and 7.19c,

$$g \sin \phi \cos \theta = \dot{\psi}(u \cos \phi \cos \theta + w \sin \theta)$$

To get manageable relationships, we note that in nearly level flight at low angles of attack and moderate angles of bank,

$$u \cos \phi \cos \theta \gg w \sin \theta$$

and

$$u \approx V$$

we have

$$\tan \phi \approx \frac{\dot{\psi} V}{g} \tag{7.30}$$

Continuing with the assumption of small α, z_B and z_W are nearly coincident so that $L = -Z$. The definition of *load factor* (n) is

$$n = \frac{L}{W} = \frac{L}{mg} \tag{7.31}$$

Then, using the body-axis Z-force equation, with $L = -Z$,

$$L = mg \cos \phi \cos \theta + m(qu - pv)$$

We now use Equations 7.19a and 7.19b, and make similar assumptions regarding the angles, yielding

$$qu - pv \approx \dot{\psi} V \sin \phi$$

$$L \approx mg \cos \phi + m \dot{\psi} V \sin \phi$$

Finally, using Equation 7.30 ($\dot{\psi} V \approx g \tan \phi$), we have

$$L \approx mg \cos \phi + mg \tan \phi \sin \phi = mg \left(\cos \phi + \frac{\sin^2 \phi}{\cos \phi} \right) = mg \sec \phi$$

This yields the approximate relationship between load factor and bank angle,

$$n \approx \sec \phi \quad \text{(Many assumptions)} \tag{7.32}$$

Level flight

Level flight simply means the aircraft is neither climbing nor descending, or $\gamma = 0$. Level flight by itself has little effect on the equations of motion, save to eliminate one variable and to ensure that the effects of changing altitude may be safely neglected. If level flight is not assumed, then the effects of altitude changes on the forces and moments (change of density, or ground effect) may require consideration.

7.4.3 The trim problem

To a pilot the term 'trim' usually means the act of adjusting trim tabs or artificial feel systems in order to eliminate the forces on the control stick (or pedals) he is required to maintain. The other meaning of 'trim' is like that in sailing in which one trims the sails so that all the forces and moments are balanced. This trim problem is the problem of determining the values of those non-zero variables such that a specified steady-state flight condition results. For example, if we specify straight steady-state flight we require that

$$C_\ell(\beta, \delta_\ell, \delta_n) = 0$$
$$C_n(\beta, \delta_\ell, \delta_n) = 0$$
$$C_Y(\beta, \delta_n) + mg \sin\phi \cos\theta / \bar{q}S = 0$$

These three equations must be satisfied simultaneously by some combination of sideslip, rolling and yawing controls, bank angle, pitch angle, velocity, and altitude. Most flight dynamics problems begin with a specification of some velocity and altitude of interest, so we may expect Mach and dynamic pressure to be given. For the rolling and yawing moment equations we may therefore specify one of the variables sideslip, rolling and yawing controls and solve for the other two. The remaining question is to find a bank angle and pitch angle that satisfies the side-force equation. One of these (normally pitch angle) may be specified. Because the functional dependencies of the force and moment coefficients are usually quite complex, the trim problem can not be solved analytically. With the linear assumption, however, closed form solutions are possible. Using the same example the rolling moment coefficient is approximated by

$$C_\ell(\beta, \hat{p}, \hat{r}, \delta_\ell, \delta_n) = C_{\ell_{Ref}} + C_{\ell_\beta}\beta + C_{\ell_p}\hat{p} + C_{\ell_r}\hat{r} + C_{\ell_{\delta_\ell}}\delta_\ell + C_{\ell_{\delta_n}}\delta_n$$

Now, if we take the reference condition to be that in which all independent variables are zero, and assume this approximation is good for a suitable range of the variables about that reference condition, then (with $p = 0$ and $r = 0$, due to straight flight):

$$C_\ell = C_{\ell_\beta}\beta + C_{\ell_{\delta_\ell}}\delta_\ell + C_{\ell_{\delta_n}}\delta_n$$

In a similar fashion we have

$$C_n = C_{n_\beta}\beta + C_{n_{\delta_\ell}}\delta_\ell + C_{n_{\delta_n}}\delta_n$$
$$C_Y = C_{Y_\beta}\beta + C_{Y_{\delta_n}}\delta_n$$

This part of the trim problem becomes

$$C_{\ell_\beta}\beta + C_{\ell_{\delta_\ell}}\delta_\ell + C_{\ell_{\delta_n}}\delta_n = 0$$
$$C_{n_\beta}\beta + C_{n_{\delta_\ell}}\delta_\ell + C_{n_{\delta_n}}\delta_n = 0$$
$$C_{Y_\beta}\beta + C_{Y_{\delta_n}}\delta_n + mg \sin\phi \cos\theta / \bar{q}S = 0$$

Equations of Motion

For example, if we now assume that ϕ, θ, and $\bar{q}$ are given, then we need to solve for β, δ_ℓ, and δ_n in the linear system

$$\begin{bmatrix} C_{\ell\beta} & C_{\ell\delta_\ell} & C_{\ell\delta_n} \\ C_{n\beta} & C_{n\delta_\ell} & C_{n\delta_n} \\ C_{Y\beta} & 0 & C_{Y\delta_n} \end{bmatrix} \begin{Bmatrix} \beta \\ \delta_\ell \\ \delta_n \end{Bmatrix} = \begin{Bmatrix} 0 \\ 0 \\ -mg\sin\phi\cos\theta/\bar{q}S \end{Bmatrix}$$

Problems

1. Define $q_0 \ldots q_3$ as the Euler parameters in the transformation from body-fixed axes to the local horizontal reference frame. Using these Euler parameters instead of Euler angles, and otherwise using the customary notation for body-axis forces and velocities, evaluate the force equation

$$\dot{u} = \frac{X}{m} + \cdots$$

2. See Table 7.1 and 7.2. Calculate the angle-of-attack and elevator setting required for the airplane to maintain steady, balanced, level flight while turning at $\phi = -30°$ of bank at the given altitude and Mach number. Ignore thrust effects and assume linear force and moment coefficient relationships:

$$C_L = C_{L_{Ref}} + C_{L_\alpha}\alpha + C_{L_{\dot\alpha}}\hat{\dot\alpha} + C_{L_q}\hat{q} + C_{L_M}\Delta M + C_{L_{\delta_e}}\Delta\delta_e$$

$$C_m = C_{m_\alpha}\alpha + C_{m_{\dot\alpha}}\hat{\dot\alpha} + C_{m_q}\hat{q} + C_{m_M}\Delta M + C_{m_{\delta_e}}\Delta\delta_e$$

Table 7.1 Twin-engine aircraft data.

Longitudinal		Lat–Dir	
$C_{L_{Ref}}$	0.40	C_{y_β}	−0.75
$C_{D_{Ref}}$	0.04	C_{ℓ_β}	−0.06
C_{L_α}	6.5	C_{n_β}	0.13
C_{D_α}	0.60	C_{ℓ_p}	−0.42
C_{m_α}	−0.72	C_{n_p}	−0.756
$C_{L_{\dot\alpha}}$	0.0	C_{ℓ_r}	0.04
$C_{m_{\dot\alpha}}$	−0.4	C_{n_r}	−0.16
C_{L_q}	0.0	$C_{\ell_{\delta_a}}$	0.06
C_{m_q}	−0.92	$C_{n_{\delta_a}}$	−0.06
C_{L_M}	0.0	$C_{y_{\delta_r}}$	0.16
C_{D_M}	−0.6	$C_{\ell_{\delta_r}}$	0.029
C_{m_M}	−0.60	$C_{n_{\delta_r}}$	−0.057
$C_{L_{\delta_e}}$	0.44		
$C_{m_{\delta_e}}$	−0.88		

Stability axes, Mach = 0.80, 40 000 ft.
From Nelson (1998).

Table 7.2 Twin-engine aircraft physical data.

Property	Value
W	38 200 lb
I_{xx}	118 773 slug · ft²
I_{yy}	135 869 slug · ft²
I_{zz}	243 504 slug · ft²
I_{xz}	5061 slug · ft²
S	542.5 ft²
b	53.75 ft
$\bar{c}$	10.93 ft

From Nelson (1998).

Note that all the independent variables should be taken as Δ from the reference value used to linearize the equations. For the given relationships the reference values of α, $\dot{\alpha}$, and q are all zero.

3. Steady, straight flight that is not symmetric is referred to as a *steady-heading sideslip*. Use the provided data and assume linear force and moment coefficient relationships:

$$C_\ell = C_{\ell_\beta}\beta + C_{\ell_p}\hat{p} + C_{\ell_r}\hat{r} + C_{\ell_{\delta_a}}\delta_a + C_{\ell_{\delta_r}}\delta_r$$
$$C_n = C_{n_\beta}\beta + C_{n_p}\hat{p} + C_{n_r}\hat{r} + C_{n_{\delta_a}}\delta_a + C_{n_{\delta_r}}\delta_r$$
$$C_Y = C_{Y_\beta}\beta + C_{Y_p}\hat{p} + C_{Y_r}\hat{r} + C_{Y_{\delta_r}}\delta_r$$

Note that all the independent variables should be taken as Δ from the reference value used to linearize the equations. For the given relationships the reference values are all zero.

(a) Let $\theta = 0$ and $\beta = 10$ deg be given. Calculate the bank angle and control inputs (ϕ, δ_a, and δ_r) in degrees that are required to maintain steady-heading sideslip.

(b) Let $\theta = 10$ deg and $\delta_r = \delta_{r_{Max}} = 20$ deg be given. Calculate the bank angle, sideslip, and lateral control input (ϕ, β, and δ_a) in degrees that are required to maintain steady-heading sideslip.

4. Assume that the right engine of a two-engined aircraft has failed, and the left engine alone is capable of maintaining level flight. Asymmetric thrust will create a yawing moment N_T that is in addition to the aerodynamic yawing moment. Let θ be given. Assume the aerodynamic force and moment dependencies given in problem 3 are valid for this aircraft. Is it generally possible for the aircraft to maintain steady, straight, symmetric flight with some combination of rolling and yawing controls? Note that the presence of asymmetric thrust changes the conclusions in the section on straight, symmetric flight (in particular, Equation 7.29b). Explain fully.

5. A multi-engine aircraft suffers an engine failure just after takeoff. In problem 4 we showed how the aircraft could be flown in order to maintain steady, straight, symmetric flight. Federal Aviation Regulations restrict the amount of bank angle that may be used in this flight condition to $\phi = \pm 5$ deg. Assume N_T is known, $\theta = 0$, and state

any other assumptions made. Show how you would determine the controls δ_ℓ and δ_n, plus the states ϕ and β, required to maintain steady, straight flight with minimum sideslip, $|\beta|_{Min}$.

6. See Table 7.1 and Table 7.2. Data are valid for stability axes. Assume the aircraft is in steady level flight, with no bank angle, but with sideslip $\beta = -15°$ (called a *skidding turn*).

   Find the lateral–directional states and controls required to trim the aircraft in this steady flight condition. Express your answers in the English system and use degrees for angular measurements. Show all steps and clearly state all assumptions made.

Reference

Nelson, R.C. (1998) *Flight Stability and Automatic Control*, 2nd edn, WCB/McGraw-Hill.

8
Linearization

8.1 General

We have derived 12 nonlinear, coupled, first order, ordinary differential equations that describe the motion of rigid aircraft in a stationary atmosphere over a flat Earth: the equations of motion. Analytical solutions to the equations of motion are obviously not forthcoming, so other means of solving them must be sought.

One such means is through numerical integration. There are several algorithms available that will allow the differential equations to be propagated forward in discrete time steps. At the beginning of each time step the entire right-hand side of each equation $\dot{x} = \ldots$ is evaluated, yielding the rate of change of x at that instant. Then, loosely, dx/dt is replaced by $\Delta x/\Delta t$ and the change Δx is approximated over the interval Δt. With sufficiently fast computers this can be done in real-time, and is the basis for flight simulation. Numerical integration will generate time-histories of the aircraft motion in response to arbitrary initial conditions and control inputs. These time-histories may then be analyzed to characterize the response in familiar terms such as its frequency and damping.

Alternatively, one may consider small inputs and variations in initial conditions applied at and about some reference (usually steady) flight condition. The advantage to this approach is that for suitably small regions about the steady condition, all the dependencies may be considered linear. This will result in 12 *linear* (though still *coupled*) ordinary differential equations for which analytical solutions are available. The process is called linearization of the equations.

Numerical integration does not require special inititial conditions or control inputs; it simply approximates what happens in real life. Linearization brings with it the assumption that all ensuing motion will be 'close' to the reference steady-state flight condition. (This assumption is sometimes wantonly disregarded and the linearized equations are treated as if they were 'the' equations of motion.)

Because linearization requires us to stay close to the steady flight condition, it is generally reasonable to neglect altitude effects in the equations of motion. Therefore, none of the three navigation equations need be considered, nor does the kinematic equation for $\dot{\psi}$ (or $\dot{\chi}$). We will therefore focus on the six motion variables and the remaining two Euler angles for linear analysis.

8.2 Taylor Series

A familiar example of the linearization of a nonlinear ordinary differential equation involves a pendulum of length ℓ, for which motion about $\theta = 0$ is described by

$$m\ell\ddot{\theta} = -mg\sin\theta$$

This second-order equation may be replaced by two coupled first-order equations by introducing the angular velocity variable ω, whence

$$\dot{\theta} = \omega$$
$$\dot{\omega} = -g\sin\theta/\ell$$

The only nonlinearity is in the term $\sin\theta$, which is normally linearized by applying the small angle approximation, $\sin\theta \approx \theta$, so

$$\dot{\theta} = \omega$$
$$\dot{\omega} = -g\theta/\ell$$

More formally, such results as the small angle approximation may be obtained from a power series representation of a function by retaining only terms through the first power of the variables. Thus the standard series expansion of $\sin\theta$ about $\theta = 0$ is

$$\sin\theta = \theta - \theta^3/3! + \theta^5/5! - \cdots$$
$$\sin\theta \approx \theta$$

In order to deal with functions of several variables it is convenient to use for the series expansion the *Taylor series* representations of the functions. For a function of a single variable, $f(x)$, about some reference value x_{Ref} we have

$$f(x) = f(x_{Ref}) + \left.\frac{df(x)}{dx}\right|_{Ref}\Delta x + \left.\frac{d^2 f(x)}{dx^2}\right|_{Ref}\frac{\Delta x^2}{2!} + \left.\frac{d^3 f(x)}{dx^3}\right|_{Ref}\frac{\Delta x^3}{3!} + \cdots$$

In this expression, $x = x_{Ref} + \Delta x$. So for example, $f(x) = \sin x$ about $x_{Ref} = 0$ yields

$$\sin x = \sin x_{Ref} + \cos x_{Ref} x - \sin x_{Ref} x^2/2! - \cos x_{Ref} x^3/3! + \cdots$$
$$= x - x^3/3! + x^5/5! - \cdots$$

Or, through the first-order term, $\sin x \approx x$.

A Taylor series for a function of n independent variables, $f(x_1, x_2, \ldots, x_n)$ about reference conditions $(x_1, x_2, \ldots, x_n)_{Ref}$, through the first-order terms is:

$$f(x_1, x_2, \ldots, x_n) \approx f(x_1, x_2, \ldots, x_n)_{Ref}$$
$$+ \left.\frac{\partial f}{\partial x_1}\right|_{Ref}\Delta x_1 + \left.\frac{\partial f}{\partial x_2}\right|_{Ref}\Delta x_2 + \cdots \quad (8.1)$$
$$+ \left.\frac{\partial f}{\partial x_n}\right|_{Ref}\Delta x_n + \overset{0}{\cancel{H.O.T.}}$$

8.3 Nonlinear Ordinary Differential Equations

To be as general as possible we will define the function to be linearized by moving everything to the left-hand side of the equation and equating it to zero. For example, the body axis kinematic equation for bank angle will be defined as

$$f_\phi(\dot{\phi}, \phi, \theta, p, q, r) \equiv \dot{\phi} - p - (q \sin\phi + r \cos\phi) \tan\theta \equiv 0$$

There are two advantages to this approach. First, the function evaluates to zero for any reference condition, and therefore the first term in the Taylor series will always vanish. Second, even though we have solved the equations of motion for the explicit derivatives of the variables, some of the forces and moments may depend on these derivatives. By defining the functions as shown, all occurences of such terms will be accounted for when derivatives are taken.

The reference conditions we will use will be those previously discussed in Section 7.4 for steady flight. Mathematically this is not necessary, but for our purposes it is. In any event the meaning of terms such as $\Delta \dot{x}$ should always be construed to mean $\Delta(dx/dt)$ and not $d(\Delta x)/dt$ or the reference conditions of $\dot{x}$ will be lost, since

$$d(\Delta x)/dt = d(x - x_{Ref})/dt = dx/dt$$

and on the other hand,

$$\Delta(dx/dt) = dx/dt - (dx/dt)_{Ref}$$

Of course, if $(dx/dt)_{Ref} = 0$ the two are the same.

8.4 Systems of Equations

The formal procedure for dealing with several equations in several variables is as follows. First, three vectors $\mathbf{x}$, $\dot{\mathbf{x}}$, and $\mathbf{u}$ are defined. The $\mathbf{x}$ vector represents each of the variables that appear as derivatives, the $\dot{\mathbf{x}}$ vector represents each of those variables as a derivative, and the $\mathbf{u}$ vector represents all of the controls. If n state variables and m controls are being considered, then $\mathbf{x}$ and $\dot{\mathbf{x}}$ are n-vectors and $\mathbf{u}$ is an m-vector. There will be n ordinary differential equations which are placed in the form $f_i(\dot{\mathbf{x}}, \mathbf{x}, \mathbf{u}) = 0, i = 1 \ldots n$. A vector-valued function $\mathbf{f}(\dot{\mathbf{x}}, \mathbf{x}, \mathbf{u})$ is formed by considering the n functions thus defined as a vector. We then write the Taylor series of $\mathbf{f}(\dot{\mathbf{x}}, \mathbf{x}, \mathbf{u})$ through the first-order terms,

$$\mathbf{f}(\dot{\mathbf{x}}, \mathbf{x}, \mathbf{u}) = \left.\frac{\partial \mathbf{f}}{\partial \dot{\mathbf{x}}}\right|_{Ref} \Delta \dot{\mathbf{x}} + \left.\frac{\partial \mathbf{f}}{\partial \mathbf{x}}\right|_{Ref} \Delta \mathbf{x} + \left.\frac{\partial \mathbf{f}}{\partial \mathbf{u}}\right|_{Ref} \Delta \mathbf{u} = 0$$

Here we have used $\mathbf{f}(\dot{\mathbf{x}}, \mathbf{x}, \mathbf{u}) = 0$ and $\mathbf{f}(\dot{\mathbf{x}}, \mathbf{x}, \mathbf{u})_{Ref} = 0$ by definition. The derivative of the n-vector of functions $\mathbf{f}(\mathbf{v})$ with respect to a p-vector $\mathbf{v}$ is a *Jacobian* and is defined as

$$\frac{\partial \mathbf{f}}{\partial \mathbf{v}} \equiv \begin{bmatrix} \partial f_1/\partial v_1 & \partial f_1/\partial v_2 & \cdots & \partial f_1/\partial v_p \\ \partial f_2/\partial v_1 & \partial f_2/\partial v_2 & \cdots & \partial f_2/\partial v_p \\ \vdots & \vdots & \vdots & \vdots \\ \partial f_n/\partial v_1 & \partial f_n/\partial v_2 & \cdots & \partial f_n/\partial v_p \end{bmatrix}$$

Thus, $\partial \mathbf{f}/\partial \mathbf{x}$ and $\partial \mathbf{f}/\partial \dot{\mathbf{x}}$ are square $n \times n$ matrices and $\partial \mathbf{f}/\partial \mathbf{u}$ is an $n \times m$ matrix. The matrix $\partial \mathbf{f}/\partial \dot{\mathbf{x}}$ is generally non-singular. We then solve for the vector $\Delta \dot{\mathbf{x}}$,

$$\Delta \dot{\mathbf{x}} = -\left[\frac{\partial \mathbf{f}}{\partial \dot{\mathbf{x}}}\right]^{-1}_{Ref} \left\{ \left[\frac{\partial \mathbf{f}}{\partial \mathbf{x}}\right]_{Ref} \Delta \mathbf{x} + \left[\frac{\partial \mathbf{f}}{\partial \mathbf{u}}\right]_{Ref} \Delta \mathbf{u} \right\} \tag{8.2a}$$

This equation is often written

$$\Delta \dot{\mathbf{x}} = A \Delta \mathbf{x} + B \Delta \mathbf{u} \tag{8.2b}$$

with the obvious meaning of A and B.

To see how this works, consider our pendulum problem, to which we add an externally applied torque M to be used as a control,

$$\dot{\theta} = \omega$$

$$\dot{\omega} = -g \sin \theta / \ell + M$$

The two derivative terms are $\dot{\theta}$ and $\dot{\omega}$ ($n = 2$) and the single control is M ($m = 1$). We therefore define

$$\mathbf{x} = \begin{Bmatrix} x_1 \\ x_2 \end{Bmatrix} \equiv \begin{Bmatrix} \theta \\ \omega \end{Bmatrix}, \quad \mathbf{u} = \{u_1\} \equiv \{M\}$$

Two scalar functions are defined to constitute the vector-valued function $\mathbf{f}(\dot{\mathbf{x}}, \mathbf{x}, \mathbf{u})$:

$$\dot{\theta} - \omega = 0 \Rightarrow f_1(\dot{\mathbf{x}}, \mathbf{x}, \mathbf{u}) \equiv \dot{x}_1 - x_2 = 0$$

$$\dot{\omega} + g \sin \theta / \ell - M = 0 \Rightarrow f_2(\dot{\mathbf{x}}, \mathbf{x}, \mathbf{u}) \equiv \dot{x}_2 + g \sin x_1 / \ell - u_1 = 0$$

$$\mathbf{f}(\dot{\mathbf{x}}, \mathbf{x}, \mathbf{u}) \equiv \begin{Bmatrix} f_1(\dot{\mathbf{x}}, \mathbf{x}, \mathbf{u}) \\ f_2(\dot{\mathbf{x}}, \mathbf{x}, \mathbf{u}) \end{Bmatrix} = \begin{Bmatrix} \dot{x}_1 - x_2 \\ \dot{x}_2 + g \sin x_1 / \ell - u_1 \end{Bmatrix} = \begin{Bmatrix} 0 \\ 0 \end{Bmatrix}$$

The derivatives are calculated and then evaluated at reference conditions $\mathbf{x}_{Ref} = \dot{\mathbf{x}}_{Ref} = 0$ and $M_{Ref} = 0$ (from which also $\Delta \mathbf{x} = \mathbf{x}$, $\Delta \dot{\mathbf{x}} = \dot{\mathbf{x}}$, and $\Delta \mathbf{u} = \mathbf{u}$):

$$\frac{\partial \mathbf{f}}{\partial \dot{\mathbf{x}}} \equiv \begin{bmatrix} \partial f_1/\partial \dot{x}_1 & \partial f_1/\partial \dot{x}_2 \\ \partial f_2/\partial \dot{x}_1 & \partial f_2/\partial \dot{x}_2 \end{bmatrix} = \begin{bmatrix} 1 & 0 \\ 0 & 1 \end{bmatrix}$$

$$\frac{\partial \mathbf{f}}{\partial \mathbf{x}} \equiv \begin{bmatrix} \partial f_1/\partial x_1 & \partial f_1/\partial x_2 \\ \partial f_2/\partial x_1 & \partial f_2/\partial x_2 \end{bmatrix} = \begin{bmatrix} 0 & -1 \\ g \cos x_1/\ell & 0 \end{bmatrix}, \quad \frac{\partial \mathbf{f}}{\partial \mathbf{x}}\bigg|_{Ref} = \begin{bmatrix} 0 & -1 \\ g/\ell & 0 \end{bmatrix}$$

$$\frac{\partial \mathbf{f}}{\partial \mathbf{u}} \equiv \begin{bmatrix} \partial f_1/\partial u_1 \\ \partial f_2/\partial u_1 \end{bmatrix} = \begin{bmatrix} 0 \\ -1 \end{bmatrix}$$

The final result becomes

$$\dot{\mathbf{x}} = \begin{bmatrix} 0 & 1 \\ -g/\ell & 0 \end{bmatrix} \mathbf{x} + \begin{bmatrix} 0 \\ 1 \end{bmatrix} \mathbf{u}$$

Linearization

This is identical to the two linearized scalar equations

$$\dot{\theta} = \omega$$
$$\dot{\omega} = -(g/\ell)\theta + M$$

The motivation for this approach proceeds from the similarity of the set of linear, ordinary differential equations in $\Delta\dot{\mathbf{x}} = A\Delta\mathbf{x} + B\Delta\mathbf{u}$ with its scalar counterpart with forcing function u, $\dot{x} = ax + bu$. Since solutions to the scalar equations are quite well known, it is reasonable to think that we can find similar solutions to the systems of equations.

The problem with this approach from our point of view is that we have eight equations with four controls to linearize, so $\partial \mathbf{f}/\partial \dot{\mathbf{x}}$ and $\partial \mathbf{f}/\partial \mathbf{x}$ will be 8×8 matrices and $\partial \mathbf{f}/\partial \mathbf{u}$ will be an 8×4 matrix. However, in the form we have derived the equations of motion, in which we have already solved for the explicit derivative term, the matrix $\partial \mathbf{f}/\partial \dot{\mathbf{x}}$ will be very nearly the identity matrix (non-unity on the diagonal and non-zero off-diagonal entries will come from force and moment dependencies on $\dot{\alpha}$, and possibly $\dot{\beta}$).

It is far simpler to linearize each equation as a scalar function of several variables and deal with $\dot{\alpha}$ and $\dot{\beta}$ dependencies as special cases. On the other hand the vector-matrix form will be very useful later, so after treating each equation as a scalar problem, we will assemble them into the form of $\Delta\dot{\mathbf{x}} = A\Delta\mathbf{x} + B\Delta\mathbf{u}$.

8.5 Examples

8.5.1 General

In order to proceed we need to specify two things:

1. The state and control variables to be used, and
2. The reference flight condition at which the equations are to be evaluated.

For purposes of discussion we will select the body axis velocities u, v, w, p, q, and r plus the two body axis Euler angles θ and ϕ for our states, and the four generic controls δ_T, δ_ℓ, δ_m, and δ_n.

Since we have picked a body-axis system it is important to state which one. The analysis is simplified somewhat by using stability axes. In that case, x_S is the projection of the velocity vector in the reference flight condition onto the plane of symmetry. Thus in the reference condition the angle between the velocity vector and the x-axis is zero, we have

$$\alpha_{Ref} = w_{Ref} = 0 \text{ (Stability axes)}$$

This is true for any reference condition if stability axes are chosen. For our reference flight condition we take steady, straight, symmetric flight. As a consequence we have for reference conditions

$$\dot{u}_{Ref} = \dot{v}_{Ref} = \dot{w}_{Ref} = \dot{p}_{Ref} = \dot{q}_{Ref} = \dot{r}_{Ref} = \dot{\theta}_{Ref} = \dot{\phi}_{Ref} = 0$$
$$v_{Ref} = \beta_{Ref} = p_{Ref} = q_{Ref} = r_{Ref} = \phi_{Ref} = 0$$

Because $v_{Ref} = w_{Ref} = 0$, and because $V_{Ref} = \sqrt{u_{Ref}^2 + v_{Ref}^2 + w_{Ref}^2}$, we also have

$$u_{Ref} = V_{Ref}$$

We assume the speed and altitude at which the analysis is to be performed have been specified, as has the climb (or descent) angle γ_{Ref}. The use of stability axes in steady, straight, symmetric flight also means that

$$\theta_{Ref} = \gamma_{Ref}$$

In the force and moment dependencies we need the definitions

$$\bar{q} = \frac{1}{2}\rho V^2$$

$$\hat{p} = \frac{pb}{2V} \qquad \hat{r} = \frac{rb}{2V}$$

$$\hat{q} = \frac{q\bar{c}}{2V} \qquad \hat{\dot{\alpha}} = \frac{\dot{\alpha}\bar{c}}{2V}$$

Because our linear velocity variables are u, v, and w, we also need, for our force and moment dependencies,

$$V = \sqrt{u^2 + v^2 + w^2}$$

$$\alpha = \tan^{-1}(w/u)$$

$$\beta = \sin^{-1}(v/\sqrt{u^2 + v^2 + w^2})$$

In taking the partial derivatives we will apply the chain rule frequently, needing to evaluate in the end some partial derivatives repeatedly. Most are zero when evaluated in steady, straight, symmetric flight (see Appendix B.1). The derivatives are as follows:

$\left.\frac{\partial \rightarrow}{\partial \downarrow}\right|_{Ref}$	V	β	α	$\hat{p}$	$\hat{q}$	$\hat{r}$	$\hat{\dot{\alpha}}$	q
u	1	0	0	0	0	0	0	$\rho_{Ref} V_{Ref}$
v	0	$\frac{1}{V_{Ref}}$	0	0	0	0	0	0
w	0	0	$\frac{1}{V_{Ref}}$	0	0	0	0	0
p	0	0	0	$\frac{b}{2V_{Ref}}$	0	0	0	0
q	0	0	0	0	$\frac{\bar{c}}{2V_{Ref}}$	0	0	0
r	0	0	0	0	0	$\frac{b}{2V_{Ref}}$	0	0
$\dot{\alpha}$	0	0	0	0	0	0	$\frac{\bar{c}}{2V_{Ref}}$	0

We will encounter derivatives relating to thrust and velocity in the force equations. The basic relationship is

$$T = \bar{q} S C_T(\hat{V}, \delta_T)$$

Recall that the functional dependency of C_T on $\hat{V}$ was introduced to allow for the fact that thrust itself may sometimes be assumed to be invariant with airspeed

Linearization

(e.g., rockets and jets). An analogous result obtains for certain engines assumed to have constant thrust-horsepower, modeled as a constant product TV. The derivations for C_{T_V} for both cases are at Appendix B.1, and result in:

$$C_{T_V} = -2C_{T_{Ref}} \qquad \text{(Constant thrust)}$$

$$C_{T_V} = -3C_{T_{Ref}} \qquad \text{(Constant thrust-horsepower)}$$

We will meet a few other non-dimensional groupings in the process of linearizing the equations of motion:

$$\hat{t} \equiv \frac{t}{\bar{c}/(2V_{Ref})} \qquad \text{(Time)}$$

$$D(\cdot) \equiv \frac{\bar{c}}{2V_{Ref}} \frac{d(\cdot)}{dt} \qquad \text{(Differentiation)}$$

$$\hat{m} \equiv \frac{m}{\rho S \bar{c}/2} \qquad \text{(Mass)}$$

$$\hat{I}_{yy} \equiv \frac{I_{yy}}{\rho S (\bar{c}/2)^3} \qquad \text{(Moments of inertia)}$$

$$\hat{I}_{xx} \equiv \frac{I_{xx}}{\rho S (b/2)^3}$$

$$\hat{I}_{zz} \equiv \frac{I_{zz}}{\rho S (b/2)^3}$$

$$\hat{I}_{xz} \equiv \frac{I_{xz}}{\rho S (b/2)^3}$$

$$\mathcal{A} \equiv b/\bar{c} \qquad \text{(Aspect ratio)}$$

We will adopt the following convention to represent partial derivatives of forces and moments with respect to their independent variables, evaluated at reference flight conditions. If X is any force or moment that is a function of y, then:

$$X_y \equiv \left.\frac{\partial X}{\partial y}\right|_{Ref}$$

This notation is by no means standard; see Section 8.6 'Customs and Conventions'.

Additional information on the process of linearizing the aircraft equations of motion may be found in Appendix B.

8.5.2 A kinematic equation

To complete the linearization of the function previously defined,

$$f_\phi(\dot{\phi}, \phi, \theta, p, q, r) \equiv \dot{\phi} - p - (q \sin\phi + r \cos\phi) \tan\theta \equiv 0$$

The linearization proceeds using Equation 8.1 as

$$f_\phi = f_{\phi Ref} + \left.\frac{\partial f_\phi}{\partial \dot\phi}\right|_{Ref} \Delta\dot\phi + \left.\frac{\partial f_\phi}{\partial \phi}\right|_{Ref} \Delta\phi + \cdots + \left.\frac{\partial f_\phi}{\partial r}\right|_{Ref} \Delta r$$

Since $f_{\phi Ref} = 0$ by definition, and since all the Δs except $\Delta\theta$ are measured from zero reference conditions, we have the linearized equation,

$$\begin{aligned}f_\phi = {}& (0) + (1)\dot\phi + [(-q\cos\phi + r\sin\phi)\tan\theta]_{Ref}\phi \\ & + [(-q\sin\phi - r\cos\phi)\sec^2\theta]_{Ref}\Delta\theta \\ & - (1)p - (\sin\phi\tan\theta)_{Ref}q - (\cos\phi\tan\theta)_{Ref}r\end{aligned}$$

Evaluating the reference conditions yields

$$\dot\phi = p + r\tan\gamma_{Ref} \qquad (8.3)$$

In terms of the non-dimensional roll and yaw rates,

$$\dot\phi = \frac{2V_{Ref}}{b}\left(\frac{pb}{2V_{Ref}} + \frac{rb}{2V_{Ref}}\tan\gamma_{Ref}\right) = \frac{2V_{Ref}}{b}(\hat p + \hat r \tan\gamma_{Ref})$$

This may be written as

$$\left(\frac{d/dt}{2V_{Ref}/b}\right)\phi = \hat p + \hat r\tan\gamma_{Ref}$$

The term in parentheses represents one way to non-dimensionalize the operator d/dt. In fact, Etkin did exactly that (Etkin and Reid, 1995), dividing time by $b/2V_{Ref}$ in the lateral–directional and by $\bar c/2V_{Ref}$ in the longitudinal equations. Here we have adopted Etkin's earlier definition (Etkin, 1972) of non-dimensionalizing time by the divisor $\bar c/2V_{Ref}$. This introduces the aspect ratio $\mathcal{A}$ into the equation, since

$$\left(\frac{d/dt}{2V_{Ref}/b}\right)\phi = \left(\frac{d/dt}{2V_{Ref}/\bar c}\right)\left(\frac{b}{\bar c}\right)\phi = \mathcal{A}D(\phi)$$

The completely non-dimensional form of the bank angle equation is therefore

$$D(\phi) = \frac{1}{\mathcal{A}}(\hat p + \hat r\tan\gamma_{Ref}) \qquad (8.4)$$

8.5.3 A moment equation

For this example we consider the body-axis rolling equation,

$$\dot p = \frac{I_{zz}}{I_D}[L + I_{xz}pq - (I_{zz} - I_{yy})qr] + \frac{I_{xz}}{I_D}[N - I_{xz}qr - (I_{yy} - I_{xx})pq]$$

Part of the linearization is easy: the explicit p, q, and r dependencies pose no problem. The rolling and yawing moments are problematic. We need to get all the dependencies

Linearization 101

down to either states, controls, or constants. Since the aerodynamic data are normally available in coefficient form, we also need to relate to those. The rolling and yawing moments are:

$$L = \bar{q}SbC_\ell(\beta, \hat{p}, \hat{r}, \delta_\ell, \delta_n)$$
$$N = \bar{q}SbC_n(\beta, \hat{p}, \hat{r}, \delta_\ell, \delta_n)$$

The only states that do not appear are θ and ϕ, and the two controls on which we have dependencies are δ_ℓ and δ_n. We therefore define the function for linearization as

$$f_p(\dot{p}, u, v, w, p, q, r, \delta_\ell, \delta_n) \equiv I_D \dot{p} - I_{zz}L - I_{xz}N$$
$$+ I_{xz}(-I_{xx} + I_{yy} - I_{zz})pq + [I_{zz}(-I_{yy} + I_{zz}) - I_{xz}^2]qr = 0$$

Before beginning the derivatives, we note that dependencies on u and w appear only through the moments L and N (either through $\bar{q}$, β, $\hat{p}$, or $\hat{r}$). We have already determined derivatives of $\bar{q}$, β, $\hat{p}$, and $\hat{r}$ with respect to u and w, and in particular only $\partial \bar{q}/\partial u$ is non-zero in the reference flight condition. However, that term will be multiplied by $C_{\ell_{Ref}}$, which is zero ($L_{Ref} = 0$). We conclude therefore that when evaluated in steady, straight, symmetric flight,

$$\partial f_p/\partial u = \partial f_p/\partial w = 0$$

The other partial derivatives go as follows:

$$\left.\frac{\partial f_p}{\partial \dot{p}}\right|_{Ref} = I_D = I_{xx}I_{zz} - I_{xz}^2$$

$$\left.\frac{\partial f_p}{\partial v}\right|_{Ref} = -I_{zz}L_v - I_{xz}N_v$$

$$\left.\frac{\partial f_p}{\partial p}\right|_{Ref} = -I_{zz}L_p - I_{xz}N_p \qquad \text{(Using } q_{Ref} = 0\text{)}$$

$$\left.\frac{\partial f_p}{\partial q}\right|_{Ref} = 0 \qquad \text{(Using } p_{Ref} = r_{Ref} = 0\text{)}$$

$$\left.\frac{\partial f_p}{\partial r}\right|_{Ref} = -I_{zz}L_r - I_{xz}N_r \qquad \text{(Using } q_{Ref} = 0\text{)}$$

$$\left.\frac{\partial f_p}{\partial \delta_\ell}\right|_{Ref} = -I_{zz}L_{\delta_\ell} - I_{xz}N_{\delta_\ell}$$

$$\left.\frac{\partial f_p}{\partial \delta_n}\right|_{Ref} = -I_{zz}L_{\delta_n} - I_{xz}N_{\delta_n}$$

We may now write out the linearization of the rolling moment equation in dimensional form as follows:

$$\dot{p} = \frac{1}{I_D}[(I_{zz}L_v + I_{xz}N_v)v + (I_{zz}L_p + I_{xz}N_p)p$$
$$+ (I_{zz}L_r + I_{xz}N_r)r + (I_{zz}L_{\delta_\ell} + I_{xz}N_{\delta_\ell})\Delta\delta_\ell$$
$$+ (I_{zz}L_{\delta_n} + I_{xz}N_{\delta_n})\Delta\delta_n] \tag{8.5}$$

If the stability axes coincide with the principal axes this simplifies to

$$\dot{p} = \frac{1}{I_{xp}}(L_v v + L_p p + L_r r + L_{\delta_\ell}\Delta\delta_\ell + L_{\delta_n}\Delta\delta_n) \quad \text{(Principal axes)}$$

The partial derivatives of the moments are in dimensional form. We may at this point take the linearized equation as-is, and evaluate each of the factors on the right-hand side at some given altitude and speed. Alternatively we may re-write the equation using the non-dimensional derivatives. The required derivatives are straightforward; for example, we have:

$$L_v = \left.\frac{\partial L}{\partial v}\right|_{Ref}$$

$$= \left.\frac{\partial[\bar{q}SbC_\ell(\beta,\hat{p},\hat{r},\delta_\ell,\delta_n)]}{\partial v}\right|_{Ref}$$

$$= Sb\left(C_\ell\left.\frac{\partial\bar{q}}{\partial v}\right|_{Ref} + \bar{q}\left.\frac{\partial C_\ell}{\partial\beta}\frac{\partial\beta}{\partial v}\right|_{Ref}\right)_{Ref}$$

$$= \left(\frac{\bar{q}_{Ref}Sb}{V_{Ref}}\right)C_{\ell_\beta}$$

In L_v, $C_{\ell_{Ref}} = 0$ in the first term and $\partial\beta/\partial v|_{Ref} = 1/V_{Ref}$ in the second.

The end result of derivatives of L and N with respect to the states and controls is that the non-zero expressions are:

$\left.\frac{\partial\to}{\partial\downarrow}\right|_{Ref}$	L	N
v	$\left(\frac{\bar{q}_{Ref}Sb}{V_{Ref}}\right)C_{\ell_\beta}$	$\left(\frac{\bar{q}_{Ref}Sb}{V_{Ref}}\right)C_{n_\beta}$
p	$\left(\frac{\bar{q}_{Ref}Sb^2}{2V_{Ref}}\right)C_{\ell_p}$	$\left(\frac{\bar{q}_{Ref}Sb^2}{2V_{Ref}}\right)C_{n_p}$
r	$\left(\frac{\bar{q}_{Ref}Sb^2}{2V_{Ref}}\right)C_{\ell_r}$	$\left(\frac{\bar{q}_{Ref}Sb^2}{2V_{Ref}}\right)C_{n_r}$
δ_ℓ	$(\bar{q}_{Ref}Sb)C_{\ell_{\delta_\ell}}$	$(\bar{q}_{Ref}Sb)C_{n_{\delta_\ell}}$
δ_n	$(\bar{q}_{Ref}Sb)C_{\ell_{\delta_n}}$	$(\bar{q}_{Ref}Sb)C_{n_{\delta_n}}$

(8.6)

Linearization

When related to the non-dimensional coefficients we have (see Appendix B.2):

$$D(\hat{p}) = \frac{1}{A\hat{I}_D}[(\hat{I}_{zz}C_{\ell_\beta} + \hat{I}_{xz}C_{n_\beta})\beta + (\hat{I}_{zz}C_{\ell_p} + \hat{I}_{xz}C_{n_p})\hat{p}$$
$$+ (\hat{I}_{zz}C_{\ell_r} + \hat{I}_{xz}C_{n_r})\hat{r} + (\hat{I}_{zz}C_{\ell_{\delta_\ell}} + \hat{I}_{xz}C_{n_{\delta_\ell}})\Delta\delta_\ell$$
$$+ (\hat{I}_{zz}C_{\ell_{\delta_n}} + \hat{I}_{xz}C_{n_{\delta_n}})\Delta\delta_n] \quad (8.7)$$

In this expression $\mathcal{A}$ is the aspect ratio and $\hat{I}_D = \hat{I}_{xx}\hat{I}_{zz} - \hat{I}_{xz}^2$. If the stability axes coincide with the principal axes,

$$D(\hat{p}) = \frac{1}{A\hat{I}_{xp}}(C_{\ell_\beta}\beta + C_{\ell_p}\hat{p} + C_{\ell_r}\hat{r} + C_{\ell_{\delta_\ell}}\delta_\ell + C_{\ell_{\delta_n}}\delta_n) \quad \text{(Principal axes)}$$

8.5.4 A force equation

For this example we take the body-axis Z-force equation,

$$\dot{w} = \frac{1}{m}(Z + T\sin\epsilon_T) + g\cos\phi\cos\theta + qu - pv$$

Define the function

$$f_w(\dot{w}, u, v, w, p, q, r, \phi, \theta, \delta_T, \delta_m, \delta_n) = \dot{w} - \frac{1}{m}(Z + T\sin\epsilon_T)$$
$$- g\cos\phi\cos\theta - qu + pv$$

Since this is a longitudinal equation, the presence of dependencies on lateral–directional independent variables v, p, r, ϕ, and δ_n is not desired. The dependencies on v and ϕ are explicit. The dependencies on v, p, r, and δ_n arise through the force dependencies, which are

$$T = \bar{q}SC_T(\hat{V}, \delta_T)$$
$$Z = \bar{q}SC_Z$$

The latter is technically complicated by the mixed system of wind- and body-axis forces (see Equation 6.1). It may be shown, however, that linearizing the correct relationship for C_Z in a *symmetric* flight conditions, yields the same result as linearizing the simplified relationship (Equation 6.2) in the same flight condition. That is, the linearization of

$$C_Z = -C_D \sin\alpha \sec\beta - C_Y \sin\alpha \tan\beta - C_L \cos\alpha$$

leaves no terms involving lateral–directional independent variables (v, p, r, ϕ, δ_ℓ, or δ_n) when evaluated at the reference flight condition, and gets the same result as setting $\beta = 0$ and linearizing

$$C_Z = -C_D \sin\alpha - C_L \cos\alpha$$

In general, however, one should never apply reference conditions prior to taking the derivatives. Applying reference conditions makes the terms constants, whose derivatives are zero.

The remaining derivatives evaluate as follows:

$$\left.\frac{\partial f_w}{\partial \dot{w}}\right|_{Ref} = 1 - Z_{\dot{w}}/m$$

$$\left.\frac{\partial f_w}{\partial u}\right|_{Ref} = -(Z_u + T_u \sin \epsilon_T)/m \qquad \text{(Using } q_{Ref} = 0\text{)}$$

$$\left.\frac{\partial f_w}{\partial w}\right|_{Ref} = -Z_w/m \qquad \text{(Using } \partial V/\partial w = 0 \text{ in } C_T\text{)}$$

$$\left.\frac{\partial f_w}{\partial q}\right|_{Ref} = -Z_q/m - V_{Ref} \qquad \text{(Using } u_{Ref} = V_{Ref}\text{)}$$

$$\left.\frac{\partial f_w}{\partial \theta}\right|_{Ref} = g \sin \gamma_{Ref} \qquad \text{(Using } \cos \phi_{Ref} = 1\text{)}$$

$$\left.\frac{\partial f_w}{\partial \delta_T}\right|_{Ref} = -T_{\delta_T} \sin \epsilon_T / m$$

$$\left.\frac{\partial f_w}{\partial \delta_m}\right|_{Ref} = -Z_{\delta_m}/m$$

The dimensional form of the linearized equation is then

$$\dot{w} = \frac{1}{m - Z_{\dot{w}}}[(Z_u + T_u \sin \epsilon_T)\Delta u + Z_w w + (Z_q + mV_{Ref})q$$
$$- mg \sin \gamma_{Ref} \Delta \theta + T_{\delta_T} \sin \epsilon_T \Delta \delta_T + Z_{\delta_m} \Delta \delta_m] \qquad (8.8)$$

Non-dimensional derivatives are evaluated in Appendix B.3, and are:

$\left.\frac{\partial \vec{\rightarrow}}{\partial \downarrow}\right|_{Ref}$	Z	T
$\dot{w}$	$-\left(\frac{\bar{q}_{Ref} S \bar{c}}{2V_{Ref}^2}\right) C_{L\dot{\alpha}}$	0
u	$\left(\frac{\bar{q}_{Ref} S}{V_{Ref}}\right)(-2C_{L_{Ref}} - M_{Ref} C_{LM})$	$\left(\frac{\bar{q}_{Ref} S}{V_{Ref}}\right)(2C_{T_{Ref}} + C_{T_V})$
w	$-\left(\frac{\bar{q}_{Ref} S}{V_{Ref}}\right)(C_{D_{Ref}} + C_{L\alpha})$	0
q	$-\left(\frac{\bar{q}_{Ref} S \bar{c}}{2V_{Ref}}\right) C_{L_q}$	0
δ_m	$-(\bar{q}_{Ref} S) C_{L_{\delta_m}}$	0
δ_T	0	$(\bar{q}_{Ref} S) C_{T_{\delta_T}}$

(8.9)

The non-dimensional form of the equation (see Appendix B.4) becomes

$$D(\alpha) = \frac{1}{2\hat{m} + C_{L_{\dot{\alpha}}}}[(-2C_W \cos\gamma_{Ref} + C_{T_V}\sin\epsilon_T - M_{Ref}C_{L_M})\Delta\hat{V}$$
$$- (C_{D_{Ref}} + C_{L_\alpha})\alpha + (2\hat{m} - C_{L_q})\hat{q} - C_W \sin\gamma_{Ref}\Delta\theta$$
$$+ C_{T_{\delta_T}}\sin\epsilon_T \Delta\delta_T - C_{L_{\delta_m}}\Delta\delta_m] \qquad (8.10)$$

8.6 Customs and Conventions

8.6.1 Omission of Δ

In the linearized equations it is customary to drop the Δ symbol whether the reference conditions are zero or not. In general any *linear* differential equation appearing in flight dynamics will have been obtained through the linearization of more complicated equations, and it may be assumed that all the variables are small perturbations from some reference. After the equations are solved it is important to remember to add the reference values to the Δs obtained in order to get the actual values.

8.6.2 Dimensional derivatives

The convention we adopted for dimensional derivatives is consistent with Etkin's usage, in which

$$X_y \equiv \left.\frac{\partial X}{\partial y}\right|_{Ref}$$

Several authors use the same notation, but include in the definition division by mass or some moment of inertia. Thus one frequently sees, for example,

$$Z_w \equiv \left.\frac{\partial Z/\partial w}{m}\right|_{Ref}, \quad L_p \equiv \left.\frac{\partial L/\partial p}{I_{xx}}\right|_{Ref}, \quad \ldots$$

In the force equations, and in the pitching moment equation, such notation marginally simplifies the expressions. However, in the rolling and yawing equations, unless the analysis is performed in principal axes, the notation actually complicates the resulting expressions.

8.6.3 Added mass

Any time a force or moment is dependent upon a state-rate (such as $\dot\alpha$ or $\dot\beta$) the result is a modification to the mass or moment of inertia factor of the derivative term in the linearized equation. Thus we had the expression $(m - Z_{\dot w})\dot w$ instead of simply $m\dot w$ in the Z-force equation. Such terms as $Z_{\dot w}$ are often referred to as *added mass parameters* (they are usually negative) for obvious reasons. In ship dynamics several such terms arise, and the intuitive description usually offered is that the terms represent the mass of water being displaced by the ship's motion.

8.7 The Linear Equations

It is important to remember that the linearization was performed at a particular reference flight condition: steady, straight, and symmetric. If any other flight condition is to be analyzed the linearization will have to be performed over again.

8.7.1 Linear equations
Dimensional longitudinal equations

$$\dot{u} = \frac{1}{m}[(X_u + T_u \cos \epsilon_T)\Delta u + X_w w - mg \cos \gamma_{Ref} \Delta\theta$$
$$+ T_{\delta_T} \cos \epsilon_T \Delta\delta_T + X_{\delta_m} \Delta\delta_m] \tag{8.11a}$$

$$\dot{w} = \frac{1}{m - Z_{\dot{w}}}[(Z_u + T_u \sin \epsilon_T)\Delta u + Z_w w + (Z_q + mV_{Ref})q$$
$$- mg \sin \gamma_{Ref} \Delta\theta + T_{\delta_T} \sin \epsilon_T \Delta\delta_T + Z_{\delta_m} \Delta\delta_m] \tag{8.11b}$$

$$\dot{q} = \frac{1}{I_{yy}}\left\{\left[M_u + \frac{M_{\dot{w}}(Z_u + T_u \sin \epsilon_T)}{m - Z_{\dot{w}}}\right]\Delta u + \left[M_w + \frac{M_{\dot{w}} Z_w}{m - Z_{\dot{w}}}\right]w$$
$$+ \left[M_q + \frac{M_{\dot{w}}(Z_q + mV_{Ref})}{m - Z_{\dot{w}}}\right]q - \left[\frac{mg M_{\dot{w}} \sin \gamma_{Ref}}{m - Z_{\dot{w}}}\right]\Delta\theta$$
$$+ \left[\frac{M_{\dot{w}} T_{\delta_T} \sin \epsilon_T}{m - Z_{\dot{w}}}\right]\Delta\delta_T + \left[M_{\delta_m} + \frac{M_{\dot{w}} Z_{\delta_m}}{m - Z_{\dot{w}}}\right]\Delta\delta_m\right\} \tag{8.11c}$$

$$\dot{\theta} = q \tag{8.11d}$$

Dimensional lateral–directional equations

$$\dot{v} = \frac{1}{m}[Y_v v + Y_p p + (Y_r - mV_{Ref})r + mg \cos \gamma_{Ref} \phi + Y_{\delta_n} \delta_n] \tag{8.12a}$$

$$\dot{p} = \frac{1}{I_D}[(I_{zz} L_v + I_{xz} N_v)v + (I_{zz} L_p + I_{xz} N_p)p$$
$$+ (I_{zz} L_r + I_{xz} N_r)r + (I_{zz} L_{\delta_\ell} + I_{xz} N_{\delta_\ell})\delta_\ell$$
$$+ (I_{zz} L_{\delta_n} + I_{xz} N_{\delta_n})\delta_n] \tag{8.12b}$$

$$\dot{r} = \frac{1}{I_D}[(I_{xz} L_v + I_{xx} N_v)v + (I_{xz} L_p + I_{xx} N_p)p$$
$$+ (I_{xz} L_r + I_{xx} N_r)r + (I_{xz} L_{\delta_\ell} + I_{xx} N_{\delta_\ell})\delta_\ell$$
$$+ (I_{xz} L_{\delta_n} + I_{xx} N_{\delta_n})\delta_n] \tag{8.12c}$$

$$\dot{\phi} = p + r \tan \gamma_{Ref} \tag{8.12d}$$

Linearization

Non-dimensional longitudinal equations

$$D(\hat{V}) = \frac{1}{2\hat{m}}[(2C_W \sin\gamma_{Ref} - M_{Ref}C_{D_M} + C_{T_V}\cos\epsilon_T)\Delta\hat{V}$$
$$+ (C_{L_{Ref}} - C_{D_\alpha})\alpha - C_W \cos\gamma_{Ref}\Delta\theta$$
$$+ C_{T_{\delta_T}}\Delta\delta_T - C_{D_{\delta_m}}\Delta\delta_m] \tag{8.13a}$$

$$D(\alpha) = \frac{1}{2\hat{m} + C_{L_\alpha}}[(-2C_W \cos\gamma_{Ref} + C_{T_V}\sin\epsilon_T - M_{Ref}C_{L_M})\hat{V}$$
$$- (C_{D_{Ref}} + C_{L_\alpha})\alpha + (2\hat{m} - C_{L_q})\hat{q} - C_W \sin\gamma_{Ref}\Delta\theta$$
$$+ \sin\epsilon_T C_{T_{\delta_T}}\Delta\delta_T - C_{L_{\delta_m}}\Delta\delta_m] \tag{8.13b}$$

$$D(\hat{q}) = \frac{1}{\hat{I}_{yy}}\{[M_{Ref}C_{m_M}$$
$$+ \frac{C_{m_{\dot{\alpha}}}(-2C_W\cos\gamma_{Ref} + C_{T_V}\sin\epsilon_T - M_{Ref}C_{L_M})}{2\hat{m} + C_{L_{\dot{\alpha}}}}\Bigg]\Delta\hat{V}$$
$$+ \left[C_{m_\alpha} - \frac{C_{m_{\dot{\alpha}}}(C_{D_{Ref}} + C_{L_\alpha})}{2\hat{m} + C_{L_{\dot{\alpha}}}}\right]\alpha + \left[C_{m_q} + \frac{C_{m_{\dot{\alpha}}}(2\hat{m} - C_{L_q})}{2\hat{m} + C_{L_{\dot{\alpha}}}}\right]\hat{q}$$
$$- \frac{C_W C_{m_{\dot{\alpha}}}\sin\gamma_{Ref}}{2\hat{m} + C_{L_{\dot{\alpha}}}}\Delta\theta + \left[C_{m_{\delta_m}} - \frac{C_{m_{\dot{\alpha}}}C_{L_{\delta_m}}}{2\hat{m} + C_{L_{\dot{\alpha}}}}\right]\Delta\delta_m \tag{8.13c}$$

$$D(\theta) = \hat{q} \tag{8.13d}$$

Non-dimensional lateral–directional equations

$$D(\beta) = \frac{1}{2\hat{m}}[C_{Y_\beta}\beta + C_{Y_p}\hat{p} + (C_{Y_r} - 2\hat{m}/\mathcal{A})\hat{r}$$
$$+ C_W \cos\gamma_{Ref}\phi + C_{Y_{\delta_n}}\Delta\delta_n] \tag{8.14a}$$

$$D(\hat{p}) = \frac{1}{\mathcal{A}\hat{I}_D}[(\hat{I}_{zz}C_{\ell_\beta} + \hat{I}_{xz}C_{n_\beta})\beta + (\hat{I}_{zz}C_{\ell_p} + \hat{I}_{xz}C_{n_p})\hat{p}$$
$$+ (\hat{I}_{zz}C_{\ell_r} + \hat{I}_{xz}C_{n_r})\hat{r} + (\hat{I}_{zz}C_{\ell_{\delta_\ell}} + \hat{I}_{xz}C_{n_{\delta_\ell}})\Delta\delta_\ell$$
$$+ (\hat{I}_{zz}C_{\ell_{\delta_n}} + \hat{I}_{xz}C_{n_{\delta_n}})\Delta\delta_n] \tag{8.14b}$$

$$D(\hat{r}) = \frac{1}{\mathcal{A}\hat{I}_D}[(\hat{I}_{xz}C_{\ell_\beta} + \hat{I}_{xx}C_{n_\beta})\beta + (\hat{I}_{xz}C_{\ell_p} + \hat{I}_{xx}C_{n_p})\hat{p}$$
$$+ (\hat{I}_{xz}C_{\ell_r} + \hat{I}_{xx}C_{n_r})\hat{r} + (\hat{I}_{xz}C_{\ell_{\delta_\ell}} + \hat{I}_{xx}C_{n_{\delta_\ell}})\Delta\delta_\ell$$
$$+ (\hat{I}_{xz}C_{\ell_{\delta_n}} + \hat{I}_{xx}C_{n_{\delta_n}})\Delta\delta_n] \tag{8.14c}$$

$$D(\phi) = \frac{1}{A}(\hat{p} + \hat{r} \tan \gamma_{Ref}) \qquad (8.14d)$$

8.7.2 Matrix forms of the linear equations

Before we place these equations in the form $\Delta \dot{\mathbf{x}} = A \Delta \mathbf{x} + B \Delta \mathbf{u}$, we note that equations involving derivatives of longitudinal states are functions only of longitudinal states and controls, and that equations involving derivatives of lateral–directional states are functions only of lateral–directional states and controls. That is, for the dimensional equations, we have differential equations for $\dot{u}$, $\dot{w}$, $\dot{q}$, and $\dot{\theta}$ that are functions of the states u, w, q, θ, and the controls δ_T and δ_m. Similarly, the differential equations for $\dot{v}$, $\dot{p}$, $\dot{r}$, and $\dot{\phi}$ are functions of the states v, p, r, ϕ, and the controls δ_ℓ and δ_n. The same holds for the non-dimensional equations except the states are $\hat{V}$, α, $\hat{q}$, and θ for the longitudinal equations, and β, $\hat{p}$, $\hat{r}$, and ϕ for the lateral–directional equations.

Thus the longitudinal equations are *decoupled* from the lateral–directional equations, and vice versa. Instead of dealing with eight equations all at once, we may break them up into two sets of four equations each. Therefore define the longitudinal state and control vectors $\mathbf{x}_{Long}$ and $\mathbf{u}_{Long}$, and the lateral–directional state and control vectors $\mathbf{x}_{LD}$ and $\mathbf{u}_{LD}$:

$$\mathbf{x}_{Long} \equiv \begin{Bmatrix} u \\ w \\ q \\ \theta \end{Bmatrix}, \quad \mathbf{u}_{Long} \equiv \begin{Bmatrix} \delta_T \\ \delta_m \end{Bmatrix} \quad \mathbf{x}_{LD} \equiv \begin{Bmatrix} v \\ p \\ r \\ \phi \end{Bmatrix}, \quad \mathbf{u}_{LD} \equiv \begin{Bmatrix} \delta_\ell \\ \delta_n \end{Bmatrix}$$

Likewise we have the non-dimensional states $\hat{\mathbf{x}}_{Long}$ and $\hat{\mathbf{x}}_{LD}$:

$$\hat{\mathbf{x}}_{Long} \equiv \begin{Bmatrix} \hat{V} \\ \alpha \\ \hat{q} \\ \theta \end{Bmatrix} \quad \hat{\mathbf{x}}_{LD} \equiv \begin{Bmatrix} \beta \\ \hat{p} \\ \hat{r} \\ \phi \end{Bmatrix}$$

Control deflections are normally defined as non-dimensional quantities (throttle from zero to one, and radians for flapping surfaces, for example), so there is no need for separate definitions of $\hat{\mathbf{u}}_{Long}$ or $\hat{\mathbf{u}}_{LD}$.

Further assumptions

For our subsequent purposes it is sufficient to further simplify the linear equations of motion with four assumptions:

1. The aircraft is in steady, straight, symmetric, level flight (SSSLF), or $\gamma_{Ref} = 0$.
2. The engine thrust-line is aligned with x_B, so $\epsilon_T = 0$.
3. The body-fixed coordinate system is aligned with the principal axes, so $I_{xz} = 0$.
4. There are no aerodynamic dependencies on $\dot{\alpha}$ (or $\dot{w}$).

While these assumptions are not necessary, they do permit us to focus on the more significant effects typically seen in the study of aircraft dynamics and control.

Linearization

Dimensional longitudinal equations (SSSLF, $\epsilon_T = 0$)

$$\dot{\mathbf{x}}_{Long} = A_{Long}\mathbf{x}_{Long} + B_{Long}\mathbf{u}_{Long} \tag{8.15a}$$

$$\mathbf{x}_{Long} \equiv \begin{Bmatrix} u \\ w \\ q \\ \theta \end{Bmatrix}, \quad \mathbf{u}_{Long} \equiv \begin{Bmatrix} \delta_T \\ \delta_m \end{Bmatrix} \tag{8.15b}$$

$$A_{Long} = \begin{bmatrix} \frac{X_u + T_u}{m} & \frac{X_w}{m} & 0 & -g \\ \frac{Z_u}{m} & \frac{Z_w}{m} & \frac{Z_q + mV_{Ref}}{m} & 0 \\ \frac{M_u}{I_{yy}} & \frac{M_w}{I_{yy}} & \frac{M_q}{I_{yy}} & 0 \\ 0 & 0 & 1 & 0 \end{bmatrix} \tag{8.15c}$$

$$B_{Long} = \begin{bmatrix} \frac{T_{\delta_T}}{m} & \frac{X_{\delta_m}}{m} \\ 0 & \frac{Z_{\delta_m}}{m} \\ 0 & \frac{M_{\delta_m}}{I_{yy}} \\ 0 & 0 \end{bmatrix} \tag{8.15d}$$

Non-dimensional longitudinal equations (SSSLF, $\epsilon_T = 0$)

$$\dot{\hat{\mathbf{x}}}_{Long} = \hat{A}_{Long}\hat{\mathbf{x}}_{Long} + \hat{B}_{Long}\mathbf{u}_{Long} \tag{8.16a}$$

$$\hat{\mathbf{x}}_{Long} \equiv \begin{Bmatrix} \hat{V} \\ \alpha \\ \hat{q} \\ \theta \end{Bmatrix} \tag{8.16b}$$

$$\hat{A}_{Long} = \begin{bmatrix} \frac{-M_{Ref}C_{D_M} + C_{T_V}}{2\hat{m}} & \frac{C_{L_{Ref}} - C_{D_\alpha}}{2\hat{m}} & 0 & \frac{-C_W}{2\hat{m}} \\ \frac{-2C_W - M_{Ref}C_{L_M}}{2\hat{m}} & \frac{-C_{D_{Ref}} - C_{L_\alpha}}{2\hat{m}} & \frac{2\hat{m} - C_{L_q}}{2\hat{m}} & 0 \\ \frac{M_{Ref}C_{m_M}}{\hat{I}_{yy}} & \frac{C_{m_\alpha}}{\hat{I}_{yy}} & \frac{C_{m_q}}{\hat{I}_{yy}} & 0 \\ 0 & 0 & 1 & 0 \end{bmatrix} \tag{8.16c}$$

$$\hat{B}_{Long} = \begin{bmatrix} \frac{C_{T_{\delta_T}}}{2\hat{m}} & \frac{-C_{D_{\delta_m}}}{2\hat{m}} \\ 0 & \frac{-C_{L_{\delta_m}}}{2\hat{m}} \\ 0 & \frac{C_{m_{\delta_m}}}{\hat{I}_{yy}} \\ 0 & 0 \end{bmatrix} \tag{8.16d}$$

Dimensional lateral–directional equations (SSSLF, $I_{xz} = 0$)

$$\dot{\mathbf{x}}_{LD} = A_{LD}\mathbf{x}_{LD} + B_{LD}\mathbf{u}_{LD} \qquad (8.17\text{a})$$

$$\mathbf{x}_{LD} \equiv \begin{Bmatrix} v \\ p \\ r \\ \phi \end{Bmatrix}, \qquad \mathbf{u}_{LD} \equiv \begin{Bmatrix} \delta_\ell \\ \delta_n \end{Bmatrix} \qquad (8.17\text{b})$$

$$A_{LD} = \begin{bmatrix} \frac{Y_v}{m} & \frac{Y_p}{m} & \frac{Y_r - mV_{Ref}}{m} & g \\ \frac{L_v}{I_{xx}} & \frac{L_p}{I_{xx}} & \frac{L_r}{I_{xx}} & 0 \\ \frac{N_v}{I_{zz}} & \frac{N_p}{I_{zz}} & \frac{N_r}{I_{zz}} & 0 \\ 0 & 1 & 0 & 0 \end{bmatrix} \qquad (8.17\text{c})$$

$$B_{LD} = \begin{bmatrix} 0 & \frac{Y_{\delta_n}}{m} \\ \frac{L_{\delta_\ell}}{I_{xx}} & \frac{L_{\delta_n}}{I_{xx}} \\ \frac{N_{\delta_\ell}}{I_{zz}} & \frac{N_{\delta_n}}{I_{zz}} \\ 0 & 0 \end{bmatrix} \qquad (8.17\text{d})$$

Non-dimensional lateral–directional equations (SSSLF, $I_{xz} = 0$)

$$\dot{\hat{\mathbf{x}}}_{LD} = \hat{A}_{LD}\hat{\mathbf{x}}_{LD} + \hat{B}_{LD}\mathbf{u}_{LD} \qquad (8.18\text{a})$$

$$\hat{\mathbf{x}}_{LD} \equiv \begin{Bmatrix} \beta \\ \hat{p} \\ \hat{r} \\ \phi \end{Bmatrix} \qquad (8.18\text{b})$$

$$\hat{A}_{LD} = \begin{bmatrix} \frac{C_{y_\beta}}{2\hat{m}} & \frac{C_{y_p}}{2\hat{m}} & \frac{C_{y_r} - 2\hat{m}/A}{2\hat{m}} & \frac{C_W}{2\hat{m}} \\ \frac{C_{\ell_\beta}}{A\hat{I}_{xx}} & \frac{C_{\ell_p}}{A\hat{I}_{xx}} & \frac{C_{\ell_r}}{A\hat{I}_{xx}} & 0 \\ \frac{C_{n_\beta}}{A\hat{I}_{zz}} & \frac{C_{n_p}}{A\hat{I}_{zz}} & \frac{C_{n_r}}{A\hat{I}_{zz}} & 0 \\ 0 & \frac{1}{A} & 0 & 0 \end{bmatrix} \qquad (8.18\text{c})$$

$$\hat{B}_{LD} = \begin{bmatrix} 0 & \frac{C_{y_{\delta_n}}}{2\hat{m}} \\ \frac{C_{\ell_{\delta_\ell}}}{\hat{I}_{xx}} & \frac{C_{\ell_{\delta_n}}}{\hat{I}_{xx}} \\ \frac{C_{n_{\delta_\ell}}}{\hat{I}_{zz}} & \frac{C_{n_{\delta_n}}}{\hat{I}_{zz}} \\ 0 & 0 \end{bmatrix} \qquad (8.18\text{d})$$

Linearization

Problems

Independent variables: body-axis velocities u, v, w, p, q, and r plus the two body axis Euler angles θ and ϕ, and their derivatives; and the four generic controls δ_T, δ_ℓ, δ_m, and δ_n. For problems that don't specify a particular airplane, assume the non-dimensional derivatives are known. The non-dimensional derivatives are those that correspond to the functional dependencies listed in Equations 6.3, for example, since C_T depends on $\hat{V}$ and δ_T, C_{T_V} and $C_{T_{\delta_T}}$ are assumed known.

1. Linearize the heading angle equation, $\dot{\psi} = (q \sin\phi + r\cos\phi) \sec\theta$. Use steady, straight, symmetric flight for reference conditions; and stability axes for body axes. The following are given: h_{Ref}, V_{Ref}, and $\theta_{Ref} = \gamma_{Ref} \neq 0$. Find just the dimensional form. Show all steps.

2. Repeat problem 1 using steady, symmetric, turning flight for reference conditions; and stability axes for body axes. The following are given: h_{Ref}, V_{Ref}, ϕ_{Ref}, and $\theta_{Ref} = \gamma_{Ref} \neq 0$. Find just the dimensional form. Show all steps.

3. In problem 2 you should have found that a change in bank angle $\Delta\phi$ does not result in a change in the turn rate, yet Equation 7.30 shows the bank angle and turn rate are directly proportional to each other. Explain.

4. Linearize the yawing moment equation using principal axes. Use steady, straight, symmetric flight for reference conditions. The following are given: h_{Ref}, V_{Ref}, α_{Ref}, and θ_{Ref}. Find just the dimensional form, and relate all dimensional derivatives to non-dimensional derivatives. Show all steps.

5. Linearize the body-axis X-force equation with $\epsilon_T = 0$. Use steady, straight, symmetric flight for reference conditions. Find just the dimensional form, and relate all dimensional derivatives to non-dimensional derivatives. Show all steps.

6. Use the data for the multi-engine aircraft aircraft given in Chapter 7, problem 6. Evaluate each term in the linearized, dimensional body-axis equation for $\dot{u}$. Data are valid for stability axes. Assume that ϵ_T and γ_{Ref}, and any derivatives not given, are zero.

7. In this problem altitude is not assumed to be constant. Using the model of a standard atmosphere in the gradient region, the following relationships are valid:

$$a = \sqrt{\gamma RT}$$

$$\rho = \rho_{SL}\left(\frac{T}{T_{SL}}\right)^{-[(g_o/rR)+1]}$$

$$T = T_{SL} + rh$$

The independent variable in these equations is altitude h, upon which depend the speed of sound a, the density ρ, and the temperature T. The constants are γ, R, ρ_{SL}, T_{SL}, r, and g_o.

Begin with the expression $M = C_m \bar{q} S \bar{c}$, where $C_m(M, \alpha, \hat{\dot{\alpha}}, \hat{q}, \hat{h}, \delta_m, \delta_T)$. For reference conditions of steady, straight, symmetric, level flight at reference conditions $C_{L_{Ref}}$, $C_{D_{Ref}}$, V_{Ref} and h_{Ref}, determine the derivative of pitching moment with respect to altitude, evaluated in reference conditions,

$$\left.\frac{\partial M}{\partial h}\right|_{Ref} = \ldots ?$$

References

Etkin, B. (1972) *Dynamics of Atmospheric Flight*, 1st edn, John Wiley & Sons, Inc.
Etkin, B. and Reid, L.D. (1995) *Dynamics of Flight: Stability and Control,* 3rd edn, John Wiley & Sons, Inc.

9
Solutions to the Linear Equations

9.1 Scalar Equations

Consider the scalar first order, linear, ordinary differential equation with constant coefficients a and b:

$$\dot{x} = ax(t) + bu(t)$$

A LaPlace transformation of both sides of the equation yields

$$sx(s) - x(t_0) = ax(s) + bu(s)$$

This is then solved for $x(s)$, with $t_0 = 0$:

$$x(s) = (s-a)^{-1}x(0) + b(s-a)^{-1}u(s)$$

The two parts on the right correspond to the *initial condition response* and the *forced response*. If $u(t) = 0$, then we have the initial condition response

$$x(s) = \frac{1}{s-a}x(0)$$

The inverse transformation yields the initial condition response in the time domain,

$$x(t) = e^{at}x(0)$$

Clearly we require $a \leq 0$ or the response will diverge. If $x(0) = 0$ then we have the forced response

$$x(s) = \frac{b}{s-a}u(s)$$

The inverse transformation of the right-hand side obviously depends upon $u(s)$. For instance, if $u(t)$ is the unit step at $t = 0$ then $u(s) = 1/s$ and

$$x(s) = \frac{b}{s(s-a)}$$

$$x(t) = \frac{b}{a}(e^{at} - 1)$$

Aircraft Flight Dynamics and Control, First Edition. Wayne Durham.
© 2013 John Wiley & Sons, Ltd. Published 2013 by John Wiley & Sons, Ltd.

The forced response is sometimes written as

$$\frac{x(s)}{u(s)} = \frac{b}{s-a}$$

The quantity on the right-hand side is identified as the *transfer function* of $u(s)$ to $x(s)$. One may also speak of $1/(s-a)$ as the transfer function from $x(0)$ to $x(s)$.

9.2 Matrix Equations

Now we perform the analogous operations with our vectors and matrices. For now we will drop the Δs in the equation, and just consider

$$\dot{\mathbf{x}} = A\mathbf{x} + B\mathbf{u}$$

Recall that $\dot{\mathbf{x}}$ and $\mathbf{x}$ are n-dimensional vectors, $\mathbf{u}$ is an m-dimensional vector, A is an $n \times n$ matrix, and B is an $n \times m$ matrix. We consider A and B to be constant matrices. The LaPlace transform of a vector is the element-by-element transform of its components, so

$$s\mathbf{x}(s) - \mathbf{x}(0) = sI\mathbf{x}(s) - \mathbf{x}(0) = A\mathbf{x}(s) + B\mathbf{u}(s)$$

$$[sI - A]\mathbf{x}(s) = \mathbf{x}(0) + B\mathbf{u}(s)$$

$$\mathbf{x}(s) = [sI - A]^{-1}\mathbf{x}(0) + [sI - A]^{-1}B\mathbf{u}(s) \qquad (9.1)$$

The time-domain response is just the element-by-element inverse LaPlace transform of $\mathbf{x}(s)$, or $\mathbf{x}(t) = \mathcal{L}^{-1}\mathbf{x}(s)$.

Now consider the state transition matrix $[sI - A]^{-1}$. It may be represented as

$$[sI - A]^{-1} = \frac{C(s)}{d(s)}$$

The $n \times n$ matrix $C(s)$ is the *adjoint* of $[sI - A]$, and each entry will be a polynomial of order $n - 1$ or less. The polynomial $d(s)$ is the determinant of $[sI - A]$ and is of order n. Thus each entry in $[sI - A]^{-1}$ is a ratio of polynomials in the complex variable s. A ratio of polynomials in which the order of the numerator is less than or equal to the order of the denominator is called *proper*, and if the numerator's order is strictly less the ratio of polynomials is called *strictly proper*. It is easy to show that each of the ratios of polynomials in $[sI - A]^{-1}$ is strictly proper (because each of the numerator polynomials is the determinant of a matrix of order $n - 1$ with first-order polynomials in s along its diagonal). Thus we may write

$$[sI - A]^{-1} = \left\{\frac{c_{ij}(s)}{d(s)}\right\}, \quad i, j = 1 \ldots n$$

The polynomial $d(s)$ is referred to as the *characteristic polynomial*, and $d(s) = 0$ is the *characteristic equation* of the system. The denominator polynomial $d(s)$ may be factored into a product of first-order polynomials in s of the form $(s - \lambda_i)$ in which λ_i, $i = 1 \ldots n$, are the roots of $d(s) = 0$, or the *eigenvalues* of A.

$$d(s) = (s - \lambda_1)(s - \lambda_2) \cdots (s - \lambda_n)$$

Solutions to the Linear Equations

If the eigenvalues are distinct then the ratio of polynomials may be further expressed as a sum of ratios in which each denominator is one of the first-order factors of the denominator polynomial. That is, for an arbitrary numerator polynomial $n(s)$ of order less than n,

$$\frac{n(s)}{d(s)} = \frac{n(s)}{(s-\lambda_1)(s-\lambda_2)\cdots(s-\lambda_n)}$$

$$= \frac{n_1}{(s-\lambda_1)} + \frac{n_2}{(s-\lambda_2)} + \cdots + \frac{n_n}{(s-\lambda_n)}$$

This is the familiar partial-fraction expansion method of solving inverse LaPlace transform problems. The inverse transform of such an expression is the sum of terms involving exponentials $n_i e^{\lambda_i t}$, $i = 1 \ldots n$:

$$\mathcal{L}^{-1}\left\{\frac{n(s)}{d(s)}\right\} = n_1 e^{\lambda_1 t} + n_2 e^{\lambda_2 t} + \cdots + n_n e^{\lambda_n t}$$

This result applies to each entry of $[sI - A]^{-1}$.

9.3 Initial Condition Response

9.3.1 Modal analysis

System modes

For the unforced, or initial-condition response ($\mathbf{u}(t) = \mathbf{u}(s) = 0$) we have

$$\dot{\mathbf{x}} = A\mathbf{x}$$

$$\mathbf{x}(s) = [sI - A]^{-1}\mathbf{x}(0)$$

Since $\mathbf{x}(0)$ is just a vector of constants, each element of $\mathbf{x}(s)$ will be a linear combination of terms like $n(s)/d(s)$, for example,

$$x_i(s) = \frac{v_{i,1}}{s-\lambda_1} + \frac{v_{i,2}}{s-\lambda_2} + \cdots + \frac{v_{i,n}}{s-\lambda_n}$$

Each element of $x_i(t)$ will therefore be a sum of terms like $v_{i,j} e^{\lambda_j t}$, or

$$x_i(t) = v_{i,1} e^{\lambda_1 t} + v_{i,2} e^{\lambda_2 t} + \cdots + v_{i,n} e^{\lambda_n t}$$

The entire vector $\mathbf{x}(t)$ may be represented as

$$\mathbf{x}(t) = \mathbf{v}_1 e^{\lambda_1 t} + \mathbf{v}_2 e^{\lambda_2 t} + \cdots + \mathbf{v}_n e^{\lambda_n t} \tag{9.2}$$

Each of the components $\mathbf{v}_i e^{\lambda_i t}$ is one of the *modes* of the system response. In practice the eigenvalues will be a mixture of real and complex numbers. The complex roots will occur in complex conjugate pairs (because A has real components), say $\mathbf{v}_i e^{\lambda_i t} + \mathbf{v}_i^* e^{\lambda_i^* t}$ (the asterisk denotes the complex conjugate). The result is, of course, a real oscillatory response. In such cases the combined pair $\mathbf{v}_i e^{\lambda_i t} + \mathbf{v}_i^* e^{\lambda_i^* t}$ is considered a single mode.

It is clearly necessary that each of the eigenvalues have negative real parts for all of the modes, and hence the system, to be stable.

Now evaluate $\dot{\mathbf{x}} - A\mathbf{x} = 0$ using Equation 9.2 for $\mathbf{x}$,

$$\dot{\mathbf{x}} = \lambda_1 \mathbf{v}_1 e^{\lambda_1 t} + \lambda_2 \mathbf{v}_2 e^{\lambda_2 t} + \cdots + \lambda_n \mathbf{v}_n e^{\lambda_n t}$$

$$A\mathbf{x} = A\mathbf{v}_1 e^{\lambda_1 t} + A\mathbf{v}_2 e^{\lambda_2 t} + \cdots + A\mathbf{v}_n e^{\lambda_n t}$$

Hence,

$$\dot{\mathbf{x}} - A\mathbf{x} = (\lambda_1 \mathbf{v}_1 - A\mathbf{v}_1)e^{\lambda_1 t} + \cdots + (\lambda_n \mathbf{v}_n - A\mathbf{v}_n)e^{\lambda_n t} = 0 \quad (9.3)$$

Since $e^{\lambda_i t}$ is never zero, Equation 9.3 requires that each of the terms in parentheses vanish independently, or

$$(\lambda_i \mathbf{v}_i - A\mathbf{v}_i) = (\lambda_i I_i - A)\mathbf{v}_i = 0, \quad i = 1 \ldots n$$

This means that the vectors $\mathbf{v}_i$ are the *eigenvectors* of A, each associated with an eigenvalue λ_i. Since the non-zero multiple (including multiplication by complex numbers) of an eigenvector is also an eigenvector, we may write the initial condition response as

$$\mathbf{x}(t) = \alpha_1 e^{\lambda_1 t} \mathbf{v}_1 + \alpha_2 e^{\lambda_2 t} \mathbf{v}_2 + \cdots + \alpha_n e^{\lambda_n t} \mathbf{v}_n \quad (9.4)$$

In Equation 9.4, for a given set of eigenvectors $\mathbf{v}_i$, the mulipliers α_i are chosen to satisfy the initial condition

$$\mathbf{x}(0) = \alpha_1 \mathbf{v}_1 + \alpha_2 \mathbf{v}_2 + \cdots + \alpha_n \mathbf{v}_n$$

If we denote each of the scalar terms $q_i(t) \equiv \alpha_i e^{\lambda_i t}$, then

$$\dot{q}_i(t) = \lambda_i \alpha_i e^{\lambda_i t} = \lambda_i q_i(t)$$

We then recognize $\alpha_i e^{\lambda_i t}$ as the initial condition response of the differential equation $\dot{q}_i(t) = \lambda_i q_i(t)$ with $\alpha_i = q_i(0)$, or $q_i(t) = q_i(0)e^{\lambda_i t}$. Now define the vector $\mathbf{q}(t) \equiv \{q_i(t)\}$, $i = 1 \ldots n$, and we may write the initial condition response

$$\mathbf{x}(t) = [\mathbf{v}_1 \mathbf{v}_2 \cdots \mathbf{v}_n]\mathbf{q}(t) \equiv M\mathbf{q}(t) \quad (9.5)$$

The modal matrix

Here we have defined the matrix M as consisting of columns which are the eigenvectors of A. M is frequently called the *modal matrix*. Note that M is in general not a direction cosine matrix, and $M^{-1} \neq M^T$. However, since we have assumed distinct eigenvalues, the eigenvectors are linearly independent and $|M| \neq 0$. Thus M^{-1} exists and $\mathbf{q}(t) = M^{-1}\mathbf{x}(t)$, so

$$\dot{\mathbf{q}}(t) = M^{-1}\dot{\mathbf{x}}(t) = M^{-1}A\mathbf{x}(t)$$
$$= M^{-1}AM M^{-1}\mathbf{x}(t) = (M^{-1}AM)M^{-1}\mathbf{x}(t)$$

Define $\Lambda \equiv M^{-1}AM$, so that

$$\dot{\mathbf{q}}(t) = \Lambda \mathbf{q}(t) \quad (9.6)$$

Solutions to the Linear Equations

Since $\dot{q}_i(t) = \lambda_i q_i(t)$, then

$$\Lambda = \begin{bmatrix} \lambda_1 & 0 & \cdots & 0 \\ 0 & \lambda_2 & \cdots & 0 \\ \vdots & \vdots & \cdots & \vdots \\ 0 & 0 & \cdots & \lambda_n \end{bmatrix} = diag\{\lambda_i\}$$

The solution of $\dot{\mathbf{q}} = \Lambda \mathbf{q}(t)$ is therefore

$$\mathbf{q}(t) = \begin{bmatrix} e^{\lambda_1 t} & 0 & \cdots & 0 \\ 0 & e^{\lambda_2 t} & \cdots & 0 \\ \vdots & \vdots & \cdots & \vdots \\ 0 & 0 & \cdots & e^{\lambda_n t} \end{bmatrix} \mathbf{q}(0)$$

Define

$$e^{\Lambda t} \equiv \begin{bmatrix} e^{\lambda_1 t} & 0 & \cdots & 0 \\ 0 & e^{\lambda_2 t} & \cdots & 0 \\ \vdots & \vdots & \cdots & \vdots \\ 0 & 0 & \cdots & e^{\lambda_n t} \end{bmatrix}$$

Whence

$$\mathbf{q}(t) = e^{\Lambda t} \mathbf{q}(0) \tag{9.7}$$

Thus, for a system with distinct eigenvalues, we may solve the initial condition response of $\dot{\mathbf{x}} = A\mathbf{x}$ with given initial conditions $\mathbf{x}(0)$ as follows:

1. Determine the eigenvalues and eigenvectors of A.
2. For the given initial conditions $\mathbf{x}(0)$, determine $\mathbf{q}(0) = M^{-1}\mathbf{x}(0)$.
3. Write down the solution $\mathbf{q}(t)$ using Equation 9.7.
4. Evaluate the solution $\mathbf{x}(t) = M\mathbf{q}(t)$.

In short, then,

$$\mathbf{x}(t) = M e^{\Lambda t} M^{-1} \mathbf{x}(0) \tag{9.8}$$

Argand diagrams

With Equation 9.8 we may calculate the time-history of each of the states in response to given initial conditions. Each real eigenvalue will contribute a mode of the form $\alpha_i e^{\lambda_i t} \mathbf{v}_i$, where each α_i is a constant (possibly zero) determined by the initial conditions and $\mathbf{v}_i$ is a constant eigenvector associated with λ_i. The time history of each state in the response of a given mode is a constant component of the vector $\alpha_i \mathbf{v}_i$ multiplied by the exponential term $e^{\lambda_i t}$. At any time, therefore, the magnitudes of the various states will be in the same ratio (as given by $\mathbf{v}_i$) as at any other time.

That is, if the initial-condition response of the ith mode is $\mathbf{x}(t) = \mathbf{v}_i e^{\lambda_i t}$, then in terms of the individual states,

$$\begin{Bmatrix} x_1(t) \\ x_2(t) \\ \vdots \\ x_n(t) \end{Bmatrix} = \begin{Bmatrix} v_{i1} \\ v_{i2} \\ \vdots \\ v_{in} \end{Bmatrix} e^{\lambda_i t} = \begin{Bmatrix} v_{i1} e^{\lambda_i t} \\ v_{i2} e^{\lambda_i t} \\ \vdots \\ v_{in} e^{\lambda_i t} \end{Bmatrix}$$

Then the ratio of the jth state to the kth at any time t is

$$\frac{x_j(t)}{x_k(t)} = \frac{v_{ij}e^{\lambda_i t}}{v_{ik}e^{\lambda_i t}} = \frac{v_{ij}}{v_{ik}}$$

Note that this does not mean that the ratio of two states is in general constant, since more than one mode may be involved. For example, if the two modes λ_1 and λ_2 are both excited, then $\mathbf{x}(t) = \mathbf{v}_1 e^{\lambda_1 t} + \mathbf{v}_2 e^{\lambda_2 t}$, and

$$\frac{x_j(t)}{x_k(t)} = \frac{v_{1j}e^{\lambda_1 t} + v_{2j}e^{\lambda_2 t}}{v_{1k}e^{\lambda_1 t} + v_{2k}e^{\lambda_2 t}} \neq \frac{v_{1j} + v_{2j}}{v_{1k} + v_{2k}}$$

If complex roots occur then an oscillatory response will result. A complex conjugate pair of roots, although representing two distinct eigenvalues, create a single mode when the roots are combined to get a real response.

Say a complex eigenvalue λ_i occurs. Then, if the entries in the A matrix are real, the conjugate eigenvalue λ_i^* will also occur.

$$\lambda_i = \sigma + j\omega, \qquad \lambda_i^* = \sigma - j\omega$$

Moreover, if the eigenvector associated with λ_i is $\mathbf{v}_i$ then the eigenvector associated with λ_i^* will be $\mathbf{v}_i^*$. Any multiple α_i of $\mathbf{v}_i$ must be accompanied by a multiple α_i^* of $\mathbf{v}_i^*$. Considering just the single mode corresponding to λ_i and λ_i^*,

$$\mathbf{x}(t) = e^{\sigma t}(\alpha_i e^{j\omega t}\mathbf{v}_i + \alpha_i^* e^{-j\omega t}\mathbf{v}_i^*)$$

In general the multiplier α_i and each of the elements of $\mathbf{v}_i$ may be a complex number. When multiplying complex numbers the polar form is preferred,

$$a + jb = Me^{j\phi}$$

$$M = \sqrt{a^2 + b^2}, \qquad \phi = \tan^{-1}\frac{b}{a}$$

Represent α_i and the kth component of eigenvector $\mathbf{v}_i$ in polar form as

$$\alpha_i = M_\alpha e^{j\phi_\alpha}, \qquad v_{ik} = M_{ik}e^{j\phi_{ik}}$$

$$\alpha_i \mathbf{v}_i = M_\alpha e^{j\phi_\alpha} \begin{Bmatrix} M_{i1}e^{j\phi_{i1}} \\ \vdots \\ M_{in}e^{j\phi_{in}} \end{Bmatrix}$$

The contribution of the eigenvalue leads to

$$\alpha_i e^{j\omega t}\mathbf{v}_i = M_\alpha \begin{Bmatrix} M_{i1}e^{j(\omega t + \phi_\alpha + \phi_{i1})} \\ \vdots \\ M_{in}e^{j(\omega t + \phi_\alpha + \phi_{in})} \end{Bmatrix}$$

Solutions to the Linear Equations

The conjugate part of the response is

$$\alpha_i^* e^{-j\omega t} \mathbf{v}_i^* = M_\alpha \begin{Bmatrix} M_{i1} e^{-j(\omega t + \phi_\alpha + \phi_{i1})} \\ \vdots \\ M_{in} e^{-j(\omega t + \phi_\alpha + \phi_{in})} \end{Bmatrix}$$

The total response for this mode may then be written as

$$\mathbf{x}(t) = e^{\sigma t} M_\alpha \begin{Bmatrix} M_{i1} \left(e^{j(\omega t + \phi_\alpha + \phi_{i1})} + e^{-j(\omega t + \phi_\alpha + \phi_{i1})} \right) \\ \vdots \\ M_{in} \left(e^{j(\omega t + \phi_\alpha + \phi_{in})} + e^{-j(\omega t + \phi_\alpha + \phi_{in})} \right) \end{Bmatrix}$$

$$= 2 e^{\sigma t} M_\alpha \begin{Bmatrix} M_{i1} \cos(\omega t + \phi_\alpha + \phi_{i1}) \\ \vdots \\ M_{in} \cos(\omega t + \phi_\alpha + \phi_{in}) \end{Bmatrix}$$

Then the kth component of $\mathbf{x}(t)$, $x_k(t)$, is

$$x_k(t) = 2 e^{\sigma t} M_\alpha M_{ik} \cos(\omega t + \phi_\alpha + \phi_{ik})$$

The magnitude of $e^{\sigma t}$ and M_α is the same for each state, as are the angles ωt and ϕ_α. Thus the relative difference between the response of two different states is contained in the magnitude M_{ik} and the phase ϕ_k, which are determined by the corresponding entries in the eigenvector. That is, if all we care about is the relationship of one state to another during an initial condition response of an oscillatory mode, all that information is contained in the eigenvector.

An *Argand diagram* is the plot in the complex plane of complex entries in an eigenvector for several different states. It is conventional to pick the eigenvector associated with the eigenvalue with the positive imaginary part, $\sigma + j\omega$. Figure 9.1 shows two such entries, corresponding to the two states x_k and x_{k+1}.

In Figure 9.1 the phase difference between the two states is the angle between v_k and v_{k+1}. The relationship between the real magnitudes of the two states is visualized as the projections of v_k and v_{k+1} onto the real axis.

The evolution of time in the response of the states in an Argand diagram may be visualized as a counter-clockwise rotation of the vectors about the origin through an angle ωt.

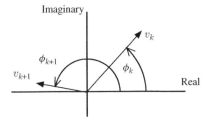

Figure 9.1 Argand diagram.

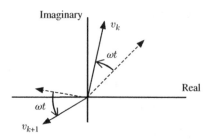

Figure 9.2 Response at time t.

Note that the counterclockise rotation occurs because we have picked the eigenvector corresponding to $\sigma + j\omega$. The eigenvector corresponding to $\sigma - j\omega$ would have to be viewed as a clockwise rotation. Thus at some later time t, the situation would be as shown in Figure 9.2.

9.4 Mode Sensitivity and Approximations

9.4.1 Mode sensitivity

In the long history of flight dynamics there have been ongoing efforts to reduce the larger problem to several of smaller order. The motivation for this effort probably began with the difficulties of analyzing large systems using only a slide rule, pencil, and paper. One major benefit to order reduction is that the smaller problems may be analyzed literally (as opposed to numerically). In many cases this analysis will show that a particular response mode is strongly dependent on a small subset of the parameters used to formulate the equations of motion. These parameters in turn are determined by the physical design of the aircraft. Thus, links between design choices and the dynamic response of an airplane may be established, enabling designers to more directly determine how well their aircraft will fly.

We have already seen some reductions in the order of the problem. Beginning with 12 rigid body equations of motion, it was observed that the geographical coordinates x_E and y_E, as well as the heading angle ψ, do not affect the dynamic response of the aircraft, and these states were ignored. The assumption of constant altitude was not so easily justifiable, but experience bears out the validity of neglecting that state. More subtly, various assumptions have been made that cause the linearized lateral–directional and longitudinal equations to be uncoupled from each other. Thus, instead of a twelfth-order system of equations, we have two separate fourth-order systems.

For each fourth-order system of linear equations there will be four eigenvalues. We will soon show that the longitudinal eigenvalues typically consist of two complex–conjugate pairs, and the lateral–directional equations consist of one complex–conjugate pair and two real roots. Thus there are two longitudinal modes and three lateral–directional modes. The next level of order reduction seeks to find dynamic systems with the same number of states as the order of the mode, that is, one state for a real eigenvalue and two for an oscillatory mode.

Solutions to the Linear Equations 121

Eigenvector analysis

One way to approach the order-reduction problem is to ask if there is a set of initial conditions that will cause only one of the modes to appear in the response. If there is, and if in this set of initial conditions some states are 'small', then maybe they can be neglected without affecting the response.

The answer to the question is fairly easy. Since the eigenvectors v_i have been assumed to be linearly independent, then if $x(0) = \alpha_i v_i$, some non-zero multiple of v_i, then all the other $\alpha_j = 0$, $j = 1 \ldots n$, $j \neq i$. This can also be shown using $q(0) = \alpha M^{-1} x(0) = \alpha M^{-1} v_i$. Since $M = [v_1 v_2 \cdots v_n]$, and $M^{-1} M = I$, $M^{-1} v_i$ is just the ith column of the identity matrix. Then $q(t) = e^{\Lambda t} q(0)$ is

$$q(t) = e^{\Lambda t} q(0) = \begin{bmatrix} e^{\lambda_1 t} & \cdots & 0 & \cdots & 0 \\ \vdots & \vdots & \cdots & \vdots & \vdots \\ 0 & \cdots & e^{\lambda_i t} & \cdots & 0 \\ \vdots & \vdots & \cdots & \vdots & \vdots \\ 0 & \cdots & \cdots & \cdots & e^{\lambda_n t} \end{bmatrix} \begin{Bmatrix} 0 \\ \vdots \\ \alpha \\ \vdots \\ 0 \end{Bmatrix} = \begin{Bmatrix} 0 \\ \vdots \\ \alpha e^{\lambda_i t} \\ \vdots \\ 0 \end{Bmatrix}$$

From this we see that only the mode corresponding to $e^{\lambda_i t}$ will be present in the response, $x(t) = \alpha_i e^{\lambda_i t} v_i$. That is: *if the initial conditions of the states are aligned with one of the eigenvectors, only the mode of the eigenvalue corresponding to that eigenvector will be present in the initial-condition response.* If the mode arises from complex roots, then we may take $x(0) = \alpha_i v_i + \alpha_i^* v_i^*$, which is a vector of real numbers.

Thus if certain of the entries in v_i are 'small' relative to all the others, then the corresponding states are not influential in the initial conditions that excite only the ith mode. In the extreme case, if an entry is zero then the corresponding state will not respond at all in that mode. In cases of complex conjugate pairs we use the analysis that led to the Argand diagram to determine relative size.

The problem with applying this information is in determining what we mean by a relatively 'small' entry. If all the states have the same units this makes some sense. In flight dynamics, however, the states are linear velocities, angular velocities, linear measurements, and angles. So, for example, we would need to decide if a change in velocity of 20 ft/s is large or small compared with a change in pitch rate of 0.2 rad/s.

Sensitivity analysis

In this analysis the question of different units does not arise, since only one state will be examined at a time. The question to be answered is whether a given mode is more sensitive to changes in some states than others. Put another way, we seek to determine whether a change in the initial condition of a single state will more strongly influence some modes than others. Thus, if a change in the initial condition of a single state, while all others are unchanged, causes one mode to respond more 'energetically' than others, then this state may be thought of as 'dominant' in the response of the mode. It is convenient to think of this sensitivity analysis as injecting energy into the system through a single state, and looking to see how this energy is then distributed to the various modes.

For our change in initial condition of the jth state we use the notation

$$\Delta \mathbf{x}_j(0) = \{x_i\}, \quad x_i = \begin{cases} 1, & i = j \\ 0, & i \neq j \end{cases} \quad (9.9)$$

This notation means there is a '1' in the jth position, and zeros elsewhere. From Equation 9.5 we can easily determine the change $\Delta \mathbf{q}_j(0)$ in $\mathbf{q}(0)$,

$$\Delta \mathbf{q}_j(0) = M^{-1} \Delta \mathbf{x}_j(0) \quad (9.10)$$

The vector $\Delta \mathbf{q}_j(0)$ has components $\{\Delta q_{j_i}\}$, $i = 1 \ldots n$. The change in the initial condition is seen in each mode according to

$$\Delta \mathbf{x}_j(0) = \Delta q_{j_1}(0) \mathbf{v}_1 + \Delta q_{j_2}(0) \mathbf{v}_2 + \cdots + \Delta q_{j_n}(0) \mathbf{v}_n \quad (9.11)$$

Equation 9.11 has all the information we need for our sensitivity analysis. The actual time-varying response is just like Equation 9.11 except for the time-varying parts,

$$\Delta \mathbf{x}_j(t) = \Delta q_{j_1}(0) e^{\lambda_1 t} \mathbf{v}_1 + \Delta q_{j_2}(0) e^{\lambda_2 t} \mathbf{v}_2 + \cdots + \Delta q_{j_n}(0) e^{\lambda_n t} \mathbf{v}_n$$

In Equation 9.11 we examine the relative values of the jth component of each of the vectors $\Delta q_i(0) \mathbf{v}_i$. For example, if we varied the initial condition of the first state ($j = 1$), then we look at the first component of each vector on the right-hand side to see how that change was distributed among the various modes.

The process of performing the sensitivity analysis is easier than it looks. Note first in Equation 9.9 that it would not have mattered whether we put zeros in the non-jth positions, since in the end we just looked at the jth component of the result. This means we can analyze all of the states simultaneously. Now note that Equation 9.10 has the effect of assigning to $\Delta \mathbf{q}_j(0)$ the jth column of the matrix M^{-1}. In analyzing all the states simultaneously we form the matrix M^{-1} and interpret its columns as the vectors $\Delta \mathbf{q}_j(0)$:

$$M^{-1} = \begin{bmatrix} \Delta \mathbf{q}_1(0) & \Delta \mathbf{q}_2(0) & \cdots & \Delta \mathbf{q}_n(0) \end{bmatrix} \quad (9.12)$$

This method of modal sensitivity analysis can be summarized as follows:

1. Calculate the eigenvalues and eigenvectors of the system, form the modal matrix M, and calculate M^{-1}.
2. Denote the rows of M as $r_1, r_2, \ldots$ and the columns of M^{-1} as $c_1, c_2, \ldots$

$$M = \begin{bmatrix} r_1 \\ r_2 \\ \vdots \\ r_n \end{bmatrix}, \quad M^{-1} = \begin{bmatrix} c_1 & c_2 & \cdots & c_n \end{bmatrix}$$

Form diagonal matrices $C_1, C_2, \ldots$ with the entries from $c_1, c_2, \ldots$ forming the diagonal.

$$c_i = \begin{bmatrix} c_{1i} \\ c_{2i} \\ \vdots \\ c_{ni} \end{bmatrix}, \quad C_i \equiv \begin{bmatrix} c_{1i} & 0 & 0 & \cdots & 0 \\ 0 & c_{2i} & 0 & \cdots & 0 \\ & & \vdots & & \\ 0 & 0 & 0 & \cdots & c_{ni} \end{bmatrix}$$

Now evaluate the $n \times n$ sensitivity matrix S, defined as

$$S = \begin{bmatrix} r_1 C_1 \\ r_2 C_2 \\ \vdots \\ r_n C_n \end{bmatrix} \qquad (9.13)$$

3. Retain just the magnitudes of the entries in S (absolute values of real numbers, magnitudes of complex numbers). Normalize each row by summing the entries, then dividing each entry by that sum.
4. Each row of S corresponds to a state, and each column corresponds to a mode. The modes are enumerated in the same order as the eigenvalues that determined the eigenvectors used to form M. Examine each column of S to assess the relative magnitudes of the numbers in that column. An entry of zero means the state in question does not respond at all in the mode, and it may be ignored in analyzing that mode. At the other extreme, if all the numbers in a column are the same, or nearly so, no states can be ignored in the analysis of that mode. Between these extremes decisions must be made. Experience has shown that states whose entry is no larger than 10% of the largest entry may be safely ignored.

9.4.2 Approximations

Once a state has been declared ignorable, it and its effects may be removed from the equations of motion to get the approximation. Mathematically we assume that the ignorable variable becomes a constant. If the state x_i is ignorable, then we take $\dot{x}_i = 0$. Now, the variable itself still appears in the remaining equations of motion, to be multiplied by the appropriate entries in those equations. The only remaining question, then, is what constant value should be assigned to the ignorable variable for use in the other equations.

There has long been a notion of 'slow' and 'fast' variables that is often applied to this sort of analysis, especially in the study of singular perturbations. The idea is that a slow ignorable variable does not change as rapidly as the variables of interest, and its initial value should be used as its constant value. Since we are dealing with perturbations in the variable, that constant value is zero. On the other hand, fast ignorable variables are thought of as having finished all their dynamics before the problem of interest has a chance to get started. In this case, the constant value is the 'steady-state' response of the ignorable variable.

The problem with slow and fast variables is that it is hard to find a rigorous definition of what these adjectives mean, or a methodology to decide whether a given variable is one or the other. Our approach will be to rank the modes from fastest to slowest according to the associated eigenvalue's real part. The larger (in magnitude) negative the real part the more quickly $e^{\sigma t}$ tends to zero. For a given mode's approximation, every state associated with faster modes will be considered a fast variable, and every state associated with slower modes will be a slow variable.

The method of constructing the approximation to a given mode therefore is:

1. Perform a sensitivity analysis and find the ignorable states for the mode in question.
2. For each ignorable state, decide if it is fast or slow according to the mode in which the state is not ignorable.

3. If an ignorable state is slow, set it to zero in the approximating equations.
4. If an ignorable state is fast, determine its 'steady-state' value in its mode, and algebraically evaluate its contribution to the approximating equations.
5. Remove the rows of the A and B matrices associated with the ignorable variable.
6. Remove the columns of A associated with the ignorable variable.
7. The remaining non-trivial equations are the approximation to the mode in question.

To illustrate this process, take a generic third-order system

$$\begin{Bmatrix} \dot{x}_1 \\ \dot{x}_2 \\ \dot{x}_3 \end{Bmatrix} = \begin{bmatrix} a_{11} & a_{12} & a_{13} \\ a_{21} & a_{22} & a_{23} \\ a_{31} & a_{32} & a_{33} \end{bmatrix} \begin{Bmatrix} x_1 \\ x_2 \\ x_3 \end{Bmatrix}$$

Assume that x_1 is dominant in mode 1, x_2 is dominant in mode 2, and x_3 is dominant in mode 3. Further assume that mode 1 is slowest and mode 3 is fastest. We are to approximate mode 2. Relative to mode 2, x_3 is fast and x_1 is slow. First we set $\dot{x}_1 = \dot{x}_3 = 0$. x_1 is slow so set $x_1 = 0$. Next we need to solve for $x_{3_{ss}}$. The equations at this point are

$$\begin{Bmatrix} 0 \\ \dot{x}_2 \\ 0 \end{Bmatrix} = \begin{bmatrix} a_{11} & a_{12} & a_{13} \\ a_{21} & a_{22} & a_{23} \\ a_{31} & a_{32} & a_{33} \end{bmatrix} \begin{Bmatrix} 0 \\ x_2 \\ x_{3_{ss}} \end{Bmatrix}$$

We evaluate $x_{3_{ss}}$ in its mode by solving the third equation for $x_{3_{ss}}$ as a function of x_2,

$$\dot{x}_3 = 0 = a_{32} x_2 + a_{33} x_{3_{ss}} \Rightarrow x_{3_{ss}} = -\frac{a_{32}}{a_{33}} x_2$$

The information is applied to the mode 2 equation to evaluate its approximation:

$$\dot{x}_2 = a_{22} x_2 + a_{23} x_{3_{ss}} = a_{22} x_2 - a_{23} \frac{a_{32}}{a_{33}} x_2 = \left(a_{22} - a_{23} \frac{a_{32}}{a_{33}} \right) x_2$$

The approximation to the second mode is therefore

$$\dot{x}_2 = \left(\frac{a_{22} a_{33} - a_{23} a_{32}}{a_{33}} \right) x_2$$

9.5 Forced Response

9.5.1 Transfer functions

For the forced response we take $\mathbf{x}(0) = 0$, and

$$\mathbf{x}(s) = [sI - A]^{-1} B \mathbf{u}(s)$$

Calculating the inverse transformation of the right-hand side obviously depends on $\mathbf{u}(t)$ which determines $\mathbf{u}(s)$. At this point $\mathbf{u}(t)$ represents the time-history of the pilot's control inputs, which assuredly are not easy to place in analytical form. We therefore resort to considering simple analytical forms of $\mathbf{u}(t)$: impulses, steps, ramps, and sinusoids.

The matrix $[sI - A]^{-1}B$ consists of ratios of polynomials, each with the characteristic polynomial $d(s) \equiv |sI - A| = d_0 + d_1 s + \cdots + d_n s^n$ as its denominator. Each entry is a transfer function that relates the jth input to the ith state. This matrix of transfer functions is often denoted as

$$[sI - A]^{-1}B \equiv G(s) = \{g_{ij}(s)\}$$

$$\mathbf{x}(s) = G(s)\mathbf{u}(s) \tag{9.14}$$

The relationship is $x_i(s) = \sum_j g_{ij}(s) u_j(s)$. Each of the $g_{ij}(s)$ may be represented as

$$g_{ij}(s) = \frac{n_{ij}(s)}{d(s)}$$

In terms of the factors of the numerator and denominator polynomials,

$$g_{ij}(s) = \frac{n_{ij}(s)}{d(s)} = \frac{k_{ij}(s - z_1)(s - z_2)\cdots(s - z_{n_z})}{(s - p_1)(s - p_2)\cdots(s - p_n)}$$

The numerator roots are called *zeros*; n_z is the number of zeros in a particular transfer function, and is generally different for each of the $g_{ij}(s)$. The denominator roots are, of course, the same as the eigenvalues, but in the analysis of transfer functions they are commonly called *poles*. If a pole and a zero are identical they may be cancelled; in that case the remaining poles are not the same as the eigenvalues.

For some given $\mathbf{u}(t)$ for which $\mathbf{u}(s)$ is known, the forced reponse may be calculated from

$$\mathbf{x}(t) = \mathcal{L}^{-1}[G(s)\mathbf{u}(s)] \tag{9.15}$$

9.5.2 Steady-state response

The steady-state reponse (if it exists) of a system to given control inputs whose LaPlace transforms are known is given by the Final Value Theorem. For a given $x_i(s)$, $u_j(s)$, and $g_{ij}(s)$,

$$\lim_{t \to \infty} x_i(t) = \lim_{s \to 0}[sx_i(s)] = \lim_{s \to 0}[s g_{ij}(s) u_j(s)]$$

Given the LaPlace transforms of each of the inputs, $u_j(s)$,

$$\lim_{t \to \infty} \mathbf{x}(t) = \lim_{s \to 0}[s\mathbf{x}(s)] = \lim_{s \to 0}[sG(s)\mathbf{u}(s)]$$

Note that the final value theorem will often give a result even if the limit does not exist. The existence, or lack thereof, usually requires some understanding of the physics of the problem being addressed.

Problems

For all problems, unless otherwise instructed MATLAB® or similar software may be used, but include print-outs of any work not done by hand.

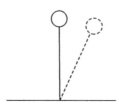

Figure 9.3 Inverted pendulum.

1. Consider an inverted pendulum (Figure 9.3) consisting of a mass m on a rigid, massless rod of length ℓ free to move as shown without friction or other resistance. Let $\ell = 32.2$ ft, $m = 1$ slug, and $g = 32.2$ ft/s^2. Let the angular displacement θ, rate $\dot{\theta}$, and acceleration $\ddot{\theta}$ be measured positive in the clockwise direction. Two forces act on the mass: its weight, and a resistance to motion (damping) $T = -8\dot{\theta}$ lb.
   (a) Determine the equation of motion for the system, $\ddot{\theta} = f(\theta, \dot{\theta})$.
   (b) Let $x_1 = \theta$ and $x_2 = \dot{\theta}$. Write the system of first-order differential equations that describe the system,
   $$\begin{Bmatrix} \dot{x}_1 \\ \dot{x}_2 \end{Bmatrix} \equiv \dot{\mathbf{x}} = \mathbf{f}(\mathbf{x}) = \begin{Bmatrix} f_1(\mathbf{x}) \\ f_2(\mathbf{x}) \end{Bmatrix}$$
   (c) Linearize the system of equations about steady-state equilibrium reference conditions, and place in the form $\dot{\mathbf{x}} = A\mathbf{x}$.
   (d) Determine the eigenvalues and eigenvectors of the linearized system.
   (e) You should have two real eigenvalues, one stable and the other unstable. In terms of the initial condition response of the linearized system, explain what the stable mode is.
2. Consider the system $\dot{\mathbf{x}} = A\mathbf{x} + B\mathbf{u}$, with
   $$A = \begin{bmatrix} -1 & -1 & 1 \\ 0 & -2 & 1 \\ 0 & -2 & 0 \end{bmatrix} \quad B = \begin{bmatrix} 1 & 1 \\ 0 & 1 \\ 0 & 1 \end{bmatrix}$$
   (a) Evaluate the state transition matrix $[sI - A]^{-1}$. Use Fedeeva's algorithm and show all steps.
   (b) Find the eigenvalues λ_1 (complex, with positive imaginary part), $\lambda_2 = \lambda_1^*$, and λ_3 of the system. The eigenvalues should be approximately $\lambda_1 \approx -1 + j1$ and $\lambda_3 \approx -4$. Find the corresponding eigenvectors, scaled such that the largest entry in each eigenvector is one. Do all calculations by hand and show all steps.
   (c) Let $M = \begin{bmatrix} \mathbf{v}_1 & \mathbf{v}_2 & \mathbf{v}_3 \end{bmatrix}$ and $\mathbf{q} = M^{-1}\mathbf{x}$. Transform the system into $\dot{\mathbf{q}} = \Lambda \mathbf{q} + M^{-1} B\mathbf{u}$.
   (d) For the initial condition response, let $\mathbf{x}(0)$ be the real part of $\mathbf{v}_1$. Find $\mathbf{q}(t)$ and $\mathbf{x}(t)$.
   (e) Perform a mode sensitivity analysis of this system. Based on those results, find a 2×2 matrix that approximates the complex mode, and a first-order system that approximates the real mode.

10
Aircraft Flight Dynamics

10.1 Example: Longitudinal Dynamics

10.1.1 System matrices

See Appendix A for data. At a particular flight condition the A-4 *Skyhawk* has the following linearized, dimensional longitudinal system and control matrices (the Δs have been dropped):

$$\dot{\mathbf{x}}_{Long} = A_{Long}\mathbf{x}_{Long} + B_{Long}\mathbf{u}_{Long}$$

$$\mathbf{x}_{Long}^T = \{u, \ \alpha, \ q, \ \theta\}$$

$$\mathbf{u}_{Long}^T = \{\delta_T, \ \delta_m\}$$

$$A_{Long} = \begin{bmatrix} -1.52 \times 10^{-2} & -2.26 & 0 & -32.2 \\ -3.16 \times 10^{-4} & -0.877 & 0.998 & 0 \\ 1.08 \times 10^{-4} & -9.47 & -1.46 & 0 \\ 0 & 0 & 1 & 0 \end{bmatrix}$$

$$B_{Long} = \begin{bmatrix} 20.5 & 0 \\ 0 & -1.66 \times 10^{-4} \\ 0 & -12.8 \\ 0 & 0 \end{bmatrix}$$

10.1.2 State transition matrix and eigenvalues

Using Fedeeva's algorithm (Appendix D) we calculate the state transition matrix $[sI - A_{Long}]^{-1}$, which results in

$$[sI - A_{Long}]^{-1} = \frac{C(s)}{d(s)} = \frac{\{c_{ij}(s)\}}{d(s)}, \ i = 1 \ldots 4, \ j = 1 \ldots 4$$

The terms $c_{ij}(s)$ in the numerator are

$$c_{11}(s) = s^3 + 2.34s^2 + 10.7s$$
$$c_{12}(s) = -2.26s^2 - 3.29s + 305$$
$$c_{13}(s) = -34.4s - 28.2$$
$$c_{14}(s) = -32.2s^2 - 75.2s - 345$$
$$c_{21}(s) = -3.16 \times 10^{-4}s^2 - 3.53 \times 10^{-4}s$$
$$c_{22}(s) = s^3 + 1.47s^2 + 2.21 \times 10^{-2}s + 3.48 \times 10^{-3}$$
$$c_{23}(s) = 0.998s^2 + 1.51 \times 10^{-2}s + 1.02 \times 10^{-2}$$
$$c_{24}(s) = 1.02 \times 10^{-2}s + 1.14 \times 10^{-2}$$
$$c_{31}(s) = 1.08 \times 10^{-4}s^2 + 3.09 \times 10^{-3}s$$
$$c_{32}(s) = -9.47s^2 - 0.144s$$
$$c_{33}(s) = s^3 + 0.892s^2 + 1.26 \times 10^{-2}s$$
$$c_{34}(s) = -3.48 \times 10^{-3}s - 9.93 \times 10^{-2}$$
$$c_{41}(s) = 1.08 \times 10^{-4}s + 3.09 \times 10^{-3}$$
$$c_{42}(s) = -9.47s - 0.144$$
$$c_{43}(s) = s^2 + 0.892s + 1.26 \times 10^{-2}$$
$$c_{44}(s) = s^3 + 2.35s^2 + 10.8s + 0.162$$

The characteristic polynomial is

$$d(s) = s^4 + 2.35s^3 + 10.76s^2 + 0.1652s + 0.0993$$

In factored form,

$$d(s) = (s + 1.17 - j3.06)(s + 1.17 + j3.06)$$
$$(s + 0.0067 - j0.096)(s + 0.0067 + j0.096)$$

Thus the eigenvalues of the system are

$$\lambda_{1,2} = -1.17 \pm j3.06$$
$$\lambda_{3,4} = -0.0067 \pm j0.096$$

The eigenvalues are in the form $\lambda = \sigma \pm j\omega_d$, in which σ is the damping term and ω_d is the damped frequency of the response. Both modes are stable (negative damping terms), although the second mode has a real part very near zero, indicating that it is only marginally stable. In standard second-order form the two modes are

$$d(s) = (s^2 + 2.34s + 10.7)(s^2 + 0.0134s + 0.00925)$$

Each of the oscillatory modes may be compared with $s^2 + 2\zeta\omega_n s + \omega_n^2$, from which we learn that the first mode has natural frequency $\omega_{n_{1,2}} = 3.27$ rad/s and damping ratio $\zeta_{1,2} = 0.357$; and the second mode has $\omega_{n_{3,4}} = 0.0962$ rad/s and $\zeta_{3,4} = 0.0696$. These results are qualitatively typical of 'conventional' aircraft: one mode is characterized by relatively large natural frequency and damping, and the other by relatively small natural frequency and damping. We may evaluate the time to half amplitude, period, and number of cycles to half amplitude associated with each of the responses (Table 10.1).

Table 10.1 Longitudinal metrics.

Metric	$\lambda_{1,2} = \lambda_{SP}$ (Short period)	$\lambda_{3,4} = \lambda_{Ph}$ (Phugoid)
$t_{1/2} = \dfrac{\ln(1/2)}{\sigma}$	0.592 s	103 s
$T = \dfrac{2\pi}{\omega_d}$	2.05 s	65.4 s
$N_{1/2} = \dfrac{t_{1/2}}{T}$	0.289	1.57

The first mode, associated with $\lambda_{1,2}$, is seen to have a relatively short period. This gives rise to the unimaginative name for this mode: it is the *short period mode*. The other mode has a much more imaginative name, the *phugoid mode*. The origin of this name is historically reported as being due to F.W. Lanchester in *Aerodonetics* (1908), who thought he had the root of the Greek word 'to fly' but erroneously picked the root of the word 'to flee'. Whatever its origins, the word *phugoid* is firmly ensconced in the argot of aeronautical engineers.

10.1.3 Eigenvector analysis

We now examine the eigenvectors of the longitudinal modes. Using software such as MATLAB®, the modal matrix M is determined to be

$$M = \begin{bmatrix} \mathbf{v}_1 & \mathbf{v}_2 & \mathbf{v}_3 & \mathbf{v}_4 \end{bmatrix}$$

$$\mathbf{v}_1 = \begin{Bmatrix} 3.66 \times 10^{-2} + j0.946 \\ 8.01 \times 10^{-2} + j5.18 \times 10^{-2} \\ -0.182 + j0.231 \\ 8.56 \times 10^{-2} + j2.68 \times 10^{-2} \end{Bmatrix}, \quad \mathbf{v}_2 = \mathbf{v}_1^*$$

$$\mathbf{v}_3 = \begin{Bmatrix} 4.90 \times 10^{-2} - j0.999 \\ -3.76 \times 10^{-6} + j3.27 \times 10^{-5} \\ 9.08 \times 10^{-6} - j2.88 \times 10^{-4} \\ -2.99 \times 10^{-3} + j1.14 \times 10^{-4} \end{Bmatrix}, \quad \mathbf{v}_4 = \mathbf{v}_3^*$$

The first two columns of M ($\mathbf{v}_{1,2}$) correspond to the short period roots, and the last two ($\mathbf{v}_{3,4}$) to the phugoid roots. In polar form, the short period eigenvector is

$$\mathbf{v}_1 = \begin{Bmatrix} 0.947 \angle 87.8 \deg \\ 9.54 \times 10^{-2} \angle 32.9 \deg \\ 0.294 \angle 128.3 \deg \\ 8.97 \times 10^{-2} \angle 17.4 \deg \end{Bmatrix}$$

From the eigenvector the Argand diagram for the short period response is drawn (Figure 10.1), with dashed lines indicating the direction of states too small to be represented.

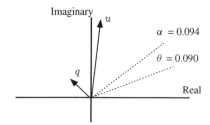

Figure 10.1 Argand diagram for the short period mode.

The Argand diagram may be used to visualize the behavior of the states relative to one another. The pitch rate leads to changes in angle-of-attack and pitch attitude by about 90 deg, reaching its maximum and minimum values about one-quarter cycle before the two angles. The undamped changes in angle-of-attack and pitch attitude are nearly the same magnitude and very close in phase. This means they will each reach their minimum and maximum values at about the same time. The undamped time histories of α and θ show this result (Figure 10.2).

If we include damping, and let the constant multipliers be unity, we have the results shown in Figure 10.3.

The damped time histories show that this particular short period mode dies out very quickly. The total response to arbitrary initial conditions will, of course, be the sum of the short period and phugoid modes. The responses associated with the phugoid, with its period of over a minute, will have barely changed during the time it takes the short period to dampen, suggesting that the short period may be viewed in isolation from the phugoid. On the other hand, the short period influences will have long since vanished from the phugoid response by the time large changes have taken place in that mode: the short period will appear as a small wrinkle at the beginning of the phugoid response.

From the Argand diagram one might be tempted to conclude that the short period mode is characterized by large changes in velocity u and pitch rate q. However, a 1 ft/s change in speed is not very large when compared with $V_{Ref} = 446.6$ ft/s. Note that if u is scaled by the reference velocity (effectively yielding $\hat{V}$) quite a different picture results, in which the short period occurs at almost constant speed, and is dominated by changes in α, θ, and q. Also note that if we use $\hat{V}$ and the non-dimensional pitch rate, $\hat{q} = q\bar{c}/2V_{Ref}$ ($\bar{c} = 10.8$ ft) then the interpretation might be that the short period is dominated by changes

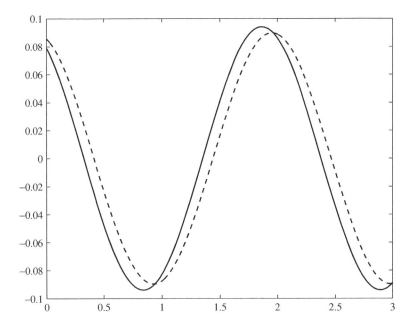

Figure 10.2 Undamped time histories, alpha and theta, time in seconds.

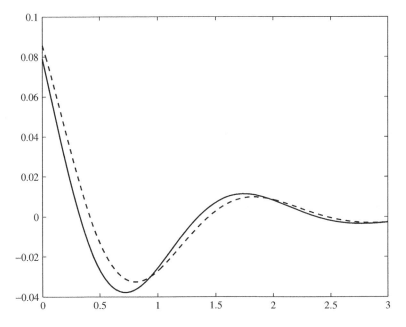

Figure 10.3 Damped time histories, alpha and theta, time in seconds.

in α and θ. Analysis such as this often depends upon knowing the right answer before beginning, in which some combination of scaling can be found to show it is right.

Now examining the phugoid response, the polar form of its eigenvector is

$$\mathbf{v}_3 = \begin{Bmatrix} 1.0 \angle -87.2 \deg \\ 3.30 \times 10^{-5} \angle 96.5 \deg \\ 2.88 \times 10^{-4} \angle -88.2 \deg \\ 3.00 \times 10^{-3} \angle 177.8 \deg \end{Bmatrix}$$

From this information the Argand diagram is drawn, as shown in Figure 10.4.

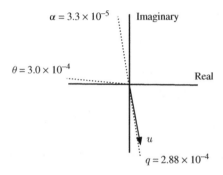

Figure 10.4 Argand diagram for the phugoid mode.

The inference from the Argand diagram is that the phugoid mode is dominated by large changes in velocity u. Even if u is scaled by $V_{Ref} = 446.6$ ft/s it is still much larger than α and q, and only slightly smaller than θ. It will turn out that this is largely true. Had we included altitude as a fifth longitudinal state we would have found its component of the phugoid eigenvector to be comparable to that of velocity and almost 180 deg out of phase with it. The phugoid mode corresponds to cyclic tradeoffs in kinetic energy (velocity) and potential energy (height), performed at nearly constant angle-of-attack. A time history of changes in velocity, Figure 10.5, shows part of this relationship.

10.1.4 Longitudinal mode sensitivity and approximations

Applying the steps described in Section 9.4.1 yields the sensitivity matrix for the longitudinal dynamics, shown in Table 10.2.

The entries in Table 10.2 are unambiguous. Each mode has two states that dominate in that mode. We should be able to approximate the short period using the states α and q, and approximate the phugoid mode using the states u and θ.

The objective of this analysis is to formulate two second-order systems, one for each mode, in terms of the stability and control parameters. Before doing that, however, we will apply the approximations to the numerical results and see how well they do.

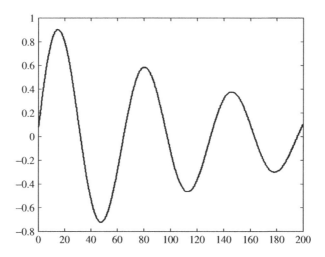

Figure 10.5 Time history of u for the phugoid, time in seconds.

Table 10.2 Longitudinal mode sensitivities.

	Short period		Phugoid	
u	0.0005	0.0005	0.4995	0.4995
α	0.4952	0.4952	0.0048	0.0048
q	0.4961	0.4961	0.0039	0.0039
θ	0.0004	0.0004	0.4996	0.4996

Short period

Numerical formulation

The short period is clearly a faster mode than the phugoid, (real part of -1.17, compared with -0.067 for the phugoid). To approximate the short period mode, we assume that $\dot{u} = \dot{\theta} = u = \theta = 0$. As a result,

$$\tilde{\mathbf{x}}_{SP} \equiv \begin{Bmatrix} \alpha \\ q \end{Bmatrix}$$

$$\tilde{A}_{SP} = \begin{bmatrix} -0.877 & 0.998 \\ -9.47 & -1.46 \end{bmatrix}$$

Denoting the short period approximation eigenvalues by $\tilde{\lambda}_{SP}$,

$$\tilde{\lambda}_{SP} = -1.17 \pm j3.06$$

This result is identical to that from the full system, and is a very good approximation indeed, at least for the current example.

Literal formulation

In terms of the parameters from which the example was derived (neglecting $Z_{\dot{w}}$ and $M_{\dot{w}}$), we have

$$\tilde{\mathbf{x}}_{SP} \equiv \begin{Bmatrix} \alpha \\ q \end{Bmatrix} \tag{10.1a}$$

$$\tilde{A}_{SP} = \begin{bmatrix} \dfrac{Z_w}{m} & \dfrac{Z_q + mV_{Ref}}{mV_{Ref}} \\ \dfrac{M_w V_{Ref}}{I_{yy}} & \dfrac{M_q}{I_{yy}} \end{bmatrix} \tag{10.1b}$$

The eigenvalues of the short period approximation are found from

$$|sI - \tilde{A}_{SP}| = s^2 + c_1 s + c_0 = 0$$

where

$$|sI - \tilde{A}_{SP}| \cong s^2 - \left(\frac{I_{yy} Z_w + mM_q}{mI_{yy}} \right) s + \frac{Z_w M_q - M_w(Z_q + mV_{Ref})}{mI_{yy}}$$

Applying this assumption to our example yields

$$\tilde{A}_{SP} = \begin{bmatrix} -0.879 & 1.00 \\ -9.77 & -1.12 \end{bmatrix}$$

$$\tilde{\lambda}_{SP} = -1.00 \pm j3.12$$

This approximation still compares favorably to $\lambda_{SP} = -1.17 \pm j3.06$, especially in terms of natural frequency (3.28 versus 3.27 rad/s) and somewhat so with respect to damping ratio (0.307 versus 0.357). Considering that $Z_{\dot{w}}$ and $M_{\dot{w}}$ are difficult to determine experimentally and thus subject to fairly large uncertainty, the approximation is probably not bad.

Phugoid

Numerical formulation

Now considering the phugoid approximation, we set $\dot{\alpha} = \dot{q} = 0$. Because α and q are associated with a faster mode, we treat them as fast variables. The procedure asks us to solve the $\dot{\alpha}$ and $\dot{q}$ for steady-state values to be used in the phugoid approximation. This yields

$$\dot{\alpha} = 0 = -(3.16 \times 10^{-4})u - 0.877\alpha_{ss} + 0.998 q_{ss}$$

$$\dot{q} = 0 = (1.08 \times 10^{-4})u - 9.47\alpha_{ss} - 1.46 q_{ss}$$

Solving this system of equations for the steady-state values of α and q as a function of u yields

$$\alpha_{ss} = -(3.295 \times 10^{-5})u, \quad q_{ss} = (2.877 \times 10^{-4})u$$

The resulting equations for $\dot{u}$ and $\dot{\theta}$ become

$$\dot{u} = -(1.51 \times 10^{-2})u - 2.26\alpha_{ss} - 32.2\theta = -(1.50 \times 10^{-2})u - 32.2\theta$$

$$\dot{\theta} = q_{ss} = (2.877 \times 10^{-4})u$$

The phugoid approximation is then

$$\tilde{\mathbf{x}}_{Ph} = \begin{Bmatrix} u \\ \theta \end{Bmatrix}$$

$$\tilde{A}_{Ph} = \begin{bmatrix} -1.50 \times 10^{-2} & -32.2 \\ 2.877 \times 10^{-4} & 0 \end{bmatrix}$$

The eigenvalues of this system are

$$\hat{\lambda}_{Ph} = -0.0075 \pm j0.096$$

Compared with the actual eigenvalues, $\lambda_{Ph} = -0.0067 \pm j0.096$, this approximation does well on the damped frequency, but overestimates the damping term by almost 15%.

Literal formulation
In terms of the dimensional stability derivatives, we have

$$\dot{\alpha} = 0 = \frac{Z_u}{mV_{Ref}}u + \frac{Z_w}{m}\alpha_{ss} + \left(\frac{Z_q}{mV_{Ref}} + 1\right)q_{ss}$$

$$\dot{q} = 0 = \frac{M_u}{I_{yy}}u + \frac{M_w V_{Ref}}{I_{yy}}\alpha_{ss} + \frac{M_q}{I_{yy}}q_{ss}$$

The derivatives Z_q and M_u are usually negligible, which results in

$$\begin{Bmatrix} \alpha_{ss} \\ q_{ss} \end{Bmatrix} = \begin{bmatrix} Z_w V_{Ref} & mV_{Ref} \\ M_w V_{Ref} & M_q \end{bmatrix}^{-1} \begin{Bmatrix} -Z_u u \\ 0 \end{Bmatrix}$$

$$= \begin{Bmatrix} -M_q \\ M_w V_{Ref} \end{Bmatrix} \frac{Z_u u}{V_{Ref}(M_q Z_w - mM_w V_{Ref})}$$

The contribution of α_{ss} appears only in the equation for $\dot{u}$, and is normally quite small (see the earlier numerical example, in which the factor of u was changed from -1.51×10^{-2} to -1.53×10^{-2}). The q_{ss} term is, however, critical to the $\dot{\theta}$ equation; without it we have $\dot{\theta} = 0$ and the system is not oscillatory at all. We therefore ignore α_{ss} but retain q_{ss}, leaving:

$$\dot{u} = \frac{X_u + T_u}{m}u - g\theta$$

$$\dot{\theta} = q_{ss} = \frac{M_w Z_u}{M_q Z_w - mM_w V_{Ref}}u$$

We have let $Z_{\dot{w}} = 0$ (since we have let $\dot{\alpha} = 0$), $\gamma_{Ref} = \epsilon_T = 0$ (they are usually quite small), leaving us with the phugoid approximation:

$$\tilde{\mathbf{x}}_{Ph} = \begin{Bmatrix} u \\ \theta \end{Bmatrix} \tag{10.2a}$$

$$\tilde{A}_{Ph} = \begin{bmatrix} \dfrac{X_u + T_u}{m} & -g \\ M_w Z_u & \\ M_q Z_w - m M_w V_{Ref} & 0 \end{bmatrix} \tag{10.2b}$$

Further simplifications

In the phugoid approximation it is normally taken that $|m M_w V_{Ref}| \gg |M_q Z_w|$, so that the system matrix becomes

$$\tilde{A}_{Ph} = \begin{bmatrix} \dfrac{X_u + T_u}{m} & -g \\ \dfrac{-Z_u}{m V_{Ref}} & 0 \end{bmatrix} \tag{10.2c}$$

The derivatives X_u and Z_u can be expressed in terms of the lift and drag of the aircraft. Moreover, since the example aircraft is a jet, $T_u = 0$. Finally, we focus on subsonic flight and therefore ignore Mach effects. The evaluation of X_u proceeds as follows:

$$X_u = \left(\dfrac{\bar{q}_{Ref} S}{V_{Ref}} \right) (2C_W \sin \gamma_{Ref} - 2C_{T_{Ref}} \cos \epsilon_T - M_{Ref} C_{D_M})$$

$$= -2 \left(\dfrac{\bar{q}_{Ref} S}{V_{Ref}} \right) C_{T_{Ref}}$$

$$= -2 \left(\dfrac{\bar{q}_{Ref} S}{V_{Ref}} \right) C_{D_{Ref}}$$

$$= -2 \left(\dfrac{D_{Ref}}{V_{Ref}} \right)$$

Similarly for Z_u:

$$Z_u = \left(\dfrac{\bar{q}_{Ref} S}{V_{Ref}} \right) (-2C_{L_{Ref}} - M_{Ref} C_{L_M})$$

$$= -2 \left(\dfrac{\bar{q}_{Ref} S}{V_{Ref}} \right) C_{L_{Ref}}$$

$$= -2 \left(\dfrac{L_{Ref}}{V_{Ref}} \right)$$

The phugoid system matrix is therefore approximated by

$$\tilde{A}_{Ph} = \begin{bmatrix} -2\dfrac{D_{Ref}}{mV_{Ref}} & -g \\ 2\dfrac{L_{Ref}}{mV_{Ref}^2} & 0 \end{bmatrix} \qquad (10.2d)$$

The characteristic polynomial of this system is

$$|sI - \tilde{A}_{Ph}| = s^2 + 2\dfrac{D_{Ref}}{mV_{Ref}}s + 2\dfrac{gL_{Ref}}{mV_{Ref}^2}$$

In the last term replace g with mg/m, and note that $L_{Ref} = W = mg$, so that

$$|sI - \tilde{A}_{Ph}| = s^2 + 2\dfrac{D_{Ref}}{mV_{Ref}}s + 2\dfrac{L_{Ref}^2}{m^2 V_{Ref}^2}$$

From this expression we see that the natural frequency of the phugoid approximation is inversely proportional to the trimmed flight speed,

$$\tilde{\omega}_{n_{Ph}} = \sqrt{2}\dfrac{L_{Ref}}{mV_{Ref}} \qquad (10.2e)$$

The damping ratio turns out to be inversely proportional to the famous aircraft performance parameter, the ratio of lift to drag (L/D):

$$\tilde{\zeta}_{Ph} = \dfrac{D_{Ref}}{\sqrt{2}L_{Ref}} = \dfrac{1}{\sqrt{2}(L/D)_{Ref}} \qquad (10.2f)$$

The phugoid is thus least well damped when the aircraft is most efficiently operated, that is, at $(L/D)_{Max}$.

The phugoid approximation given by Equation 10.2c can be derived by treating the aircraft as a point-mass that possesses the lift and drag characteristics of the aircraft. Then, by analyzing the kinetic and potential energy of the system, Equation 10.2c results. Note that the assumption that got us to that equation could be reached by letting $M_q = 0$. When the aircraft is treated as a point-mass, there are no pitch dynamics, so the two approaches are equivalent.

The phugoid approximation is a much-studied problem, and there are many methods to deal with it. For an interesting comparison of the several approximation methods, see Pradeep (1998).

10.1.5 Forced response

We now turn to the forced response. The matrix of transfer functions is easily evaluated from $[sI - A]^{-1}B \equiv G(s) = \{g_{ij}(s)\}$.

$$G(s) = \dfrac{\begin{bmatrix} (20.5s^3 + 48.0s^2 + 220s) & (44.2s + 363) \\ (-6.49 \times 10^{-3}s^2 - 7.26 \times 10^{-3}s) & (-12.8s^2 - 0.194s - 0.131) \\ (2.22 \times 10^{-3}s^2 + 0.0634s) & (-12.8s^3 - 11.5s^2 - 0.162s) \\ (2.22 \times 10^{-3}s + 0.0634) & (-12.8s^2 - 11.5s - 0.162) \end{bmatrix}}{(s + 1.17 \pm j3.06)(s + 0.0067 \pm j0.096)} \qquad (10.3)$$

This result is more useful with the numerator factored to show the zeros:

$$G(s) = \frac{\begin{bmatrix} 20.5s\,(s+1.17\pm j3.06) & 44.2(s+0.820) \\ -6.49\times 10^{-3}s(s+1.12) & -12.8(s+0.0076\pm j0.101) \\ 2.22\times 10^{-3}s(s+28.6) & -12.8s(s+0.877)(s+0.0143) \\ 2.22\times 10^{-3}(s+28.6) & -12.8(s+0.877)(s+0.0143) \end{bmatrix}}{(s+1.17\pm j3.06)(s+0.0067\pm j0.096)}$$

Two interesting results may be directly observed from the matrix of transfer functions. First, with respect to the influence of the throttle on changes in speed,

$$g_{11}(s) = \frac{u(s)}{\delta_T(s)} = \frac{20.5s\,(s+1.17\pm j3.06)}{(s+1.17\pm j3.06)(s+0.0067\pm j0.096)}$$

$$= \frac{20.5s}{s+0.0067\pm j0.096}$$

The short period mode has been cancelled by an identically placed pair of zeros in the numerator. This cancellation means that changes in throttle setting will have *no* effect on the short period dynamics. The reason for this is that we have modeled the thrust as acting straight along the aircraft x-axis, hence it will create no pitching moment or off-axis forces.

The transfer function of the pitching moment controller (elevator) to angle-of-attack has a similar result, but in this case it is the phugoid mode that is (almost) canceled:

$$g_{22}(s) = \frac{\alpha(s)}{\delta_m(s)} = \frac{-12.8(s+0.0076\pm j0.101)}{(s+1.17\pm j3.06)(s+0.0067\pm j0.096)}$$

$$\approx \frac{-12.8}{s+1.17\pm j3.06}$$

The matrix of transfer functions may be used to evaluate steady-state longitudinal responses to control inputs. There are three considerations we must observe:

1. The input and response must be 'small',
2. There must be no coupling of the response with lateral–directional modes, and
3. The steady-state conditions must exist.

The first two requirements mean that we want to stay within the range of validity of the assumptions made in linearizing the equations of motion. Because the equations are linear, the magnitudes of inputs and responses may be uniformly scaled, so if a test input results in large response, we simply reduce inputs and responses by the same factor. The second consideration, no coupling, will be satisfied for the longitudinal equations of motion in straight, symmetric flight because we have assumed away any dependence on lateral–directional variables: there is no mechanism to create sideslip, roll or yaw rates, or bank angle changes. Finally, we will have to apply common sense to determine whether the inputs result in true steady-state conditions. For instance, if the analysis shows steady-state values of pitch rate and pitch angle, we should dismiss the results since this is not possible. In addition, of course, we require that the system be stable, otherwise the steady-state solution will not exist.

Throttle input

In response to a step throttle input of magnitude 0.1, the steady-state responses of the longitudinal states are

$$\mathbf{x}_{Long}(\infty) = \lim_{s \to 0}(s)\frac{\begin{Bmatrix} 20.5s\,(s+1.17 \pm j3.06) \\ -6.49 \times 10^{-3}s(s+1.12) \\ 2.22 \times 10^{-3}s(s+28.6) \\ 2.22 \times 10^{-3}(s+28.6) \end{Bmatrix}}{s^4 + 2.35s^3 + 10.76s^2 + 0.1652s + 0.0993}\left(\frac{0.1}{s}\right)$$

$$\mathbf{x}_{Long}(\infty) = \begin{Bmatrix} 0 \\ 0 \\ 0 \\ 0.0638\ rad \end{Bmatrix} = \begin{Bmatrix} \Delta u_{SS} \\ \alpha_{SS} \\ q_{SS} \\ \theta_{SS} \end{Bmatrix}$$

The steady-state values will be obtained only after the short-period and phugoid modes have subsided.

This result may at first seem counter-intuitive. One might think that increasing the throttle should make the aircraft go faster, but instead the speed and angle-of-attack return to the trim value and the aircraft ends up in a climb. Recall, however, the steady-state requirement $M + M_T = 0$. The functional dependency of the pitching moment coefficient (neglecting altitude dependency) is $C_m(M, \alpha, \hat{\dot{\alpha}}, \hat{q}, \delta_m)$. In our example $M_T = 0$. There is no Mach dependency, no change in pitching moment control, and $\dot{\alpha}$ and q are zero in steady, straight flight. Therefore if $C_m(M, \alpha, \hat{\dot{\alpha}}, \hat{q}, \delta_m) = 0$ in the reference condition, and in the steady-state condition, the sole remaining variable, α, must be unchanged. Moreover, from $\dot{\alpha} = 0 \Rightarrow \dot{w} = 0$, every term in the numerator of the $\dot{w}$ equation is zero in the steady state except $Z_u \Delta u$, which must vanish as well:

$$Z_u \Delta u = 0 \Rightarrow \Delta u = 0$$

Elevator input

In response to a -1 deg step input in the elevator (TEU), the steady-state responses of the longitudinal states are

$$\mathbf{x}_{Long}(\infty) = \lim_{s \to 0}(s)\frac{\begin{bmatrix} 44.2\,(s+0.820) \\ -12.8(s+0.0076 \pm j0.101) \\ -12.8s(s+0.877)(s+0.0143) \\ -12.8(s+0.877)(s+0.0143) \end{bmatrix}}{s^4 + 2.35s^3 + 10.76s^2 + 0.1652s + 0.0993}\left(\frac{-0.01745}{s}\right)$$

$$\mathbf{x}_{Long}(\infty) = \begin{Bmatrix} \Delta u_{SS} \\ \alpha_{SS} \\ q_{SS} \\ \theta_{SS} \end{Bmatrix} = \begin{Bmatrix} -6.37\ \text{ft/s} \\ 0.0231\ rad \\ 0 \\ 0.0282\ rad \end{Bmatrix}$$

The elevator input has resulted in the aircraft seeking a new trim airspeed and angle-of-attack. The new value of α_{SS} is relative to the stability-axis value of zero. Since, in wings-level flight, we have $\gamma = \theta - \alpha$, we see that the aircraft is also climbing at an angle of $\gamma_{SS} = 0.292 \deg$.

10.2 Example: Lateral–Directional Dynamics

10.2.1 System matrices

See Appendix A for data. At a particular flight condition the A-4 *Skyhawk* has the following linearized, dimensional lateral–directional system and control matrices (the Δs have been dropped):

$$\dot{\mathbf{x}}_{LD} = A_{LD}\mathbf{x}_{LD} + B_{LD}\mathbf{u}_{LD}$$

$$\mathbf{x}_{LD} = \begin{Bmatrix} \beta \\ p \\ r \\ \phi \end{Bmatrix} \quad \mathbf{u}_{LD} = \begin{Bmatrix} \delta_\ell \\ \delta_n \end{Bmatrix}$$

$$A_{LD} = \begin{bmatrix} -0.248 & 0 & -1 & 0.072 \\ -23.0 & -1.68 & 0.808 & 0 \\ 13.5 & -0.0356 & -0.589 & 0 \\ 0 & 1 & 0 & 0 \end{bmatrix}$$

$$B_{LD} = \begin{bmatrix} 0 & 0.0429 \\ 17.4 & -21.9 \\ 4.26 & 0.884 \\ 0 & 0 \end{bmatrix}$$

10.2.2 State transition matrix and eigenvalues

Using Fedeeva's algorithm (Appendix D) we calculate the state transition matrix $[sI - A_{Long}]^{-1}$, which results in

$$[sI - A_{Long}]^{-1} = \frac{C(s)}{d(s)} = \frac{\{c_{ij}(s)\}}{d(s)}, \quad i = 1 \ldots 4, \; j = 1 \ldots 4$$

The terms $c_{ij}(s)$ in the numerator are

$$c_{11}(s) = s^3 + 2.27s^2 + 1.02s$$
$$c_{12}(s) = 0.108s + 0.0425$$
$$c_{13}(s) = -s^2 - 1.68s + 0.0582$$
$$c_{14}(s) = 0.0720s^2 + 0.164s + 0.0735$$
$$c_{21}(s) = -23.0s^2 - 2.64s$$
$$c_{22}(s) = s^3 + 0.837s^2 + 13.6s$$
$$c_{23}(s) = 0.808s^2 + 23.2s$$

$$c_{24}(s) = -1.66s - 0.190$$
$$c_{31}(s) = 13.5s^2 + 23.5s$$
$$c_{32}(s) = -0.0356s^2 - 8.81 \times 10^{-3}s + 0.972s$$
$$c_{33}(s) = s^3 + 1.93s^2 + 0.417s + 1.66$$
$$c_{34}(s) = 0.972s + 1.69$$
$$c_{41}(s) = -23.0s - 2.64$$
$$c_{42}(s) = s^2 + 0.837s + 13.6$$
$$c_{43}(s) = 0.808s + 23.2$$
$$c_{44}(s) = s^3 + 25.2s^2 + 15.1s + 23.8$$

The characteristic polynomial is

$$d(s) = s^4 + 2.52s^3 + 15.s^2 + 25.s + 0.190$$
$$= (s + 0.340 \pm j3.70)(s + 1.83)(s + 7.51 \times 10^{-3})$$
$$= (s^2 + 0.679s + 13.8)(s + 1.83)(s + 7.51 \times 10^{-3})$$

The eigenvalues of the system are

$$\lambda_{1,2} = -0.340 \pm j3.70$$
$$\lambda_3 = -1.83$$
$$\lambda_4 = -7.51 \times 10^{-3}$$

The complex roots give rise to a stable oscillatory mode with $\omega_n = 3.71$ rad/s and $\zeta = 0.0914$. The two real roots are both stable (damped exponentials), but λ_4 is very small and nearly unstable. This distribution of eigenvalues is fairly typical of 'conventional' aircraft, and the fourth eigenvalue is sometimes slightly unstable. For reasons that will become more clear following the modal analysis, the oscillatory mode is called the *Dutch roll*, the real root greatest in magnitude is the *Roll mode*, and the last root is the *Spiral mode*. The associated metrics are given in Table 10.3.

Table 10.3 Lateral-directional metrics.

Metric	$\lambda_{1,2} = \lambda_{DR}$ (Dutch roll)	$\lambda_3 = \lambda_R$ (Roll mode)	$\lambda_4 = \lambda_S$ (Spiral mode)
$t_{1/2} = \dfrac{\ln(1/2)}{\sigma}$	2.04 s	0.379 s	92.3 s
$T = \dfrac{2\pi}{\omega}$	1.70 s	–	–
$N_{1/2} = \dfrac{t_{1/2}}{T}$	1.2	–	–

10.2.3 Eigenvector analysis

The eigenvectors associated with these modes are as follows. With

$$M = \begin{bmatrix} \mathbf{v}_1 & \mathbf{v}_1^* & \mathbf{v}_3 & \mathbf{v}_4 \end{bmatrix}$$

$$\mathbf{v}_1 = \begin{Bmatrix} -0.0601 + j0.126 \\ -0.495 - j0.651 \\ 0.450 + j0.245 \\ -0.162 + j0.149 \end{Bmatrix} \quad \mathbf{v}_3 = \begin{Bmatrix} -0.00480 \\ -0.878 \\ 0.0269 \\ 0.479 \end{Bmatrix} \quad \mathbf{v}_4 = \begin{Bmatrix} 0.00305 \\ -0.00749 \\ 0.0711 \\ 0.997 \end{Bmatrix}$$

Only one Argand diagram is necessary. For the Dutch roll mode, we have the relationships shown in Figure 10.6. The Dutch roll mode appears to have significant components of each of the four lateral–directional states, suggesting (from this analysis) that there is no reasonable approximation. From the Argand diagram we see relatively large roll and yaw rates that are nearly 180 deg out of phase with each other. This means, roughly, that the aircraft will be simultaneously rolling one way and yawing the other. Since the bank angle is about 90 deg out of phase with both roll and yaw rates, we surmise that, for example, as the wings pass through the level position the aircraft will be rolling to the right while yawing to the left, both at near maximum rates.

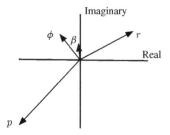

Figure 10.6 Argand diagram of the Dutch roll.

In Figure 10.7, moving from bottom to top, the aircraft begins with its wings level and a slight amount of positive sideslip. At this instant the yaw rate is positive (nose right) and the roll rate is negative (left wing down). One-quarter cycle later the bank angle has reached its maximum left wing down position and sideslip is near its maximum negative value. Roll and yaw rates are near or just past zero. Half-cycle into the time history the aircraft is again wings-level, but now with some negative sideslip, negative yaw rate (nose left) and positive roll rate (right wing down). The roll continues to the right and the yaw to the left, until the three-quarters cycle position at the top of the figure. Here the bank angle has reached its maximum right wing down position and sideslip is near its maximum positive values. One-quarter cycle later the aircraft is again in the position in the bottom figure, although the amplitudes of all four states will have diminished due to damping. In ice-skating a Dutch roll is executed by gliding with the feet parallel and pressing alternately on the edges of each foot; the similarity of the skater's motion to that shown by the aircraft gives rise to the name of this mode.

Other useful information may be gleaned from the Dutch roll eigenvector. As we shall later see, the ratio of peak roll angle to sideslip angle is of importance in determining

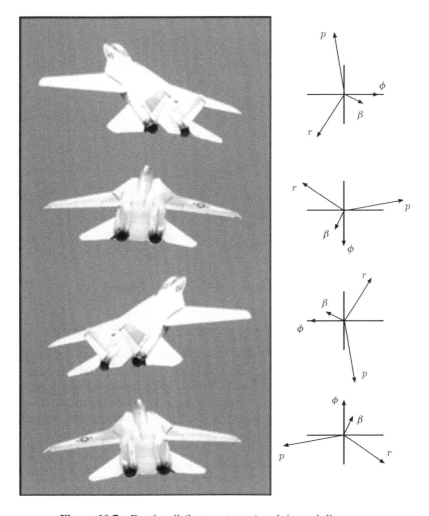

Figure 10.7 Dutch roll (bottom to top) and Argand diagrams.

the flying qualities of an aircraft: loosely, how easy and pleasant it is to fly. This ratio is determined from the magnitudes of the ϕ and β components of the eigenvector, and for our example it is

$$\left|\frac{\phi}{\beta}\right| = \frac{0.220}{0.140} = 1.57$$

Thus the magnitude of the roll angle is more than half again as great as that of sideslip; in supersonic flight this relationship is often reversed, giving rise to a 'snaking' motion. Except in aircraft with highly swept wings it is fairly easy for a pilot to observe the relationship between ϕ and β during the Dutch roll. The change in heading angle $\Delta\psi$ is almost exactly the negative of the sideslip angle β during the Dutch roll. As the aircraft rolls and yaws the left wing tip will trace a path which, for this example, looks like Figure 10.8.

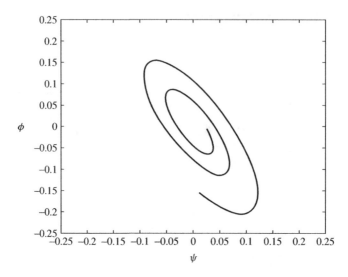

Figure 10.8 Path traced by the left wing tip.

10.2.4 Lateral–directional mode sensitivity and approximations

Results of the mode sensitivity analysis for the lateral–directional modes are shown in Table 10.4. The results are unambiguous and show that β and r are dominant in the Dutch roll mode, p in the roll mode, and ϕ in the spiral mode. So far as the speed of the three modes is concerned, the roll mode is fastest with a real eigenvalue of -1.83 s^{-1}, followed by the Dutch roll with a real part of -0.34 s^{-1}, followed by the real spiral root of -7.51×10^{-3} s^{-1}.

Table 10.4 Lateral–directional mode senstivities.

	Dutch roll		Roll	Spiral
β	0.4931	0.4931	0.0135	0.0003
p	0.0207	0.0207	0.9545	0.0041
r	0.4506	0.4506	0.0385	0.0604
ϕ	0.0147	0.0127	0.0522	0.9184

Roll mode

Numerical formulation

The roll rate p is dominant in the roll mode. The roll mode is almost purely motion about the body x-axis. The roll mode is the fastest of the lateral–directional modes, so we let $\dot{\beta} = \dot{r} = \dot{\phi} = \beta = r = \phi = 0$, leaving simply

$$\dot{p} = -1.68p$$

This simple first-order system has eigenvalue $\tilde{\lambda}_R = -1.68$, which compares favorably with the actual eigenvalue, $\lambda_R = -1.83$.

Aircraft Flight Dynamics

Literal formulation

The roll mode approximation is simply

$$\tilde{x}_R = \{p\} \tag{10.4a}$$

$$\dot{p} = \frac{L_p}{I_{xx}} p \tag{10.4b}$$

We may include the $\dot{\phi}$ equation, $\dot{\phi} = p$, without changing the roll-mode eigenvalue.

$$\tilde{x}_R = \begin{Bmatrix} p \\ \phi \end{Bmatrix}$$

$$\tilde{A}_R = \begin{bmatrix} L_p/I_{xx} & 0 \\ 1 & 0 \end{bmatrix}$$

This roll-mode approximation treats the aircraft as if it were constrained to motion about the x-axis only, and the only moment that remains is due to the roll-rate damping, L_p. Since only the roll of the aircraft is involved the mode is naturally called the roll mode, or sometimes the *roll-subsidence mode*.

Dutch roll

Numerical formulation

The Dutch roll is dominated by the variables β and r. It is slower than the roll mode but faster than the spiral mode. For the slower mode we take $\dot{\phi} = \phi = 0$. The roll mode is faster so we take $\dot{p} = 0$ but treat p as p_{ss}. This leads to

$$\dot{p} = 0 = -23.0\beta - 1.68 p_{ss} + 0.808r \Rightarrow p_{ss} = -13.69\beta + 0.4810r$$

Substituting this result into the equations for $\dot{\beta}$ and $\dot{r}$ effects only $\dot{r}$, since $\dot{\beta}$ does not depend on p:

$$\dot{r} = 13.5\beta - 0.0356 p_{ss} - 0.590r = 14.0\beta - 0.607r$$

The Dutch roll approximation (for this example) then becomes

$$\tilde{x}_{DR} = \begin{Bmatrix} \beta \\ r \end{Bmatrix}$$

$$\tilde{A}_{DR} = \begin{bmatrix} -0.248 & -1 \\ 14.0 & -0.607 \end{bmatrix}$$

$$\tilde{\lambda}_{DR} = -0.428 \pm j3.74$$

Compared with the actual eigenvalues, $\lambda_{DR} = -0.340 \pm j3.70$, this approximation appears reasonable. The estimated damping ratio ($\tilde{\zeta}_{DR} = 0.114$) captures the poorly damped characteristics of the Dutch roll ($\zeta_{DR} = 0.0914$). The natural and damped frequencies of the estimate and actual Dutch roll are very nearly the same.

Literal formulation

The literal expression is quite messy unless we neglect I_{xz}. A survey of stability and control data for various aircraft suggests I_{xz} is usually small, so we do assume $I_{xz} = 0$. The expression for p_{ss} becomes

$$p_{ss} = -\frac{L_v}{L_p}v - \frac{L_r}{L_p}r = -\frac{V_{Ref}L_v}{L_p}\beta - \frac{L_r}{L_p}r$$

When this expression is substituted into the equations for $\dot{\beta}$ and $\dot{r}$,

$$\dot{\beta} = \left(\frac{Y_v L_p - Y_p L_v}{mL_p}\right)\beta + \left(\frac{Y_r L_p - Y_p L_r}{mV_{Ref}L_p} - 1\right)r$$

$$\dot{r} = \left(\frac{V_{Ref}(N_v L_p - N_p L_v)}{I_{zz}L_p}\right)\beta + \left(\frac{N_r L_p - N_p L_r}{I_{zz}L_p}\right)r$$

Most stability and control data take $Y_p = Y_r = 0$, so that

$$\dot{\beta} = \left(\frac{Y_v}{m}\right)\beta - r$$

$$\dot{r} = \left(\frac{V_{Ref}(N_v L_p - N_p L_v)}{I_{zz}L_p}\right)\beta + \left(\frac{N_r L_p - N_p L_r}{I_{zz}L_p}\right)r$$

The Dutch roll approximation is then given by:

$$\tilde{\mathbf{x}}_{DR} = \begin{Bmatrix} \beta \\ r \end{Bmatrix} \tag{10.5a}$$

$$\tilde{A}_{DR} = \begin{bmatrix} \dfrac{Y_v}{m} & -1 \\ \dfrac{V_{Ref}(L_p N_v - L_v N_p)}{I_{zz}L_p} & \dfrac{L_p N_r - L_r N_p}{I_{zz}L_p} \end{bmatrix} \tag{10.5b}$$

In Equation 10.5b, further simplification is possible by making assumptions about the relative magnitudes of the stability derivatives involved. For the example aircraft being used, $|L_p N_r| \gg |L_r N_p|$, and $|L_p N_v| \gg |L_v N_p|$. A suitable approximation for the *Skyhawk's* Dutch roll mode at the given flight condition is therefore

$$\tilde{A}_{DR} = \begin{bmatrix} \dfrac{Y_v}{m} & -1 \\ \dfrac{V_{Ref}N_v}{I_{zz}} & \dfrac{N_r}{I_{zz}} \end{bmatrix}$$

The eigenvalues of this approximation are $\lambda_{DR} = -0.419 \pm j3.67$, not much different from the previous approximation, $\tilde{\lambda}_{DR} = -0.428 \pm j3.74$. Note that this approximation to the approximation is equivalent to taking $p_{ss} = 0$. This ambiguous effect of the roll rate equation may be due to the fact that the roll mode is only about five times faster than the Dutch roll.

Aircraft Flight Dynamics

Spiral mode

Numerical formulation

The spiral mode eigenvector shows a dominance in bank angle. Unlike the roll mode, however, there is no accompanying large roll rate. When viewed in the context of the long time scale associated with this mode, we may visualize the aircraft with some bank angle, slowly rolling back towards the wings level position. If this eigenvalue had been positive then the interpretation would have been of slowly increasing bank angle.

For the spiral mode approximation we let $\dot{\beta} = \dot{p} = \dot{r} = 0$. The steady-state values of β, p, and r are calculated from the roll and Dutch roll modes,

$$\begin{Bmatrix} 0 \\ 0 \\ 0 \end{Bmatrix} = \begin{bmatrix} -0.248 & 0 & -1 & 0.072 \\ -23.0 & -1.68 & 0.808 & 0 \\ 13.5 & -0.0356 & -0.590 & 0 \end{bmatrix} \begin{Bmatrix} \beta_{ss} \\ p_{ss} \\ r_{ss} \\ \phi \end{Bmatrix}$$

$$\begin{bmatrix} -0.248 & 0 & -1 \\ -23.0 & -1.68 & 0.808 \\ 13.5 & -0.0356 & -0.590 \end{bmatrix} \begin{Bmatrix} \beta_{ss} \\ p_{ss} \\ r_{ss} \end{Bmatrix} = \begin{Bmatrix} -0.072 \\ 0 \\ 0 \end{Bmatrix} \phi$$

The only one of the steady-state values needed is p_{ss}, which is determined to be

$$p_{ss} = -0.0081\phi$$

The spiral-mode approximation then becomes

$$\dot{\phi} = -0.0081\phi$$

The eigenvalue associated with this equation is $\tilde{\lambda}_S = -0.0081$, which is close to the actual value $\lambda_S = -0.00751$.

Literal formulation

First solving for p_{ss} (with $I_{xz} = 0$),

$$p_{ss} = \frac{-\begin{vmatrix} \frac{V_{Ref} L_v}{I_{xx}} & \frac{L_r}{I_{xx}} \\ \frac{V_{Ref} N_v}{I_{zz}} & \frac{N_r}{I_{zz}} \end{vmatrix}}{\begin{vmatrix} \frac{Y_v}{m} & \frac{Y_p}{mV_{Ref}} & \left(\frac{Y_r}{mV_{Ref}} - 1\right) \\ \frac{V_{Ref} L_v}{I_{xx}} & \frac{L_p}{I_{xx}} & \frac{L_r}{I_{xx}} \\ \frac{V_{Ref} N_v}{I_{zz}} & \frac{N_p}{I_{zz}} & \frac{N_r}{I_{zz}} \end{vmatrix}} \left(\frac{g}{V_{Ref}}\right)\phi$$

Again taking $Y_p = Y_r = 0$,

$$p_{ss} = \frac{-\begin{vmatrix} \frac{V_{Ref} L_v}{I_{xx}} & \frac{L_r}{I_{xx}} \\ \frac{V_{Ref} N_v}{I_{zz}} & \frac{N_r}{I_{zz}} \end{vmatrix}}{\begin{vmatrix} \frac{Y_v}{m} & 0 & -1 \\ \frac{V_{Ref} L_v}{I_{xx}} & \frac{L_p}{I_{xx}} & \frac{L_r}{I_{xx}} \\ \frac{V_{Ref} N_v}{I_{zz}} & \frac{N_p}{I_{zz}} & \frac{N_r}{I_{zz}} \end{vmatrix}} \left(\frac{g}{V_{Ref}}\right)\phi$$

$$\tilde{x}_S = \{\phi\} \tag{10.6a}$$

$$\dot{\phi} = \frac{g(L_v N_r - L_r N_v)}{\frac{Y_v}{m}(L_p N_r - L_r N_p) - V_{Ref}(L_v N_p - L_p N_v)} \phi \tag{10.6b}$$

Further simplifications
In Equation 10.6b, further simplification may be possible by making assumptions about the relative magnitudes of the stability derivatives involved. For instance, for the example aircraft being used, the denominator term $V_{Ref}(L_v N_p - L_p N_v)$ is more than 16 times as large as the other term, and $|L_p N_v| \gg |L_v N_p|$. A suitable approximation for the *Skyhawk's* spiral mode at the given flight condition is:

$$\dot{\phi} = \frac{g(L_v N_r - L_r N_v)}{V_{Ref} L_p N_v} \phi$$

When the terms in this approximation are calculated, the eigenvalue that results is $\tilde{\lambda}_S = -0.00776$, which is (coincidentally) closer to the actual value $\lambda_S = -0.00751$ than the full approximation.

10.2.5 Forced response

We now turn to the forced response. Evaluating the matrix of transfer functions $[sI - A_{LD}]^{-1}B \equiv G(s)$, we have

$$\mathbf{x}_{LD} = \begin{Bmatrix} \beta \\ p \\ r \\ \phi \end{Bmatrix} \quad \mathbf{u}_{LD} = \begin{Bmatrix} \delta_\ell \\ \delta_n \end{Bmatrix}$$

$$G(s) = \frac{\begin{bmatrix} -4.26(s+1.41)(s-0.165) & 0.0429(s-22.3)(s+3.76)(s+0.243) \\ 17.4s(s+0.517 \pm j4.36) & -21.9(s+0.425 \pm j3.54) \\ 4.26(s+2.52)(s+0.368 \pm j1.45) & 0.884(s-1.89)(s+2.68 \pm j2.17) \\ 17.4(s+0.517 \pm j4.36) & -21.9(s+0.425 \pm j3.54) \end{bmatrix}}{(s+0.340 \pm j3.70)(s+1.83)(s+0.00751)}$$

Note that there are no pole-zero cancellations (with the possible exception of considering the free s in three of the numerators as cancelling the spiral mode). The interpretation of this observation is that, due to the great amount of cross-coupling between the rolling and yawing moment controls, application of either control will excite all three lateral–directional modes. Application of only the rolling moment control (in this case, aileron) will not generate pure roll, but will also excite the other two modes (primarily the Dutch roll). Pilots adapt to this phenomenon by learning to accompany aileron inputs with simultaneous rudder inputs.

There is no point in investigating the steady-state response of the aircraft to step inputs in rudder and in aileron. Any change in bank angle will incline the lift vector away from the vertical and will cause immediate coupling with the longitudinal modes. Application of the rolling moment controller (aileron) will certainly change the bank angle, and so will the yawing moment controller (rudder) through first the change in sideslip, then through a dihedral effect. The significance of this may be observed by considering a step input in aileron, which will, among other consequences, cause the aircraft to roll continuously. During part of the roll the lift vector will be pointing downward, causing the flight path to curve until the aircraft is accelerating in a steep descent.

Problems

For all problems, unless otherwise instructed MATLAB® or similar software may be used, but include print-outs of any work not done by hand.

1. An airplane's short-period response is approximated by:

$$\tilde{x}_{SP} \equiv \begin{Bmatrix} \alpha \\ q \end{Bmatrix}$$

$$\tilde{A}_{SP} = \begin{bmatrix} \dfrac{Z_w}{m} & \dfrac{Z_q + mV_{Ref}}{mV_{Ref}} \\ \dfrac{M_w V_{Ref}}{I_{yy}} & \dfrac{M_q}{I_{yy}} \end{bmatrix} \quad \tilde{B}_{SP} = \begin{bmatrix} \dfrac{Z_{\delta_m}}{m} & \dfrac{Z_{\delta_F}}{m} \\ \dfrac{M_{\delta_m}}{I_{yy}} & 0 \end{bmatrix}$$

The control δ_F represents wing trailing-edge flaps. The airplane is cruising in steady, straight, symmetric, level flight at 10 000 ft ($\rho = 0.001\,755\,56\,\text{slug/ft}^3$, $a = 1077.4\,\text{ft/s}$) at $Mach = 0.8$. The following data are given (any derivatives not listed may be assumed to be zero):

$m = 1000\,\text{slugs}$ $S = 400\,\text{ft}^2$
$\bar{c} = 11.52\,\text{ft}$ $I_{yy} = 124{,}000\,\text{slug}\cdot\text{ft}^2$
$C_{L_\alpha} = 5.73$ $C_{L_{\delta_m}} = 0.36$ $C_{L_{\delta_F}} = 1.80$
$C_{D_{Ref}} = 0.0198$ $C_{D_\alpha} = 0.36$
$C_{m_q} = -5.69$ $C_{m_\alpha} = -0.08$ $C_{m_{\delta_m}} = -0.50$

(a) Find the approximate short-period natural frequency, damped frequency, damping ratio, period, and time to half amplitude.
(b) Assume the flaps are deflected in a $+5\,\text{deg}$ step input at $t = 0$. Find the step elevator deflection δ_m that must be simultaneously applied so that the pitch rate returns to zero. What are the corresponding values of $\Delta\alpha(\infty)$ and $\Delta\theta(\infty)$?

2. Analyze the lateral–directional dynamics of the multi-engine aircraft in Chapter 7 problems using the data given. Data are valid for stability axes. Assume the aircraft is in steady, straight, symmetric, gliding flight with the engine off ($T = 0$).

3. Our Dutch roll approximation for the *Skyhawk* was based on:

$$\tilde{x}_{DR} = \begin{Bmatrix} \beta \\ r \end{Bmatrix}, \quad \tilde{A}_{DR} = \begin{bmatrix} -0.248 & -1 \\ 14.0 & -0.607 \end{bmatrix}$$

(a) Apply the linearized $\dot{\psi}$ equation to the Dutch roll approximation to find a third-order system in which the states are β, r, and ψ.
(b) Calculate the eigenvalues of your answer to part 3a, and calculate the eigenvector associated with the real eigenvalue. If you have more than one real eigenvalue, select the one that is nearest in magnitude to zero.
(c) Explain the physical significance of this eigenvalue and eigenvector.
(d) Determine the magnitude and phase relationships between β and ψ in the Dutch roll. Express the phase relationship in degrees.

4. We derived expressions for approximating the phugoid natural frequency and damping ratio for an aircraft with constant thrust, T:

$$\tilde{\omega}_{nPh} = \sqrt{2}\frac{L_{Ref}}{mV_{Ref}} \quad \tilde{\zeta}_{Ph} = \frac{1}{\sqrt{2}(L/D)_{Ref}}$$

Find expressions for approximating the phugoid natural frequency and damping ratio for an aircraft with constant thrust-horsepower $TV = T_{Ref}V_{Ref}$).

5. Given the linearized lateral–directional equations of motion of an aircraft in SSSLF, stability axes, with $V_{Ref} = 733$ ft/s, and

$$\mathbf{x}_{LD} = \begin{Bmatrix} v \\ p \\ r \\ \phi \end{Bmatrix}$$

$$A_{LD} = \begin{bmatrix} -0.0294 & 0 & -731.1882 & 32.1740 \\ -0.0033 & -1.5857 & 0.2777 & 0 \\ 0.0012 & 0.0423 & -0.1842 & 0 \\ 0 & 1.0000 & 0 & 0 \end{bmatrix}$$

Using the Dutch roll approximation derived above, find the time between maximum sideslip angle and maximum yaw rate.

References

Lanchester, F.W. (1908) *Aerodynamics; Flight; Stability of Airplanes*, A. Constable, London.
Pradeep, S. (1998) A century of phugoid approximations, *Aircraft Design*, 1(2), 89–104.

11

Flying Qualities

11.1 General

This chapter provides a very brief overview of the subject of flying qualities, and presents examples just sufficient to relate the results of the previous chapters to some of the quantitative measures of flying qualities. Fuller treatments of flying qualities may be found in the excellent works by Cooper and Harper, for example Harper (1986), and by Hodgkinson (1999), among many others.

The author appears to be in a minority in preferring the term 'flying qualities' to 'handling qualities' in this discussion. The bias can be traced back several years to the author's days in a branch of flight test called FQ&P, which meant Flying Qualities and Performance. Old habits die hard.

The term 'flying qualities' means those qualities or characteristics of an aircraft which govern the ease and precision with which a pilot is able to perform the tasks required in support of an aircraft role. Performance of these tasks implies control of the aircraft states: airspeed, altitude, pointing angle (Euler angles), and so on. Some tasks may require very precise control over the states: for example, formation flying, in-flight refueling, carrier landings.

The nonlinearity and coupling of states and controls within the state equations of motion means that control of a single state cannot be accomplished without affecting others as well. Moreover, there is no direct control available for most of the states, meaning they must be controlled indirectly through other states. An obvious example of the latter is the control of bank angle, which is accomplished by varying the roll rate, which results from changing the rolling moment through the ailerons. In this example coupling occurs in various ways: a non-zero bank angle creates an unbalanced side component of gravity which tends to generate sideslip; the aileron deflection, plus a non-zero roll rate, may create yawing moments; and the non-zero roll rate may create 'gyroscopic' forces and moments in all axes, and inertia coupling.

To learn to fly is to learn to control desired states while minimizing undesired changes in other states. All conventional aircraft have the same equations of motion, and they differ dynamically primarily in their aerodynamics, geometry, mass, and moments of inertia. Having mastered flying techniques in one aircraft, transition to another aircraft requires the same basic skills but with compensation for the variations in dynamics between the aircraft. The pilot's ability to compensate for variations in aircraft dynamics allows him to fly different airplanes with a variety of characteristics with little change in task performance, but the workload required by the pilot to achieve a given level of performance is greatly affected by the aircraft dynamics. Another measure of workload is the degree to which the pilot has to compensate for inadequacies of the aircraft.

The ultimate measure of an airplane's flying qualities is the pilot's opinion. For many reasons, however, it is desirable to have engineering-type measures of an aircraft's flying qualities.

- The primary purpose of engineering measures is to provide design guidelines that will help assure good flying qualities.
- Secondarily, such measures often become the basis for contractural specifications when aircraft are purchased or modified.

A large part of research into flying qualities aims to determine concrete numbers based on design parameters that can be used to determine whether a given aircraft will or will not have good flying qualities.

11.1.1 Method

Since the flying qualities of an aircraft are bound up in its dynamic responses, the obvious place to start looking for flying qualities' measures is in the equations of motion. A great amount of information regarding an aircraft's nonlinear dynamics may be gleaned from its linearized dynamics. Linearized dynamics are characterized by many parameters, such as the natural frequency and damping ratio of the short period mode. A baseline of sorts may be established by comparing the linearized dynamics of various different aircraft with pilots' opinions of those aircraft. A more precise experimental method seeks to determine correlations between values of dynamical parameters and a pilot's opinion of how his workload in performing some well-defined task is affected by the different dynamic responses.

The experimental method requires:

1. A means to vary certain parameters independently of others.
2. The selection of a suitable task.
3. The definition of desired and adequate performance.
4. A pilot evaluation for each combination.

The comparison of flying qualities of existing aircraft fails the first requirement, since two different aircraft will generally differ in many more ways than one. The answer to this problem lies in the use of simulators. Since even a very high-fidelity ground based simulator is still just a simulator, not an airplane, very sophisticated in-flight simulators are often employed. For an excellent overview, see Weingarten (2005). These aircraft have the capability of using feedback control systems to artificially vary certain parameters. A feedback control method similar to that used in variable stability aircraft will be the subject of Section 13.2.

Selection of suitable tasks is accomplished by examining the intended operational use of an aircraft and selecting those areas in which poor flying qualities are most likely to have a great effect. For example, a normal, straight-in approach to landing may be fairly easy to perform in a given aircraft. If, however, the approach is in instrument conditions and requires a large correction in line-up just prior to touchdown, a different picture may emerge if the aircraft has poor Dutch roll or roll mode characteristics.

The definitions of desired and adequate performance are required to help quantify the pilot's opinion. Continuing with the example of a late line-up correction, desired performance might be defined as landing within ten feet of centerline, and adequate performance as being within 25 feet. Various rating scales have been devised for quantifying the pilot's opinion; the most widely employed is the Cooper–Harper rating scale, shown in Figure 11.1. Following performance of a task, the pilot begins at the bottom left and follows the decision tree to establish a numerical rating. The rating is often called the Handling Qualities Rating (HQR), and less often the Pilot Rating (PR) or Cooper–Harper Rating (CHR). The rating varies from 1 (pilot compensation not a factor) through 4 (moderate pilot compensation) for desired performance; then from 5 (considerable pilot compensation) through 7 (maximum tolerable pilot compensation) for adequate performance; and finally from 8 to 10 if controllability is in question.

Pilot opinion is, of course, a matter of opinion. Different pilots with different skills, backgrounds, experience, and expectations may differ in their assessment of the compensation required to perform a certain task. The way around this problem is to statistically evaluate the ratings of several pilots. Figure 11.2 (Chalk *et al.*, 1969) shows a representative result, in which the Dutch roll frequency and damping ratio, for different values

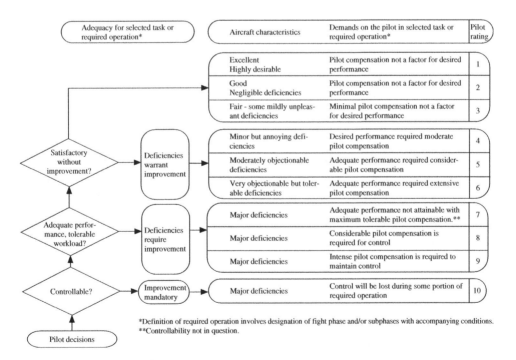

Figure 11.1 Cooper–Harper rating scale.

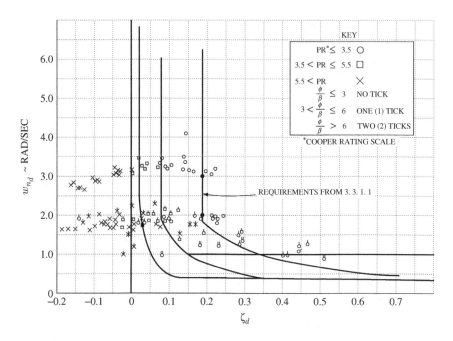

Figure 11.2 Dutch roll example (scanned from original).

of the ratio $|\phi/\beta|$, have been varied. It is clear at a glance that there has been a wide variance in the pilots' opinions of the workload required to perform the specified task.

For example, in Figure 11.2 there is a small circle with a single vertical tick near the coordinates $\zeta_d = -0.15$ and $\omega_{n_d} = 2.0$ rad/s, generally to the left of a veritable sea of × marks. The circle signifies that one pilot found, at worst, 'Moderately objectionable deficiencies' in the airplane in his performance of the task at hand, possibly a line-up correction on landing approach. The × marks reflect pilots whose opinions of the flying qualities were poor, and resulted in high HQR (here 'PR') ratings. A negative ζ_d means the Dutch roll is unstable, and those pilots used words like 'Adequate performance not attainable with maximum tolerable pilot compensation' to describe the airplane.

Also shown in Figure 11.2 are lines that divide the results up into various areas. These areas correspond to different levels of flying qualities for a particular class of aircraft in various flight phases. There are three levels of flying qualities that are related to the adequacy of the aircraft to perform its mission. The correspondence to levels of flying qualities and HQRs is (loosely):

Level 1 – HQR 1-3.5
Level 2 – HQR 3.5-6.5
Level 3 – HQR 6.5-10

Aircraft are not designed normally to display Level 2 or 3 flying qualities. These measures come into play when considering failure modes and effects, and tell the airplane designer how much the flying qualities are allowed to deteriorate after the failure of some system, based on the probability of the occurence of that failure.

Sometimes airplanes do enter service with worse than Level 1 flying qualities in some specific area, almost always accompanied by explicit requirements for additional pilot training or experience to permit mission performance. For example, night aircraft carrier landings in some airplanes may be restricted to pilots with scores or more of daytime landings.

11.1.2 Specifications and standards

In the field of military aircraft procurement the original specification was MIL-F-8785, 'Military Specification, Flying Qualities of Piloted Airplanes'. The last version of this specification was 8785C (MIL-F-8785C 1980), released in 1980. Contracts for new aircraft procurement included this specification by reference, and all of its provisions were applicable unless specifically waived in the contract. The specification contains both numerical and subjective requirements.

The specifications are based on extensive flight and simulator testing. The results were converted into numerical requirements using the language developed throughout this book. Each of MIL-F-8785B (Chalk *et al.*, 1969) and MIL-F-8785C (Moorhouse and Woodcock, 1982) was accompanied by a 'Background Information and User Guide' (BIUG) in which the experiments were described and rationale given for the criteria that resulted. These guides are indispensible for the conduct of any serious research into flying qualities. The reader of this book will find many familiar terms and transfer functions there.

The numerical requirements of 8785C generally are stated in terms of a linear mathematical description of the airplane. Certain factors, for example flight control system nonlinearities and higher-order characteristics or aerodynamic nonlinearities, can cause the aircraft response to differ significantly from that of the linear model. In that case *equivalent classical systems* (i.e., equivalent to the linear models we have developed) which have responses most closely matching those of the actual aircraft are identified. Then those numerical requirements which are stated in terms of linear system parameters (such as frequency, damping ratio, and modal phase angles) apply to the parameters of that equivalent system rather than to any particular modes of the actual higher-order system.

In several instances throughout the specification, subjective terms, such as 'objectionable flight characteristics', 'realistic time delay', 'normal pilot technique' and 'excessive loss of altitude or buildup of speed', have been employed to permit latitude where absolute quantitative criteria might be unduly restrictive.

The one-size-fits-all specifications of 8785C often resulted in conflicting requirements. Much of aircraft design involves trade-offs, and a point may easily be reached at which satisfying competing requirements will entail much greater development and manufacturing costs. For that reason and many others the specifications were replaced by standards. The current flying qualities standard is MIL-STD-1797A (1990). 1797A requires the procuring agency and the manufacturer to agree on the specifications based on guidance provided as an appendix to the standard. The guidance in 1797A reiterates in large part the specifications of 8785C with changes of the imperative 'shall' to 'should'.

The appendix to 1797A provides guidance for each section of the standard usually based on 8785C. For example, the standard for coupled roll–spiral oscillations says 'A coupled roll–spiral mode will be permitted only if it has the following characteristics', followed by a blank to be filled in for the specific airplane being addressed. The guidance for that standard presents a slightly modified version of the standard from 8785C and labels it 'Recommended minimum values of roll–spiral damping coefficient. $\zeta_{RS}\omega_{RS}$'. Following that is a discussion that mirrors that found in the BIUG for 8785C.

A major change from the specifications of 8785C to the guidance of 1797A is found in the section regarding longitudinal flying qualities, wherein a very large body of results based on more current research has been added. This section runs to nearly 100 pages and describes alternative ways of analyzing the phenomenon.

For pedagogic purposes, the linear system requirements of 8785C provide concrete numbers that are fairly easily assessed using the analysis methods from Chapter 9. In any event, an understanding and appreciation of the requirements of 8785C will not be wasted when using 1797A.

11.2 MIL-F-8785C Requirements

Selected sections of 8785C are reproduced in the following. For the purposes of cross-referencing, section, figure, and table numbers in parentheses are those found in the original document.

11.2.1 General

(1.3) Classification of airplanes. For the purpose of this specification, an airplane shall be placed in one of the following Classes:

Class I: Small, light airplanes

Class II: Medium weight, low-to-medium maneuverability airplanes

Class III: Large, heavy, low-to-medium maneuverability airplanes

Class IV: High-maneuverability airplanes

(1.4) Flight Phase Categories. Flight Phases descriptive of most military airplane missions are:

Category A: Those nonterminal Flight Phases that require rapid maneuvering, precision tracking, or precise flight-path control. Some of the flight phases in this Category are air-to-air combat (CO), ground attack (GA), in-flight refueling (receiver) (RR), and close formation flying (FF).

Category B: Those nonterminal Flight Phases that are normally accomplished using gradual maneuvers and without precision tracking, although accurate flight-path control may be required. Some of the flight phases in this Category are climb (CL), cruise (CR), and descent (D).

Category C: Terminal Flight Phases are normally accomplished using gradual maneuvers and usually require accurate flight-path control. Included in this Category are takeoff (TO), catapult takeoff (CT), approach (PA), wave-off/go-around (WO), and landing (L).

(1.5) Levels of flying qualities. Where possible, the requirements have been stated in terms of the values of the stability or control parameter being specified. Each value is a minimum condition to meet one of three Levels of acceptability related to the ability to complete the operational missions for which the airplane is designed. The Levels are:

Level 1: Flying qualities clearly adequate for the mission Flight Phase.

Level 2: Flying qualities adequate to accomplish the mission Flight Phase, but some increase in pilot workload or degradation in mission effectiveness, or both, exists.

Level 3: Flying qualities such that the airplane can be controlled safely, but pilot workload is excessive or mission effectiveness is inadequate, or both. Category A Flight Phases can be terminated safely, and Category B and C Flight Phases can be completed.

Those requirements of 8785C that are amenable to linear analysis are summarized below. For a more complete description of the requirements see the corresponding paragraphs in the source.

11.2.2 Longitudinal flying qualities

(3.2.1.1) Longitudinal static stability

For Levels 1 and 2 there shall be no tendency for airspeed to diverge aperiodically when the airplane is disturbed from trim with the cockpit controls fixed and with them free. For Level 3 the time to double amplitude must be greater than 6 seconds.

(3.2.1.2) Phugoid stability

The long-period oscillations which occur when the airplane seeks a stabilized airspeed following a disturbance shall meet the following requirements:

Level 1 – ζ_{Ph} at least 0.04
Level 2 – ζ_{Ph} at least 0
Level 3 – T_2 at least 55 seconds

(3.2.1.3) Flight-path stability

Flight-path stability is defined in terms of flight-path-angle change where the airspeed is changed by the use of pitch control only (throttle setting not changed by the crew). For the landing approach Flight Phase, the curve of flight-path angle versus true airspeed shall have a local slope at $V_{o_{Min}}$ which is negative or less positive than:

Level 1 – 0.06 deg/knot
Level 2 – 0.15 deg/knot
Level 3 – 0.25 deg/knot

The thrust setting shall be that required for the normal approach glide path at $V_{o_{Min}}$. The slope of the curve of flight-path angle versus airspeed at 5 knots slower than $V_{o_{Min}}$ shall not be more than 0.05 deg/knot more positive than the slope at $V_{o_{Min}}$, as illustrated by Figure 11.3.

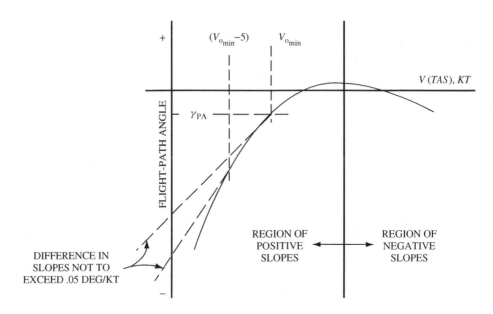

Figure 11.3 Flight path stability criteria.

(3.2.2.1.1) Short-period frequency and acceleration sensitivity
The equivalent short-period undamped natural frequency, $\omega_{n_{SP}}$, shall be within the limits shown on *figures 1, 2, and 3* (Figure 11.4, Figure 11.5, Figure 11.6). If suitable means of directly controlling normal force are provided, the lower bounds on $\omega_{n_{SP}}$ and n/α of *figure 3* (Figure 11.6) may be relaxed if approved by the procuring activity.

(3.2.2.1.2) Short-period damping
The equivalent short-period damping ratio, ζ_{SP}, shall be within the limits of *table IV* (Table 11.1).

11.2.3 Lateral–directional flying qualitities

(3.3.1) Lateral–directional mode characteristics

(3.3.1.1) Lateral–directional oscillations (Dutch roll)
The frequency, ω_{n_d}, and damping ratio, ζ_d, of the lateral–directional oscillations following a yaw disturbance input shall exceed the minimum values in *table VI* (Table 11.2). The requirements shall be met in trimmed and in maneuvering flight with cockpit controls fixed and with them free, in oscillations of any magnitude that might be experienced in operational use. If the oscillation is nonlinear with amplitude, the requirement shall apply to each cycle of the oscillation. In calm air residual oscillations may be tolerated only if the amplitude is sufficiently small that the motions are not objectionable and do not

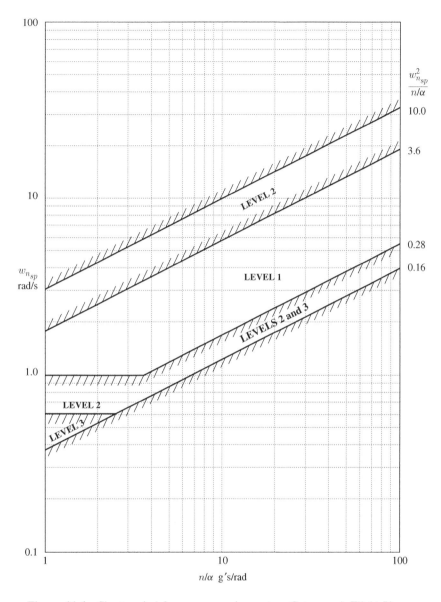

Figure 11.4 Short-period frequency requirements – Category A Flight Phases.

impair mission performance. For Category A Flight Phases, angular deviations shall be less than ±3 mils.

(3.3.1.2) Roll mode
The roll-mode time constant, τ_R, shall be no greater than the appropriate value in *table VII* (Table 11.3).

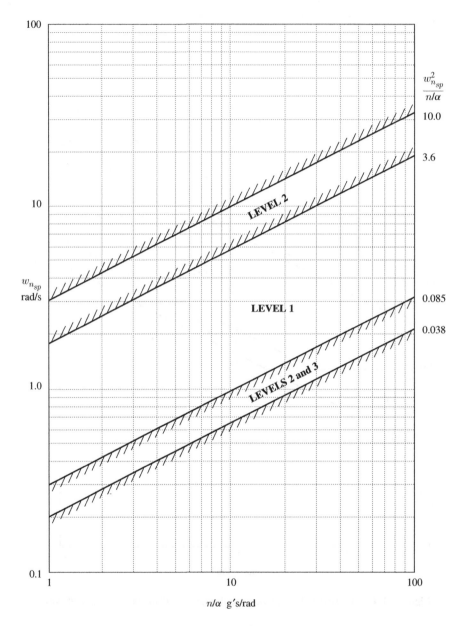

Figure 11.5 Short-period frequency requirements – Category B Flight Phases.

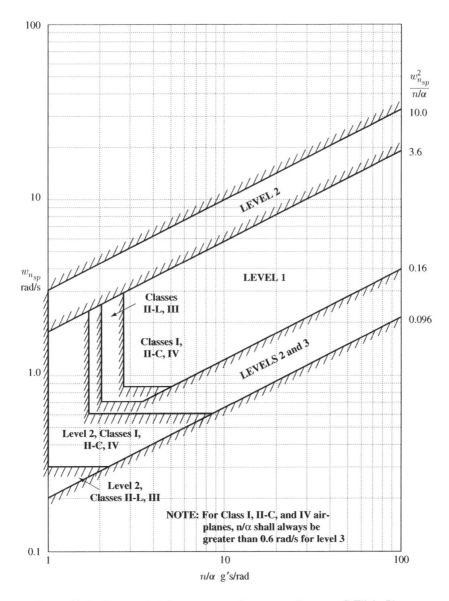

Figure 11.6 Short-period frequency requirements – Category C Flight Phases.

(3.3.1.3) Spiral stability

The combined effects of spiral stability, flight-control-system characteristics and rolling moment change with speed shall be such that following a disturbance in bank of up to 20 deg, the time for the bank angle to double shall be greater than the values in *table VIII* (Table 11.4). This requirement shall be met with the airplane trimmed for wings-level, zero-yaw-rate flight with the cockpit controls free.

Table 11.1 *(TABLE IV)* Short-period damping ratio limits.

Level	Category A & C Flight Phase		Category B Flight Phase	
	Minimum	Maximum	Minimum	Maximum
1	0.35	1.30	0.30	2.00
2	0.25	2.00	0.20	2.00
3	0.15	–	0.15	–

Table 11.2 *(TABLE VI)* Minimum Dutch roll frequency and damping.

Level	Flight Phase category	Class	Min $\zeta_d{}^a$	Min $\zeta_d \omega_{n_d}{}^a$ rad/s	Min ω_{n_d} rad/s
	A (CO and GA)	IV	0.4	–	1.0
	A	I, IV	0.19	0.35	1.0
		II, III	0.19	0.35	0.4^b
1	B	All	0.08	0.15	0.4^b
	C	I, II-C, IV	0.08	0.15	1.0
		II-L, III	0.08	0.10	0.4^b
2	All	All	0.02	0.05	0.4^b
3	All	All	0	0	0.4^b

^{a}The governing damping requirement is that yielding the larger value of ζ_d, except that ζ_d of 0.7 is the maximum required for Class III.
^{b}Class III airplanes may be excepted from the minimum ω_{n_d} requirement, subject to approval by the procuring activity, if the requirements of *3.3.2* through *3.3.2.4.1*, *3.3.5*, and *3.3.9.4* are met. When $\omega_{n_d}|\phi/\beta|_d$ is greater than 20 (rad/s)^2, the minimum $\zeta_d \omega_{n_d}$ shall be increased above the $\zeta_d \omega_{n_d}$ minimums listed above by:

Level 1 – $\Delta \zeta_d \omega_{n_d} = 0.014(\omega_{n_d}|\phi/\beta|_d - 20)$
Level 2 – $\Delta \zeta_d \omega_{n_d} = 0.009(\omega_{n_d}|\phi/\beta|_d - 20)$
Level 3 – $\Delta \zeta_d \omega_{n_d} = 0.004(\omega_{n_d}|\phi/\beta|_d - 20)$
with ω_{n_d} in rad/s.

Table 11.3 *(TABLE VII)* Maximum roll-mode time constant (seconds).

Flight Phase category	Class	Level		
		1	2	3
A	I, IV	1.0	1.4	
	II, III	1.4	3.0	
B	All	1.4	3.0	10
C	I, II-C, IV	1.0	1.4	
	II-L, III	1.4	3.0	

Table 11.4 *(TABLE VIII)* Spiral stability – minimum time to double amplitude (seconds).

Flight Phase category	Level 1	Level 2	Level 3
A and C	12	8	4
B	20	4	4

(3.3.1.4) Coupled roll–spiral oscillation

For Flight Phases which involve more than gentle maneuvering, such as CO and GA, the airplane characteristics shall not exhibit a coupled roll–spiral mode in response to the pilot roll control commands. A coupled roll–spiral mode will be permitted for Category B and C Flight Phases provided the product of frequency and damping ratio exceeds the following requirements:

Level	$\zeta_{RS}\omega_{n_{RS}}$, rad/s
1	0.5
2	0.3
3	0.15

(3.3.4) Roll-control effectiveness

Roll performance in terms of a bank angle change in a given time, ϕ_t, is specified in *table IXa* (Table 11.5) for Class I and Class II airplanes, in *3.3.4.1* for Class IV airplanes, and in *3.3.4.2* for Class III airplanes. For rolls from banked flight, the initial condition shall be coordinated, that is, zero lateral acceleration. The requirements apply to roll commands to the right and to the left, initiated both from steady bank angles and from wings-level flight except as otherwise stated. Inputs shall be abrupt, with time measured from the

Table 11.5 *(TABLE IXa)* Roll performance for Class I and II airplanes.

		\multicolumn{6}{c	}{Time to achieve the following bank angle change (seconds)}
		Category A		Category B		Category C	
Class	Level	60	45	60	45	30	25
I	1	1.3		1.7		1.3	
	2	1.7		2.5		1.8	
	3	2.6		3.4		2.6	
II-L	1		1.4		1.9	1.8	
	2		1.9		2.8	2.5	
	3		2.8		3.8	3.6	
II-C	1		1.4		1.9		1.0
	2		1.9		2.8		1.5
	3		2.8		3.8		2.0

initiation of control force application. The pitch control shall be fixed throughout the maneuver. Yaw control pedals shall remain free for Class IV airplanes for Level 1, and for all carrier-based airplanes in Category C Flight Phases for Levels 1 and 2; but otherwise, yaw control pedals may be used to reduce sideslip that retards roll rate (not to produce sideslip which augments roll rate) if such control inputs are simple, easily coordinated with roll control inputs and consistent with piloting techniques for the airplane class and mission. For Flight Phase TO, the time required to bank may be increased proportional to the ratio of the rolling moment of inertia at takeoff to the largest rolling moment of inertia at landing, for weights up to the maximum authorized landing weight.

(3.3.4.1) Roll performance for Class IV airplanes
Roll performance in terms of ϕ_t for Class IV airplanes is specified in *table IXb* (Table 11.6). Additional or alternate roll performance requirements are specified in *3.3.4.1.1* and *3.3.4.1.2*; these requirements take precedence over *table IXb* (Table 11.6). Roll performance for Class IV airplanes is specified over the following ranges of airspeeds:

Speed range symbol	Equivalent airspeed range For level 1	For levels 2 and 3
VL	$V_{0_{Min}} \le V \le V_{Min} + 20 kts$	$V_{Min} \le V \le V_{Min} + 20 kts$
L	$V_{Min} + 20 kts^{a} \le V < 1.4 V_{Min}$	$V_{Min} + 20 kts \le V < 1.4 V_{Min}$
M	$1.4 V_{0_{Min}} < V \le 0.7 V_{Max}^{b}$	$1.4 V_{Min} \le V < 0.7 V_{Max}$
H	$0.7 V_{Max}^{(2)} \le V \le V_{0_{Max}}$	$0.7 V_{Max} \le V \le V_{Max}$

^{a}Or $V_{0_{Min}}$, whichever is greater
^{b}Or $V_{0_{Max}}$, whichever is less

Table 11.6 *(TABLE IXb)* Roll performance for Class IV airplanes.

Level	Speed range	Category A (30)	Category A (50)	Category A (90)	Category B (90)	Category C (30)
1	VL	1.1			2.0	1.1
	L	1.1			1.7	1.1
	M			1.3	1.7	1.1
	H		1.1		1.7	1.1
2	VL	1.6			2.8	1.3
	L	1.5			2.5	1.3
	M			1.7	2.5	1.3
	H		1.3		2.5	1.3
3	VL	2.6			3.7	2.0
	L	2.0			3.4	2.0
	M			2.6	3.4	2.0
	H		2.6		3.4	2.0

(3.3.4.1.1) Roll performance in Flight Phase CO

Roll performance for Class IV airplanes in Flight Phase CO is specified in *table IXc* (Table 11.7) in terms of ϕ_t for 360 deg rolls initiated at 1g, and in *table IXd* (Table 11.8) for rolls initiated at load factors between $0.8n_0(-)$ and $0.8n_0(+)$.

Table 11.7 *(TABLE IXc)* Flight Phase CO roll performance in 360 deg rolls.

Level	Speed range	30 deg	90 deg	180 deg	360 deg
1	VL	1.0			
	L		1.4	2.3	4.1
	M		1.0	1.6	2.8
	H		1.4	2.3	4.1
2	VL	1.6			
	L	1.3			
	M		1.3	2.0	3.4
	H		1.7	2.6	4.4
3	VL	2.5			
	L	2.0		2.3	4.1
	M		1.7	3.0	
	H		2.1		

Table 11.8 *(TABLE IXd)* Flight Phase CO roll performance.

Level	Speed range	30 deg	50 deg	90 deg	180 deg
1	VL	1.0			
	L		1.1		
	M			1.1	2.2
	H		1.0		
2	VL	1.6			
	L	1.3			
	M			1.4	2.8
	H		1.4		
3	VL	2.5			
	L	2.0			
	M			1.7	3.4
	H		1.7		

(3.3.4.1.2) Roll performance in Flight Phase GA
The roll performance requirements for Class IV airplanes in Flight Phase GA with large complements of external stores may be relaxed from those specified in *table IXb* (Table 11.6), subject to approval by the procuring activity. For any external loading specified in the contract, however, the roll performance shall be not less than that in *table IXe* (Table 11.9) where the roll performance is specified in terms of ϕ_t for rolls initiated at load factors between $0.8n_0(-)$ and $0.8n_0(+)$. For any asymmetric loading specified in the contract, roll control power shall be sufficient to hold the wings level at the maximum load factors specified in *3.2.3.2* with adequate control margin *(3.4.10)*.

Table 11.9 *(TABLE IXe)* Flight Phase GA roll performance.

		Time to achieve the following bank angle change			
Level	Speed range	30 deg	50 deg	90 deg	180 deg
1	VL	1.5			
	L		1.7		
	M			1.7	3.0
	H		1.5		
2	VL	2.8			
	L	2.2			
	M			2.4	4.2
	H		2.4		
3	VL	4.4			
	L	3.8			
	M			3.4	6.0
	H		3.4		

Problems

1. Refer to the data for the A-4 *Skyhawk* (a Class IV aircraft) (Appendix A). Evaluate the *Skyhawk's* compliance with as many provisions of MIL-F-8785C as possible. Assume the aircraft is in the ground attack (GA) role.

References

Anon. (1980) Military Specification, Flying Qualities of Piloted Airplanes, Department of Defense, MIL-F-8785C, Washington, D.C., November 1980.

Anon. (1990) Flying Qualities of Piloted Aircraft, Department of Defense, MIL-STD-1797A, Washington, D.C., January 1990.

Chalk, C.R. et al. (1969) *Background Information and User Guide for MIL-F-8785B(ASG), "Military Specification–Flying Qualities of Piloted Airplanes"*, Technical Report AFFDL-TR-69-72, Air Force Flight Dynamics Laboratory, Wright-Patterson Air Force Base, Ohio, August 1969.

Harper, R.P. and Cooper, G.E. (1986) Handling qualities and pilot evaluation. *Journal of Guidance, Control and Dynamics*, **9** (5), 515–529.

Hodgkinson, J. (1999) *Aircraft Handling Qualities*, AIAA Inc. and Blackwell Science Ltd.

Moorhouse, D.J. and Woodcock, R.J. (1982) *Background Information and User Guide for MIL-F-8785C, "Military Specification–Flying Qualities of Piloted Airplanes"*, Technical Report AFWAL-TR-81-3109, Air Force Wright Aeronautical Laboratories, Wright-Patterson Air Force Base, Ohio, July 1982.

Weingarten, N.R. (2005) History of in flight simulation & flying qualities research at Calspan. *AIAA Journal of Aircraft*, **42** (2), March/April.

12

Automatic Flight Control

We now turn to the problem of how to modify an airplane's dynamic response characteristics in order to give it good flying qualities. In this discussion, we are speaking of the flying qualities of the 'bare airframe', meaning the airplane without any artificial measures to improve its response characteristics.

Sometimes an airplane may have undesirable flying qualities because of intentional decisions that were made during the design phase. For instance, there are certain performance advantages to relaxing the static stability of an airplane by moving the center of gravity aft, but this action will degrade its longitudinal flying qualities. The poor flying qualities may result from trade-offs in other areas as well, such as poor directional stability that results from reducing the size of the vertical stabilizer to accomodate shipboard storage.

For whatever reason, sometimes airplanes have flying qualities that need to be improved, and the solution is often an electronic feedback system (called a Stability Augmentation System, or SAS). An SAS may be designed to modify the parameters that define the airplane's flying qualities.

Conceptually an SAS is simple: it senses the response of the airplane (pitch rate, sideslip angle, etc.) and deflects control surfaces (*control effectors*) to modify the response it is sensing. Difficulties arise because the pilot may be attempting to deflect the control surfaces for his own purposes, and will not appreciate having his inputs countermanded. Various mechanical and electronic means have been devised to handle this problem.

The SAS may be integrated with other electronics to yield a Control Augmentation System (CAS), or an autopilot. A CAS changes the meaning of a pilot's inputs from the command of one response to that of another. For instance, longitudinal stick deflection usually commands (in the short term at least) a pitch rate, but it could also be treated as a command for a certain load factor, or g-command. Autopilots range from a simple altitude-hold, which frees the pilot to think of other things, to systems that can take off and land the airplane, even on aircraft carriers. They perform by sensing additional states of the airplane, and deflecting the control effectors as necessary to achieve some goal.

The far end of the spectrum of automatic flight control systems is the fly-by-wire (FBW) system. In FBW systems, the pilot's inputs to the cockpit controllers (*control inceptors*) are not directly connected to the control effectors, but rather are intercepted by a computer that interprets the pilot's intentions and commands the control inceptors to achieve them. Such systems often have an emergency mode, which will bypass the computer via a direct electrical link.

This chapter will serve as an introduction to automatic flight control systems, and will show that all the previous analysis is sufficient to design rudimentary stability augmentation systems. A few new concepts will be needed, so we start with very simple systems.

For a detailed treatment of control system design, D'Azzo and Houpis (1988) has always served the author well. For applications to aircraft, Stevens and Lewis (1992) is an excellent work.

12.1 Simple Feedback Systems

12.1.1 First-order systems

There aren't many first-order systems in the study of flight dynamics. The dynamics of the hydraulic actuators that power the control surface deflections are often modeled as first-order systems. Simple electrical networks used to smooth or filter electrical signals may be first-order. There are, of course, the roll-mode and spiral-mode approximations, so the study of first-order systems will be useful in itself.

Given a first order ordinary differential equation with forcing function,

$$\dot{x} = ax(t) + bu(t) \tag{12.1}$$

This process can be represented by the block diagram shown in Figure 12.1. The blocks and connections in this diagram don't have real-world counterparts – for instance, there is no device in nature that integrates signals – but analysis of the diagram yields the given Equation 12.1.

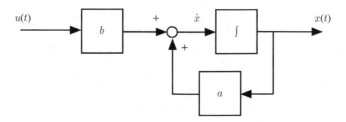

Figure 12.1 First-order system.

The single eigenvalue of this system (with subscript OL to represent 'Open Loop') is easily determined to be:

$$\lambda_{OL} = a \tag{12.2}$$

The block diagram is simpler in transfer function form. The forced response becomes

$$x(s) = \frac{b}{s-a} u(s) \tag{12.3}$$

The block diagram for the transfer function representation of this system is shown in Figure 12.2.

Figure 12.2 Transfer function representation, first-order system.

Now consider a feedback scheme in which the state $x(t)$ is measured, the measured value amplified by a factor k, and used to modify the input such that the input $u(t)$ is now the sum of $kx(t)$ and a new (reference) signal $r(t)$. In terms of the complex variable s, $u(s) = kx(s) + r(s)$ as shown in Figure 12.3.

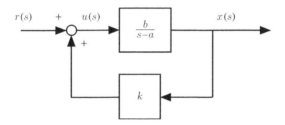

Figure 12.3 Closed-loop system.

Before proceeding, it is important to keep in mind what Figure 12.3 is telling us.

- The reference r is perhaps an electrical signal, a voltage in a circuit that is proportional to some physical device, such as a control inceptor.
- The state x represents our measurement of the state in question, such as a vane that deflects with changes in angle-of-attack to measure α. It would probably be better to call this signal something different than x, and it often is given the notation y and called an output. For now though we'll keep it simple and assume that we have measured the state x and have a signal proportional to it.
- The feedback block with gain k could be an electrical amplifier. The amplified state is summed with the reference signal; circuits that add voltages are quite common. Note that here we have indicated by the + sign that the feedback is to be added to the reference signal, and not the negative feedback some might expect. There is no problem with positive feedback when we have control of the sign of the gain k.
- The connection between this summed voltage and the control effector u is a bit more complicated than the straight connection shown here; the voltage is applied, for example, to a solenoid on a hydraulic actuator that moves the actuator ram to the right position. While on the subject of control effectors, always bear in mind that in the real world they have physical limits. If your system commands an elevator deflection greater than the elevator can travel the pilot won't thank you.

- Finally, the box with the transfer function represents a physical system. To the extent that this physical system is described by a first order ordinary differential equation, and that the values of a and b are accurately known, then this model will do for our purposes.

With that homily aside, the block diagram using transfer functions is shown in Figure 12.3.

The equation of motion of the closed-loop (CL) system becomes

$$\dot{x} = ax(t) + b[kx(t) + r(t)] = (a + bk)x(t) + br(t) \tag{12.4}$$

The closed-loop eigenvalue is then

$$\lambda_{CL} = a + bk \tag{12.5}$$

In other words, by proper choice of k the system eigenvalue may be assigned arbitrarily. Keep in mind the previous discussion: the input $u(t)$ is the sum of $kx(t)$ and $r(t)$, and while k may appear to be arbitrary, there are limits on what u can be.

The same result can be arrived at using the transfer function. We have

$$x(s) = \frac{b}{s-a} u(s) = \frac{b}{s-a}[kx(s) + r(s)]$$

$$\left(1 - \frac{bk}{s-a}\right) x(s) = \frac{b}{s-a} r(s)$$

$$\frac{s-a-bk}{s-a} x(s) = \frac{b}{s-a} r(s)$$

$$x(s) = \frac{b}{s-(a+bk)} r(s) \tag{12.6}$$

The block diagram of the closed loop system is shown in Figure 12.4.

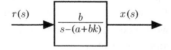

Figure 12.4 Transfer function representation, closed-loop first-order system.

12.1.2 Second-order systems

The two second-order systems of immediate interest to us are the short-period and Dutch-roll approximations. Thousands of pages of published findings based on tens of thousands of hours of research and experimentation have gone into understanding the requirements for good longitudinal flying qualities, with the short-period being the chief phenomenon of interest. The same is true to a lesser extent for the role of the Dutch roll in lateral–directional flying qualities. One only needs to look at Figure 10.7 and think of the airplane as an airliner with a queasy passenger in the cabin to appreciate that the Dutch roll warrants serious consideration.

Open-loop eigenvalues

A simple mass-spring-damper system (Figure 12.5) is used to illustrate closed-loop control of second-order systems.

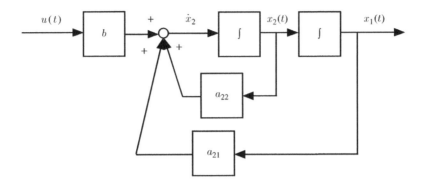

Figure 12.5 Second-order system.

Here, x_1 is the position and x_2 is the velocity, with $\dot{x}_1 = x_2$. The system is represented generically by Equation 12.7,

$$\dot{x}_1 = x_2(t)$$
$$\dot{x}_2 = a_{21}x_1(t) + a_{22}x_2(t) + bu(t) \qquad (12.7)$$

The parameter a_{21} is related to the spring in the system (proportional to displacement x_1), and the parameter a_{22} to the damping (proportional to velocity x_2).

We assume the system has complex eigenvalues, $\lambda_{OL} = \sigma \pm j\omega$, so that the response is oscillatory. The characteristic equation of this system is easily verified to be $s^2 - a_{22}s - a_{21} = 0$. We may relate the parameters a_{21} and a_{22} to the system natural frequency ω_n and damping ratio ζ as shown in Equations 12.8 and 12.9.

$$s^2 - a_{22}s - a_{21} = s^2 + 2\zeta\omega_n s + \omega_n^2 = 0 \qquad (12.8)$$

$$\omega_n = \sqrt{-a_{21}}$$

$$\zeta = \frac{-a_{22}}{2\sqrt{-a_{21}}} \qquad (12.9)$$

Using a little algebra, the real and imaginary parts of the system eigenvalues, σ and ω, are related to the natural frequency, damping ratio, and system parameters:

$$\lambda_{OL} = \sigma \pm j\omega = -\zeta\omega_n \pm j\omega_n\sqrt{1-\zeta^2} \qquad (12.10)$$

$$\sigma = -\zeta\omega_n = \frac{a_{22}}{2}$$

$$\omega = \omega_n\sqrt{1-\zeta^2} = \frac{1}{2}\sqrt{-a_{22}^2 - 4a_{21}} \qquad (12.11)$$

In terms of the transfer functions, the state transition matrix is

$$[sI - A]^{-1}B = \begin{bmatrix} s & -1 \\ -a_{21} & s - a_{22} \end{bmatrix}^{-1} \begin{bmatrix} 0 \\ b \end{bmatrix}$$

$$= \frac{1}{s^2 - a_{22}s - a_{21}} \begin{bmatrix} s - a_{22} & 1 \\ a_{21} & s \end{bmatrix} \begin{bmatrix} 0 \\ b \end{bmatrix}$$

$$= \frac{1}{s^2 - a_{22}s - a_{21}} \begin{bmatrix} b \\ bs \end{bmatrix}$$

The transfer functions are therefore

$$\frac{x_1(s)}{u(s)} = \frac{b}{s^2 - a_{22}s - a_{21}}$$

$$\frac{x_2(s)}{u(s)} = \frac{bs}{s^2 - a_{22}s - a_{21}}$$

From these relationships (and from $\dot{x}_1 = x_2$) we have

$$\frac{x_1(s)}{x_2(s)} = \frac{x_1(s)}{u(s)} \frac{u(s)}{x_2(s)} = \frac{1}{s}$$

The block diagram in the LaPlace domain then becomes as shown in Figure 12.6.

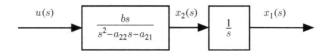

Figure 12.6 Transfer-function representation, second-order system.

A pole-zero map of the transfer function $x_1(s)/u(s)$ is shown in Figure 12.7.

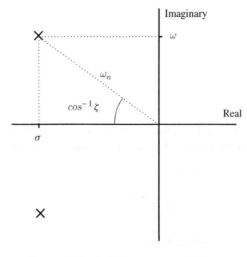

Figure 12.7 Pole-zero map, $x_1(s)/u(s)$.

Position feedback

Now consider feedback of the position variable, $u(t) = k_1 x_1(t) + r(t)$. This feedback does not affect the kinematic equation $\dot{x}_1 = x_2$, but changes the acceleration $\dot{x}_2$ as follows:

$$\dot{x}_2 = a_{21} x_1(t) + a_{22} x_2(t) + b[k_1 x_1(t) + r(t)]$$
$$= (bk_1 + a_{21}) x_1(t) + a_{22} x_2(t) + br(t)$$

Position feedback therefore affects only the spring parameter. The characteristic polynomial becomes $s^2 - a_{22} s - (a_{21} + bk_1)$.

In the LaPlace domain, we have $u(s) = k_1 x_1(s) + r(s)$, with the block diagram shown in Figure 12.8.

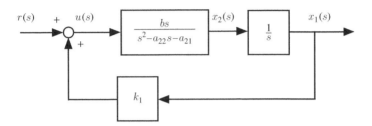

Figure 12.8 Block diagram, position feedback.

The closed-loop transfer function is easily determined,

$$x_1(s) = \frac{b}{s^2 - a_{22} s - a_{21}} u(s)$$
$$= \frac{b}{s^2 - a_{22} s - a_{21}} [k_1 x_1(s) + r(s)]$$

Simplifying,

$$\frac{x_1(s)}{r(s)} = \frac{b}{s^2 - a_{22} s - (bk_1 + a_{21})}$$

Since $\sigma = a_{22}/2$, just as it was in the open-loop system, position feedback does not change the damping term (real part of the eigenvalue) of the mass–spring–damper system. Therefore, as k_1 is varied, the roots (eigenvalues) will move vertically in the complex plane, as shown in Figure 12.9.

Rate feedback

With rate feedback we have $u(t) = k_2 x_2(t) + r(t)$. The acceleration $\dot{x}_2$ then becomes:

$$\dot{x}_2 = a_{21} x_1(t) + a_{22} x_2(t) + b[k_2 x_2(t) + r(t)]$$
$$= a_{21} x_1(t) + (bk_2 + a_{22}) x_2(t) + br(t) \tag{12.12}$$

Rate feedback therefore affects only the damping parameter. The characteristic polynomial becomes $s^2 - (bk_1 + a_{21}) s - a_{21}$. The transfer function and block diagram of this

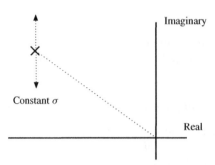

Figure 12.9 Effect of position feedback.

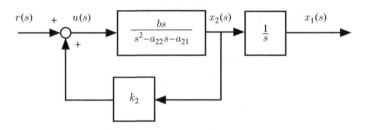

Figure 12.10 Block diagram, rate feedback.

system are given in Equation 12.13 and Figure 12.10.

$$\frac{x_1(s)}{r(s)} = \frac{b}{s^2 - (bk_2 + a_{22})s - a_{21}} \qquad (12.13)$$

The unchanged term now is $\omega_n = \sqrt{-a_{21}}$. Therefore, as k_2 is varied, the roots will move in a circular arc about the origin, as shown in Figure 12.11.

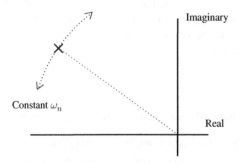

Figure 12.11 Effect of rate feedback.

Automatic Flight Control

By combining position and rate feedback, as shown in Figure 12.12, the transfer function becomes

$$\frac{x_1(s)}{r(s)} = \frac{b}{s^2 - (bk_2 + a_{22})s - (bk_1 + a_{21})} \tag{12.14}$$

Thus, the eigenvalues of the mass–spring–damper system may be placed in any arbitrary position.

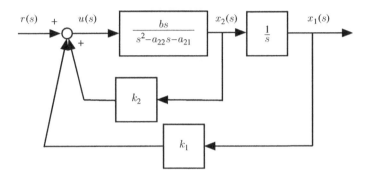

Figure 12.12 Block diagram, position and rate feedback.

12.1.3 A general representation

A more general representation of a multi-variable feedback control system is shown in Figure 12.13. In the figure, $\mathbf{x}(s)$, $\mathbf{u}(s)$, and $\mathbf{r}(s)$ are vectors, and $G(s)$ and $K(s)$ are matrices.

With reference to Figure 12.13,

$$\mathbf{u}(s) = K\mathbf{x}(s) + \mathbf{r}(s)$$

$$\mathbf{x}(s) = GK\mathbf{x}(s) + G\mathbf{r}(s) \tag{12.15}$$

$$\mathbf{x}(s) = [I - GK(s)]^{-1} G\mathbf{r}(s) \tag{12.16}$$

The matrix $G(s)$ is a matrix of transfer functions, the same as derived in Section 9.5.1.

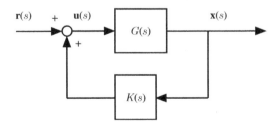

Figure 12.13 Generic multi-input multi-output system.

Application to the mass–spring–damper

The following assignments relate Figure 12.13 to the mass–spring–damper system:

$$\mathbf{x}(s) = \begin{Bmatrix} x_1(s) \\ x_2(s) \end{Bmatrix}$$

$$\mathbf{u}(s) = \{u(s)\}$$

$$\mathbf{r}(s) = \{r(s)\}$$

$$G(s) = \begin{bmatrix} \frac{b}{s^2 - a_{22}s - a_{21}} \\ \frac{bs}{s^2 - a_{22}s - a_{21}} \end{bmatrix}$$

$$K(s) = \begin{bmatrix} k_1(s) & k_2(s) \end{bmatrix} \qquad (12.17)$$

With these assignments, we have

$$\mathbf{x}(s) = \begin{bmatrix} \frac{b}{s^2 - (bk_2 + a_{22})s - (bk_1 + a_{21})} \\ \frac{bs}{s^2 - (bk_2 + a_{22})s - (bk_1 + a_{21})} \end{bmatrix} \mathbf{r}(s) \qquad (12.18)$$

12.2 Example Feedback Control Applications

Unless otherwise noted, in discussing the aircraft dynamics angles are measured in radians, distance in feet, and time in seconds.

12.2.1 Roll mode

We derived and presented the dimensional lateral–directional equations in Section 8.7.2, and then derived the first-order approximation for the dynamics of the roll mode in Section 10.2.4. Extending the approximation to include the inputs δ_ℓ and δ_n from Equation 8.17d we have

$$I_{xx}\dot{p} = L_p p + L_{\delta_\ell}\delta_\ell + L_{\delta_n}\delta_n$$

We will need only the primary roll controller δ_ℓ, and consider it to be the airplane's ailerons. We therefore let $\delta_n = 0$ so that the roll mode of the aircraft is approximated by

$$I_{xx}\dot{p} = L_p p + L_{\delta_a}\delta_a \qquad (12.19)$$

Since we have defined the roll rate and roll moment to both be positive in the same direction, and since rate damping resists the motion, $L_p < 0$. Also, since we have defined positive aileron deflection such that the right aileron is down and the left is up, positive aileron will generate a negative rolling moment.

$$L_p < 0, \; L_{\delta_a} < 0$$

With substitutions

$$x = p, \; u = \delta_a, \; a = L_p/I_{xx}, \; b = L_{\delta_a}/I_{xx} \qquad (12.20)$$

The roll-mode approximation fits our generic first-order system equation,

$$\dot{x} = ax(t) + bu(t)$$

The transfer function from $u(s)$ to $x(s)$ is

$$\frac{x(s)}{u(s)} = \frac{b}{s-a} \qquad (12.21)$$

If $u(t)$ is a step input of magnitude m_a then $u(s) = m_a/s$ and

$$x(s) = \frac{m_a b}{s(s-a)}$$

$$x(t) = -\frac{m_a b}{a}(1 - e^{at}) \qquad (12.22)$$

With the substitutions in Equation 12.20 and the roll-mode time constant defined as

$$\tau_r = -\frac{1}{a} = -\frac{I_{xx}}{L_p} \qquad (12.23)$$

The roll rate in response to a positive unit step aileron input is

$$p(t) = -\frac{m_a L_{\delta_a}}{L_p}(1 - e^{-t/\tau_r}) \qquad (12.24)$$

From Equation 12.24 the steady-state roll rate is seen to be

$$p_{ss} = -\frac{m_a L_{\delta_a}}{L_p} \qquad (12.25)$$

We note that our sign conventions have resulted in $\tau_r > 0$ and, if m_a is positive, $p_{ss} < 0$ (a positive aileron deflection generates a negative roll rate).

There are two flying qualities specifications related to the roll mode that may be addressed.

Roll-mode time constant

The importance of the roll-mode time constant on the pilot's ability to easily and precisely control the airplane may not at first be apparent. It comes to play in maneuvers in which the airplane bank angle must be precisely controlled, such as during formation flying (wherein everything must be precisely controlled). During landing approach the pilot may need to make several changes to the alignment of the flight path with the runway, and this is done by setting a certain bank angle, waiting for the heading to change, and then removing the bank angle.

Recall that the meaning of the roll-mode time constant is such that when $t = \tau_r$ $p = 0.632 p_{ss}$, when $t = 2\tau_r$ $p = 0.865 p_{ss}$, and so on. Thus the roll-mode time constant is a measure of how quickly the roll rate achieves its steady-state value.

Flying qualities research (and perhaps common sense) concludes that pilots prefer a small roll-mode time constant so that the steady state is quickly attained, and the lateral stick (or yoke) is perceived as *rate controller*. In other words, an input by the pilot seems to result almost immediately in a steady roll rate, and removing the input results in a quick cessation of roll rate.

Figure 12.14 was taken from the U.S. Naval Test Pilot School's Flight Test Manual (USNTPS, 1997). The intent of the figure is to illustrate that small τ_r allows the pilot to generate and remove roll rate quickly, allowing for precise bank angle control.

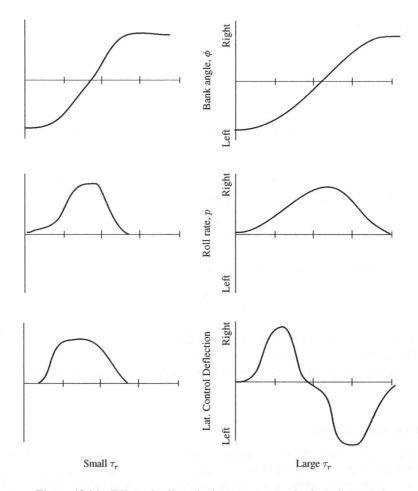

Figure 12.14 Effect of roll-mode time constant on bank angle control.

The lower right graph in the figure suggests that with a large (slow) τ_r the pilot must *anticipate* the desired bank angle and not only remove the control input, but actually reverse it in order to stop the roll rate at the desired bank angle. Note that after the pilot's input is removed (just before the second time tick) the roll rate continues to increase prompting the pilot to apply input in the opposite direction.

Now examine the use of roll-rate feedback to improve (decrease) τ_r. We take the system from Figure 12.3 and substitute our variables into $u(t) = kx(t) + r(t)$ yielding our *control law*:

$$\delta_a(t) = k_p p(t) + r(t)$$
$$\delta_a(s) = k_p p(s) + r(s) \qquad (12.26)$$

Here, we interpret the reference signal $r(t)$ as the pilot's input (stick or yoke). The closed-loop transfer functions from pilot input (r) to the roll rate (p) is

$$\frac{x(s)}{r(s)} = \frac{b}{s - (a + bk_p)} = \frac{p(s)}{r(s)} = \frac{L_{\delta_a}/I_{xx}}{s - (L_p/I_{xx} + k_p L_{\delta_a}/I_{xx})}$$

The augmented roll-mode time constant $\tau_{r_{aug}}$ is

$$\tau_{r_{aug}} = -\frac{1}{a + bk_p} = -\frac{I_{xx}}{L_p + k_p L_{\delta_a}}$$

Since we desire $\tau_{r_{aug}} < \tau_r$, with $L_p < 0$ and $L_{\delta_a} < 0$

$$|L_p + k_p L_{\delta_a}| > |L_p|$$

so that $k_p > 0$.

It therefore appears that by measuring the roll rate (perhaps with a gyroscope mounted to the airframe), then amplifying it with attention to the polarity of the signal relative to the reference signal (to make it negative), and linking the whole affair up to the aileron actuators as shown, we may suitably modify the roll-mode time constant.

Time to change bank angle

The unaugmented airplane

Other roll-mode specifications are expressed in terms of the time required to change the bank angle by a certain amount, which will be influenced by the roll-mode time constant and the maximum rate of roll. Let us look further into the bank angle change requirements to see what may be the effect of our tinkering with the roll-mode time constant.

We need first to clarify what we mean by 'change the bank angle'. The Test Pilot School figure above (Figure 12.14) tells us what it means there: the change in ϕ. But one need only look back to Equation 7.4 to see that this won't do. We therefore qualify the meaning of a change in bank angle to mean the change in the angle ϕ with $\theta \approx 0$, which allows us to write

$$\dot{\phi} = p(t)$$

One need specify as well how the change in bank angle is to be accomplished. Can we get a running jump at it, not starting the clock until the roll rate has spooled up a bit? Also, how do we define when the change has been attained, when we pass through it, or do we have to stop the roll when the airplane reaches the new bank angle?

The answer is neither of those. The test is begun at a fixed bank angle, usually wings level, but it may be done from a non-zero bank angle (in turning flight in other words). The pilot then applies the cockpit control in as nearly a step input as he is able, using two hands if necessary, and the clock is started. When the airplane rolls *through* the specified change the clock is stopped, and no attempt is made to stop the roll at the new angle. (Moreover, the tests are performed rolling in both directions to look for worst cases. The worst case for propeller engine airplanes depends on whether the airplane is British or American, since the propellers go in different directions.)

With this description of the performance of the time-to-bank test we can use the $\dot{p}$ equation with a step input (Equations 12.22–12.24). Since we are trying to change the bank angle as quickly as possible, we let $m_a = m_{a_{Max}}$, meaning maximum deflection in either direction. The LaPlace transform of $\dot{\phi}$ is $s(\phi(s) - \phi(0))$ so, with a few substitutions,

$$s(\phi(s) - \phi(0)) = p(s) = \frac{m_{a_{Max}} L_{\delta_a}/I_{xx}}{s(s - L_p/I_{xx})} = \frac{p_{ss}/\tau_r}{s(s + 1/\tau_r)}$$

$$\phi(s) = \phi(0) + \frac{p_{ss}/\tau_r}{s^2(s + 1/\tau_r)}$$

$$\phi(t) = \phi(0) + p_{ss}\tau_r \left(t/\tau_r - 1 + e^{-t/\tau_r}\right)$$

With this we can calculate the change in bank angle, $\Delta\phi$, in response to a step input in aileron of magnitude $m_{a_{Max}}$, in a specified time t_{Spec}, to be

$$\Delta\phi = p_{ss}\tau_r \left(t_{Spec}/\tau_r - 1 + e^{-t_{Spec}/\tau_r}\right) \tag{12.27}$$

In Equation 12.27 $p_{ss} = -m_a L_{\delta_a}/L_p$ (Equation 12.25). If the $\Delta\phi$ that results is greater than that specified to be accomplished in t_{Spec}, the airplane satisfies the requirement.

The effect of roll-rate feedback

Now go back and examine the results of our modification of the roll-mode time constant. When considering the unaugmented airplane we did not distinguish between the pilot's input and the aileron deflection, assuming a fixed correspondence between them. That is, the aileron deflection δ_a is related to the pilot's control input r by a constant factor. Without loss of generality assume a one-to-one correspondence between them, and scale them both so that the maximum deflection is one,

$$\delta_a(t) = r(t), \quad -1 \leq r \leq 1$$

Recall that our sign convention is such that positive aileron deflection generates a negative rolling moment, so a positive pilot control deflection does too. Thinking of a center-stick configuration, left stick is positive and right stick is negative.

As we have formulated the feedback controller,

$$\delta_a(t) = r(t) + k_p p(t), \quad -1 \leq r \leq 1$$

Before analyzing the effects of roll-rate feedback mathematically, let us think our way through the problem. Beginning our time-to-bank test from a fixed bank angle, $p(0) = 0$

and $\delta_a(0) = r(0)$. An instant later the pilot applies a step input to the stick, so $r(0^+) = 1$. The positive aileron deflection generates a negative moment and subsequent negative roll rate. Since $k_p > 0$ the product $k_p p(t)$ grows to a negative value. Thus, even though the stick is held firmly against its limits, the ailerons do not stay fixed at their limits, and decrease as the negative roll rate grows.

Let's look more closely at what we've done. We derived an expression for the augmented roll-mode time constant,

$$\tau_{r_{aug}} = -\frac{I_{xx}}{L_p + k_p L_{\delta_a}}$$

We took $k_p > 0$ so that $\tau_{r_{aug}} < \tau_r$. Also, we previously derived the transfer function with roll-rate feedback

$$\frac{p(s)}{r(s)} = \frac{L_{\delta_a}/I_{xx}}{s - (L_p/I_{xx} + k_p L_{\delta_a}/I_{xx})}$$

For a step lateral stick input at $t = 0$ of magnitude 1 the steady-state roll rate (augmented) $p_{ss_{aug}}$ is

$$p_{ss_{aug}} = \frac{L_{\delta_a}}{L_p + k_p L_{\delta_a}}$$

Keeping track of the signs of the parts,

$$\left|\frac{L_{\delta_a}}{L_p + k_p L_{\delta_a}}\right| < \left|\frac{L_{\delta_a}}{L_p}\right|$$

$$|p_{ss_{aug}}| < |p_{ss}| \tag{12.28}$$

$$\Delta\phi_{aug} = p_{ss_{aug}} \tau_{r_{aug}} \left(t_{Spec}/\tau_{r_{aug}} - 1 + e^{-t_{Spec}/\tau_{r_{aug}}}\right) \tag{12.29}$$

With the smaller steady-state roll rate and smaller time constant, the change in bank angle at a specified time following a maximum step input is always less than the unaugmented airplane.

We can better visualize the problem by looking at a time history representative of our system before and after feedback. Our example aircraft, the A-4 *Skyhawk*, is a poor candidate for roll-rate feedback since its roll-mode time constant is 0.55 s and satisfies the 1.0 s specification comfortably. We will use a very simple system, $\dot{x} = -x(t) + u(t)$ first with $u(t) = r(t)$ and then with $u(t) = r(t) + kx(t)$, $k < 0$. The input $r(t)$ will be a unit step at $t = 0$.

$$x(t) = 1 - e^{-t} \text{ (unaugmented)}$$

$$x(t) = \frac{1}{1-k}\left(1 - e^{-t(1-k)}\right) \text{ (augmented)}$$

Time histories were generated using MATLAB® and are shown in Figure 12.15. The upper curve is unaugmented, the lower is augmented using $k = -1$. The augmented response reaches steady state more quickly, but this decreased time constant is made possible only by reducing the steady-state value. In other words, it gets there faster because it doesn't have as far to go.

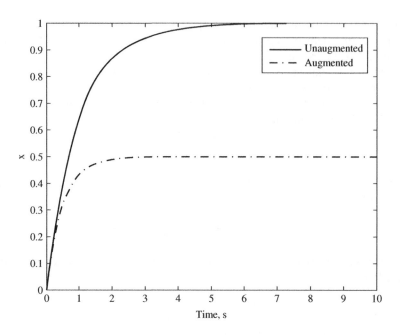

Figure 12.15 Time-histories, with and without roll-rate feedback.

12.2.2 Short-period mode

Formulation

The short-period approximation is based on the consideration of just the states α and q, and the pitching moment control δ_e (the elevator):

$$\tilde{\mathbf{x}}_{SP} = \begin{Bmatrix} \alpha \\ q \end{Bmatrix}$$

In terms of the dimensional derivatives, neglecting $\dot{\alpha}$ dependency, and ignoring changes in lift due to elevator deflection (effectively $Z_{\delta_m} = 0$) the system matrices are:

$$\tilde{A}_{SP} = \begin{bmatrix} \dfrac{Z_w}{m} & \dfrac{Z_q + mV_{Ref}}{mV_{Ref}} \\ \dfrac{M_w V_{Ref}}{I_{yy}} & \dfrac{M_q}{I_{yy}} \end{bmatrix}$$

$$\tilde{B}_{SP} = \begin{bmatrix} 0 \\ \dfrac{M_{\delta_m}}{I_{yy}} \end{bmatrix}$$

The terms in $\tilde{A}_{SP}$ or $\tilde{B}_{SP}$ are denoted generically as

$$\tilde{A}_{SP} = \begin{bmatrix} a_{11} & a_{12} \\ a_{21} & a_{22} \end{bmatrix}$$

$$\tilde{B}_{SP} = \begin{bmatrix} 0 \\ b \end{bmatrix}$$

The state transition matrix is easily evaluated,

$$[sI - \tilde{A}_{SP}]^{-1} = \frac{\begin{bmatrix} s - a_{22} & a_{12} \\ a_{21} & s - a_{11} \end{bmatrix}}{s^2 - (a_{11} + a_{22})s + (a_{11}a_{22} - a_{12}a_{21})}$$

The open-loop matrix of transfer functions is therefore:

$$G(s) = \begin{bmatrix} \alpha(s)/\delta_e(s) \\ q(s)/\delta_e(s) \end{bmatrix} = \frac{\begin{bmatrix} ba_{12} \\ bs - ba_{11} \end{bmatrix}}{s^2 - (a_{11} + a_{22})s + (a_{11}a_{22} - a_{12}a_{21})}$$

The short-period mode as a mass–spring–damper

To see how the short-period mode relates to a mass–spring–damper system, we apply $\mathbf{u}(t) = K\mathbf{x}(t) + \mathbf{r}(t)$, from which we showed

$$\mathbf{x}(s) = [I - GK(s)]^{-1} G\mathbf{r}(s)$$

Let $K = \begin{bmatrix} k_1 & k_2 \end{bmatrix}$. Then

$$G(s) = \frac{\begin{bmatrix} ba_{12} \\ bs - ba_{11} \end{bmatrix}}{d(s)}$$

where

$$d(s) = s^2 - (a_{11} + a_{22} - k_2 b)s + (a_{11}a_{22} - a_{12}a_{21} - k_1 ba_{12} + k_2 ba_{11})$$

If $k_1 \neq 0$ and $k_2 = 0$ (α feedback only) the closed-loop characteristic polynomial becomes

$$d(s) = s^2 - (a_{11} + a_{22})s + (a_{11}a_{22} - a_{12}a_{21} - k_1 ba_{12}) \qquad (12.30)$$

In this case σ is left unchanged, as it was for position feedback in the mass–spring–damper.

However, if $k_2 \neq 0$ and $k_1 = 0$ (pitch rate feedback only) the closed-loop characteristic polynomial becomes

$$d(s) = s^2 - (a_{11} + a_{22} - k_2 b)s + (a_{11}a_{22} - a_{12}a_{21} + k_2 ba_{11}) \qquad (12.31)$$

Now both σ and ω_n are affected, which is not the same as rate-feedback for the mass–spring–damper. Even so, most airplanes respond to pitch-rate feedback very much like a mass–spring–damper. We can illustrate this with data from our example aircraft, the A-4 *Skyhawk*. The numerical values for the short-period approximation yielded

$$\tilde{A}_{SP} = \begin{bmatrix} -0.877 & 0.9978 \\ -9.464 & -1.46 \end{bmatrix} \quad \tilde{B}_{SP} = \begin{bmatrix} 0 \\ -12.85 \end{bmatrix} \qquad (12.32)$$

The states are α and q, and the controls are the throttle (which we ignore with a 0 in the $\tilde{B}_{SP}$ matrix) and δ_e.

With pitch-rate feedback (Equation 12.31),

$$d(s) = s^2 + (2.34 - 12.9k_2)\, s + (10.7 + 11.3k_2)$$

MATLAB® was used to generate the root-locus for this system with results shown in upper graph of Figure 12.16 (negative gain was specified for the feedback). While the natural frequency clearly decreases, it is obvious that the dominant effect of pitch-rate feedback is to change the damping ratio, ζ.

Let us verify our observations regarding α feedback for our example aircraft. Using Equation 12.30 we have

$$d(s) = s^2 + (2.34)\, s + (10.7 + 12.8k_1)$$

Again using MATLAB® to generate a root-locus plot, the lower graph of Figure 12.16 results. As expected, the A-4 responds to α feedback the same as a mass–spring–damper

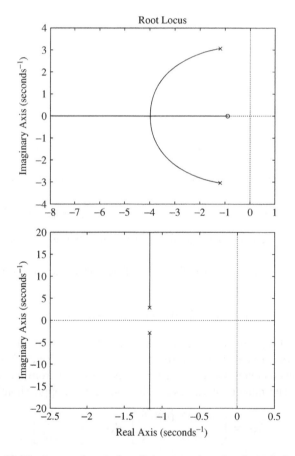

Figure 12.16 Locus of roots for pitch-rate and angle-of-attack feedback.

does to position feedback. With a combination of α and q feedback, the short-period roots may be arbitrarily placed. If the short-period approximation is used to select the gains, then the effect of those gains should be verified using the full set of longitudinal equations. Of course, the effect of the feedback scheme on the control effectors should also be determined.

Control effector considerations

The control effector limiting encountered in our investigation of roll-rate feedback is in a way peculiar to the roll mode. Pilots seldom have need to apply full longitudinal control inceptor deflection (stick or yoke) or full rudder deflection, with a few notable exceptions. Lateral maneuvering is quite different; in tactical maneuvering, full lateral inceptor deflection applied in step inputs is quite common.

The few notable exceptions wherein full longitudinal inceptor deflection is used include several air-show maneuvers, the intentional performance of stalls, recovery from stalls and spins, low-speed air-combat maneuvering, and other similar occasions that are performed at low dynamic pressure. The most common control limiting encountered in such flight regimes result from stick limiters and pushers, mechanical devices intended to prevent the pilot from entering stalls and spins. They activate primarily in response to high angle-of-attack, not pitch rate.

The most likely scenarios of interest to us are windup turns and wings-level pullups from dives. Analysis of a windup turn is complicated by the coupling of the lateral–directional states (recall Equation 5.7 where all the components of ω_B are non-zero). It is very unlikely that a pilot would encounter full stick deflection during a level pullup, but it could happen if the situation was dire.

We will consider that our *Skyhawk* pilot is recovering from a steep dive. The wings are level and he has slowly and smoothly applied increasing aft stick (so that at any time the flight condition is approximately steady). The airspeed bleeds off because of the high induced drag, until finally he runs out of aft stick, coincidentally just when passing through $M = 0.4$ near sea-level (the conditions for which our data are valid). We will ignore the *Skyhawk's* aerodynamically actuated leading-edge slats, as well as the fact that they are as likely as not to deploy asymmetrically.

Unaugmented aircraft

We first examine the unaugmented aircraft with a mechanical link from stick to elevator. Assume a 1:1 relationship between longitudinal stick (δ_{LS}) and elevator so that, measured in the same units,

$$\delta_e = \delta_{LS}$$

Since we are considering a pullup we want the elevator deflection to be trailing edge up (TEU) which is a negative deflection according to our sign conventions. Therefore aft stick will be called negative as well.

Using the short-period approximation the transfer function of interest is

$$\frac{q(s)}{\delta_e(s)} = \frac{-12.85s - 11.27}{s^2 + 2.337s + 10.72}$$

The negative value of the transfer function arises because of our sign conventions. Then, for $\delta_{LS} = -\delta_{LS_{Max}}$,

$$q_{SS} = 1.05\delta_{LS_{Max}} \text{ s}^{-1}$$

Augmented aircraft

The short-period roots of the example A-4 *Skyhawk* were compared with the Level 1 flying qualities requirements of a Class IV airplane in the ground-attack role (GA, Category A). From Table 11.1 the short-period damping ratio limits are found to be $0.35 \leq \zeta_{SP} \leq 1.30$. We approximate the value of n/α as follows:

$$\frac{n}{\alpha} = \frac{L/W}{\alpha} = \frac{\bar{q}SC_{L_\alpha}\alpha}{W\alpha} = \frac{\bar{q}SC_{L_\alpha}}{W}$$

Using the given properties of the example aircraft, this results in $n/\alpha = 12.1$ g/rad. Reference to Table 11.4 yields $1.8 \leq \omega_{n_{SP}} \leq 6.6$ rad/s. The actual short-period characteristics of the airplane were previously determined to be $\omega_{n_{SP}} = 3.27$ rad/s and $\zeta_{SP} = 0.357$. The natural frequency is thus well within the specification requirement, but the damping is very close to the lower Level 1 limit. Let us then apply pitch-rate feedback to improve the short-period damping.

Using Equations 12.32 and MATLAB® functions `rlocus` and `rlocfind` it was found that a feedback gain of $k_q = 0.222 \text{ s}^{-1}$ yielded roots $\tilde{\lambda}_{SP} = -2.59 \pm j2.55 \text{ s}^{-1}$, or $\tilde{\zeta}_{SP} = 0.72$.

Our reference signal is δ_{LS} and the control law is

$$\delta_e(t) = \delta_{LS}(t) - 0.222q(t)$$

The closed-loop transfer function is

$$\frac{q(s)}{\delta_{LS}(s)} = \frac{-12.85s - 11.27}{s^2 + 5.189s + 13.22}$$

$$q_{SS} = 0.852\delta_{LS_{Max}} \text{ s}^{-1}$$

Thus, rate feedback in this simple case has reduced the steady-state pitch rate by almost 20%. Whether or not this degradation in steady-state pitch rate would be a problem depends on how well the airplane performs with respect to other specifications and, of course, whether or not the pilots like it.

12.2.3 Phugoid

Flight-path control using thrust

In this design example we will meet the idea of *dynamic elements* in the control system design. To-date we have only considered static gains, that is, multiplying the signal by a number in the block diagram. Dynamic elements are devices, computer algorithms, or electrical networks that are described by their own dynamical equations. The specific dynamic element we will look at is an integrator, whose output is (or approximates) the integral of its input. Integral networks, or filters, are useful to remove lingering errors in the system – the longer an error is present, the greater the output of the integrator. If the reader has an electrical engineering background, we are talking about a charging capacitor.

The particular design chosen for this example has a real and very serious background. There have been several airplane accidents involving failure of some or all of the primary flight control surfaces, often resulting in loss of aircraft and often of life too. In one notable instance a DC-10 suffered loss of pitch control but the pilots managed to sufficiently control flight-path angle by varying engine thrust such that they were able to guide the airplane to an airfield in Idaho, USA. The aircraft broke up following a very hard landing, but more than half the airplane occupants survived.

For our phugoid feedback controller, we turn to a proposed method to permit control of an airplane using thrust alone. Burken et al. (1996) addressed that DC-10 accident and others as motivation for the design and flight test of a longitudinal emergency control system using thrust modulation on an MD-11 airplane. The dynamics through which this controller operates is the phugoid mode, and the primary control inceptor is the throttle.

Burken et al. (2000) subsequently patented the idea. The patent shows two configurations. With some changes in notation Figure 12.17 duplicates the primary design. Let us examine some of the features of this system to simplify it for application to our example aircraft.

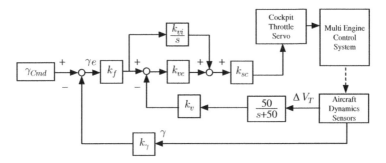

Figure 12.17 'Emergency Control Aircraft System Using Thrust Modulation', U.S. Patent 6,041,273.

Our model (Appendix A) of the A-4 *Skyhawk*, has naive dynamics for its engine. This consists simply of a throttle that ranges from zero to one and an engine that responds proportionally, without delay, with thrust from zero to 11,200 lb. Therefore the three blocks in the figure labeled k_{sc}, 'Cockpit Throttle Servo', and 'Multi Engine Control System' will be replaced by the appropriate entry in B_{Long} and δ_T. Likewise, we have modeled no aircraft sensors, so we will assume perfect measurement of the states.

The block in Figure 12.17 that acts on the sensed ΔV_T is a low-pass, or simple lead filter. It is an electrical network or digital filter that removes high frequencies, in this case it attenuates signals of 50 rad/s by 3 db and higher frequencies by greater amounts, smoothing the signal. Since we have assumed perfect measurement of airspeed, there is no noise and the filter is not required.

The block in Figure 12.17 labled k_v is the velocity feedback gain, and its selection determines the inner-loop dynamics. The block labled k_γ is the flight-path feedback gain.

Because it is compared with a signal proportional to the commanded flight path angle, γ_{Cmd} it is normally set to $k_\gamma = 1$ so that the output of the summing node, γ_e, is an actual error. The loop gain that deals with γ feedback is taken up in the feedforward gains.

Thus, we remove the block for k_γ, and our design will choose k_v and k_f, or its equivalent.

With these changes our simplified block diagram for flight-path angle control through throttle is shown in Figure 12.18.

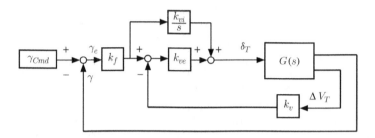

Figure 12.18 'Emergency Control Aircraft System Using Thrust Modulation', simplified.

The block $G(s)$ represents the three transfer functions required:

$$G(s) = \begin{bmatrix} \dfrac{u(s)}{\delta_T(s)} \\ \dfrac{\alpha(s)}{\delta_T(s)} \\ \dfrac{\theta(s)}{\delta_T(s)} \end{bmatrix}$$

Before proceeding we take a brief detour to see how to select feedforward gains using MATLAB®. The MATLAB® command `rlocus` expects the first argument to be a single-input single-output system, and creates a root locus corresponding to a gain in a negative feedback loop between output and input, k_{fb} in Figure 12.19.

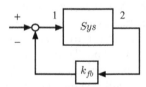

Figure 12.19 Configuration MATLAB® expects.

In Figure 12.19 the block named *Sys* may be the series connection of a number of elements, everything in the actual problem that lies between the numbers 1 and 2 in the diagram. We place no label on the input on the left since that information is not part of the problem MATLAB® is solving – it is something that we add to the design after MATLAB® has told us what to use for k_{fb}.

What we have with feedforward is simplified in Figure 12.20.

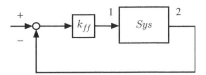

Figure 12.20 Feedforward.

We put the feedforward block in the feedback portion by simply moving the summing point around as shown in Figure 12.21.

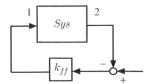

Figure 12.21 Feedforward block in the feedback loop.

There is no negative summing block between k_{ff} and point 1 of *Sys*, but it works out because k_{ff} is still inverted. In other words, MATLAB® does not know or care where the negation occurs, so long as it is in the loop. The main thing to be careful of is to make sure that everything between points 1 and 2 is correct.

With that detour behind us, we proceed. The block k_{vi}/s is an integrator. The integrator is quite common in control system design, and is usually discussed as a component of a 'Proportional-Integral-Derivative' (PID) controller. An excellent introduction to PID controllers may be found at Langton (2006), another book in Wiley's *Aerospace Series*. Here we have just the proportional and integral parts, so this network is referred to as a *PI controller*. k_{ve} is the proportional gain, and k_{vi} the integral gain.

Now, the inner-loop feedback of ΔV_T is amplified twice in Figure 12.18, first by k_v and then by k_{ve}. Let us combine those two operations into one. Likewise, γ is subject to both the gain k_f and the proportional part of the PI filter k_{ve}, so we will let k_{ve} absorb the function of k_f. The result is as shown in Figure 12.22. These changes simplify the problem without losing the essential elements, and permit us to isolate the inner loop so that k_v can be selected without worrying about the algebraic effect of subsequent choices for other gains.

One last change, and then we'll get on with gain selection. The two blocks labeled 'PI Filter' in Figure 12.22 are replaced with three blocks in Figure 12.23. The reader should verify that the transfer function of the PI filter has not changed.

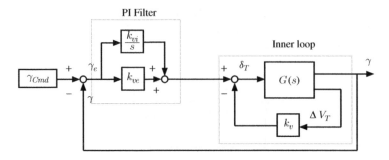

Figure 12.22 k_f eliminated and inner loop isolated.

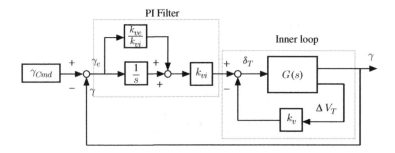

Figure 12.23 Flight path control, design block diagram.

The reason for the change in the PI filter is that in Figure 12.22 there are two gains to be selected. In Figure 12.23 we will pick a value for k_{ve}/k_{vi} (actually, we will pick a value for k_{vi}/k_{ve}) and then use root locus to pick k_{vi}. The ratio k_{vi}/k_{ve} has significance, in that the transfer function of the two blocks in parallel, k_{ve}/k_{vi} and $1/s$, is

$$\frac{1}{s} + \frac{k_{ve}}{k_{vi}} = \frac{1 + s\frac{k_{ve}}{k_{vi}}}{k_{vi}}$$

So that a zero is introduced into the loop at

$$s = -\frac{k_{vi}}{k_{ve}}$$

The terms can be a little confusing. The ratio k_{ve}/k_{vi} is a gain in the box so marked in Figure 12.23. The zero that results from that gain is at $-k_{vi}/k_{ve}$ in the complex plane. Thus when we say the zero was placed at -0.025 we mean that $k_{ve}/k_{vi} = 40$.

The design will proceed as follows:

1. Perform root locus for k_v.
2. Select a value for the transmission zero $s = -k_{vi}/k_{ve}$.
3. Perform a root locus for k_{vi}.
4. Repeat as necessary.

Automatic Flight Control

To tell MATLAB® what outputs to use, we create a C_{Long} matrix that makes the first output the same as the first state u, and the second output the fourth state θ minus the second state α, both converted from radians into degrees for ease of interpreting the time histories.

$$C_{Long} = \begin{bmatrix} 1 & 0 & 0 & 0 \\ 0 & -180/\pi & 0 & 180/\pi \end{bmatrix}$$

Inner loop

The phugoid roots of the example aircraft were previously determined to be $\hat{\lambda}_{Ph} = -0.0067 \pm j0.096$, yielding the damping ratio $\zeta_{Ph} = 0.078$ and natural frequency $\omega_{n_{Ph}} = 0.096$. We note in passing that the only applicable flying qualities specification is the Level 1 requirement $\zeta_{Ph} \geq 0.04$, so the aircraft is well within the specification.

Now, modeling just the part of Figure 12.23 called 'Inner loop', we find the appropriate transfer function is (as it was in Section 10.1):

$$\frac{u(s)}{\delta_T(s)} = \frac{20.5s^3 + 47.9s^2 + 219.8s}{s^4 + 2.352s^3 + 10.76s^2 + 0.1652s + 0.09928}$$

The root locus is shown in Figure 12.24.

From the root locus we pick the roots $-0.060 \pm j0.066$ with $k_v = 6.09 \times 10^{-3}$ s^{-1}. The phugoid now has a damping ratio $\zeta = 0.727$ and natural frequency $\omega = 0.0962$ rad/s.

Outer loop

Before picking the location of the transmission zero we check to see what effect the closing of the outer loop is going to have, given our selection for the inner loop. Then we will add the PI filter and see what effect the transmission zero has.

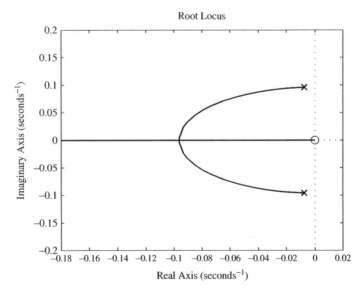

Figure 12.24 Root locus, velocity feedback.

We first close the inner loop using the MATLAB® command feedback. The system thus created is the box in dashed lines in Figure 12.23 labeled 'Inner loop'. From this we have two transfer functions, one corresponding to u and the one we select that corresponds to γ.

$$\frac{\gamma(s)}{\gamma_e(s)} = \frac{0.3712s^2 + 0.5421s + 3.624}{s^4 + 2.352s^3 + 10.76s^2 + 0.1652s + 0.09928}$$

This transfer function is valid for the inner loop with $k_v = 6.09 \times 10^{-3}$ s^{-1}.

A root locus was made for this transfer function, effectively ignoring the PI filter. The result is shown in Figure 12.25 and in closeup in Figure 12.26. Figure 12.25 shows that we can drive the system unstable if we increase the outer-loop gain too high. The closeup in Figure 12.26 shows that the outer-loop gain acts to increase the natural frequency while decreasing the damping.

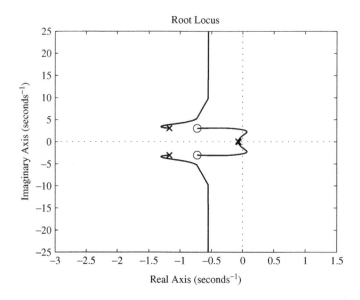

Figure 12.25 Root locus, outer loop.

To see what would happen if we backed off a bit on the inner-loop gain, we did that step again, selecting roots on the real axis at -0.185 and -0.050 (gain $k_v = 0.0108$). Then, repeating the outer-loop root locus, Figure 12.27 resulted.

Now the outer-loop gain moves the inner-loop roots back to a point roughly midway between their starting locations, where they split and become a complex conjugate pair. The locus of that pair look encouraging for selecting good natural frequency and damping.

Transmission zero
We begin this step with $k_v = 0.0108$. The inner-loop transfer function to γ (the input is not labeled–it is at the output of k_{vi} in Figure 12.23) for this gain is

$$g_{inner} = \frac{0.3712s^2 + 0.5421s + 3.624}{s^4 + 2.573s^3 + 11.28s^2 + 2.543s + 0.09928} \quad (12.33)$$

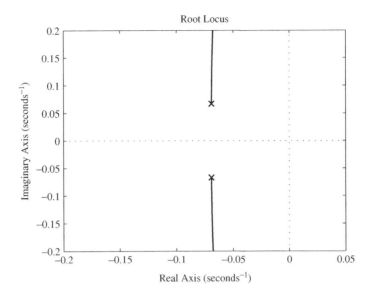

Figure 12.26 Root locus, outer loop zoomed.

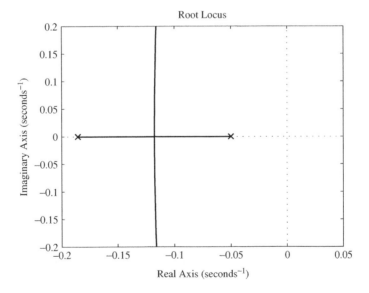

Figure 12.27 Root locus, outer loop again.

To see the effect of the placement of the PI filter zero on the outer-loop closure, different values were tried: one at -0.25 (to the left of the inner-loop pole at -0.185), one at -0.025 (to the right of the inner-loop pole at -0.05), and one at -0.12 (midway between the two inner-loop poles). For each selection of the zero the forward path was defined beginning just to the right of the k_{vi} block in Figure 12.23 and ending just to its left.

For example, with the zero at $k_{vi}/k_{ve} = 0.120$,

$$\frac{1}{s} + \frac{k_{ve}}{k_{vi}} = \frac{8.333s + 1}{s} \qquad (12.34)$$

The feedforward loop for the selection of k_{vi} is the product of Equations 12.33 and 12.34,

$$g_{fwd} = \frac{3.093s^3 + 4.888s^2 + 30.74s + 3.624}{s^5 + 2.573s^4 + 11.28s^3 + 2.543s^2 + 0.09928s}$$

It was found that placing the zero anywhere to the left of the inner-loop pole at -0.05 results in that pole joining the PI pole at zero to form complex roots near the origin. The root loci that resulted from the zero at -0.25 and -0.12 are shown in Figure 12.28 and Figure 12.29, respectively.

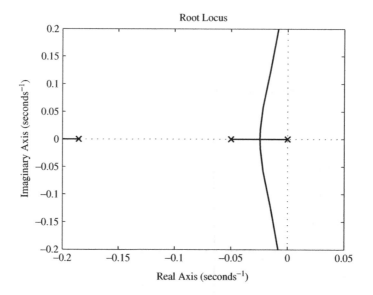

Figure 12.28 Root locus, zero at -0.25.

Placing the zero to the right of the inner-loop pole at -0.05 allows us to approximately regain the poles from the outer-loop closure without the PI filter. A zero placement at -0.040 was selected and root locus performed to determine k_{vi}. The final placement is shown in Figure 12.30.

The gains that resulted (units of 1/deg) are

$$k_v = 5.08 \ k_{vi} = 0.00222, \ k_{ve} = 0.0555$$

The Dutch roll roots are virtually unchanged, at $\lambda_{DR} = -1.17 \pm j3.06$. The phugoid is at $\lambda_{Ph} = -0.099 \pm j0.102$. The root due to the PI filter is at -0.0358.

Using those gains, the response to a 1° step input in γ_{Cmd} was generated by combining the elements into a single transfer function and using the MATLAB® command step. The results are shown in Figure 12.31.

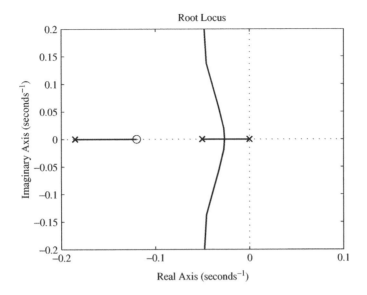

Figure 12.29 Root locus, zero at -0.12.

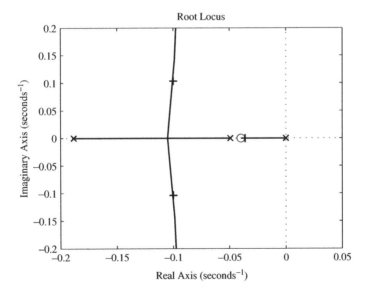

Figure 12.30 Final selection of k_{vi}.

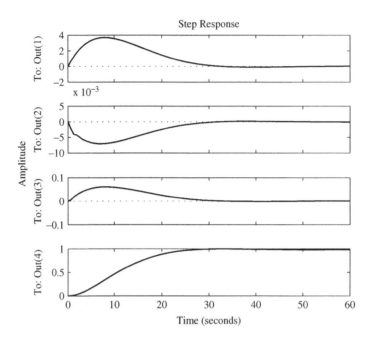

Figure 12.31 Following a 1° step input $\gamma_{C_{md}}$. Time histories are Δu (ft/s), α (deg), q (deg/s), and γ (deg). The scale of the α plot is $\times 10^{-3}$.

The flight-path angle γ appears to have a time constant of roughly 15 seconds. The pilot is clearly going to have to think well ahead as he commands different flight path angles. It is not likely that aircraft carrier landings will be performed using this design.

12.2.4 Coupled roll–spiral oscillation

For our last feedback control design challenge we turn to a rather unusual aircraft with a very interesting history: the Northrop M2-F2 (Figure 12.32). The M2-F2 was a lifting-body technology demonstrator flown in support of re-entry vehicle research. The shape of the M2-F2 grew out of theoretical and wind-tunnel studies of horizontal-landing vehicles that could withstand the environment associated with reentry from earth orbit; see for example Pyle and Swanson (1967).

The M2-F2 first flew in July 1966 and crashed less than a year later (film footage of the crash was used during the opening credits of the television movie and series *The Six-Million Dollar Man*). Of the many bad flying qualities associated with the aircraft its lateral–directional characteristics required especially extensive pilot compensation. The vehicle was prone to lateral pilot-induced oscillations (PIO), a phenomenon with many causes but often strongly linked to sluggish and delayed responses to pilot control inputs. The pilot's control inputs become 180° out of phase with the response of the aircraft, causing oscillations of increasing amplitude.

Data from different sources were used to approximate the M2-F2's lateral–directional dynamics. When linearized, Equation 12.35 resulted. The states are β, p, r, and ϕ; and

Figure 12.32 M2-F2 (and F-104 chase plane). Source: NASA.

the controls δ_a and δ_r. All angular measures are in degrees.

$$A_{LD} = \begin{bmatrix} -0.4183 & 0 & -1.0 & 0.0919 \\ -115.0 & -0.885 & 1.18 & 0 \\ 8.50 & 0.136 & -0.794 & 0 \\ 0 & 1.0 & 0 & 0 \end{bmatrix} \quad B_{LD} = \begin{bmatrix} 0.0214 & 0.0306 \\ 13.0 & 9.0 \\ -2.17 & -5.13 \\ 0 & 0 \end{bmatrix} \quad (12.35)$$

$$\lambda_{DR} = -0.993 \pm j2.77 (\zeta_{DR} = 0.337, \omega_{n_{DR}} = 2.94 \text{ rad/s})$$
$$\lambda_{RS} = -0.0559 \pm j0.927 (\zeta_{RS} = 0.0601, \omega_{n_{RS}} = 0.929 \text{ rad/s}) \quad (12.36)$$

Note that the M2-F2 used a sign convention for positive aileron deflection that is opposite of what we have used, as evinced by the positive number in $B_{LD}(2, 1)$.

The eigenvalues of this system are interesting. The Dutch roll is present, but the roll and spiral modes have combined to form a new oscillatory mode, the 'coupled roll–spiral oscillation' (subscript RS) which was mentioned in Section 11.2.3. The coupled roll–spiral mode is sometimes called the 'lateral phugoid'. The specifications for this mode are repeated here:

Level	$\zeta_{RS}\omega_{n_{RS}}$, rad/s
1	0.5
2	0.3
3	0.15

The M2-F2 with $\zeta_{RS}\omega_{n_{RS}} = 0.0559$ rad/s fails at any level, and this may help explain the PIO tendency of the aircraft.

As an interesting historical note, the specifications for the coupled roll–spiral mode were determined in part using data from piloted re-entry vehicles, including the M2-F2 (Chalk et al., 1969). The analysis was performed by Cornell Aeronautical Laboratory, Inc., later to become Calspan, famous for many things including variable-stability airplanes used to determine flying qualities criteria.

Before we begin, let us perform a sensitivity analysis of A_{LD} for the M2-F2 (Equation 12.35). We introduced the sensitivity analysis as a tool for approximating the various modes, but it can as well show us which states are likeliest to have the greater effect on a mode when fed back. Using the m-file in Section E.6.2,

$$S_{M2F2LD} = \begin{bmatrix} 0.4233 & 0.4233 & 0.0767 & 0.0767 \\ 0.1186 & 0.1186 & 0.3814 & 0.3814 \\ 0.2567 & 0.2567 & 0.2433 & 0.2433 \\ 0.1031 & 0.1031 & 0.3969 & 0.3969 \end{bmatrix} \quad (12.37)$$

The first two columns of S_{M2F2LD} correspond to the Dutch roll, and the third and fourth to the coupled roll–spiral mode. The first row shows the relative influence of β in the modes, the second of p, the third of r, and the fourth of ϕ.

There is not a single state, or even a pair of states that dominate either of the two modes. Any feedback system will likely consist of careful tuning of the feedback of two or three states.

The relatively large effect of ϕ in the coupled roll-spiral mode, inherited from its role in the spiral mode, suggests that it may be a candidate for feedback. The only thing the Euler angles have to do with the dynamics of our airplane is through the orientation of the gravity vector. The mechanism through which ϕ acts on the spiral mode is to place the gravity vector to one side or the other of the plane of symmetry, which results in an acceleration in that direction. Loosely speaking, that acceleration will generate sidewards velocity, hence sideslip β, then a rolling moment through the dihedral effect. The problem with such twice-removed effects (sideslip then moment) is that they introduce zeros into the transfer functions, not always in good places. We will examine this shortly.

The matrix of transfer functions corresponding to this system was created in MATLAB®. Normally one does not try to cure lateral–directional modes with bank angle feedback, but given the rather different configuration of the M2-F2, along with the appearance of the coupled roll–spiral mode, all possibilities were considered.

Several of the transfer functions had zeros in the right-half plane; with two of the open-loop roots so near the imaginary axis it seems likely that any feedback at all would send these roots straight toward the zeros, decreasing the damping and then becoming unstable for low levels of gain.

For example, the aileron to yaw rate transfer function has a pole-zero map as shown in Figure 12.33.

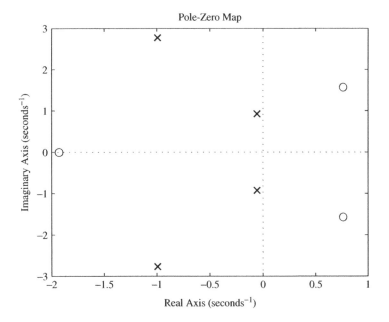

Figure 12.33 Pole-zero map, $r(s)/\delta_a(s)$.

A root locus plot (Figure 12.34) of rudder to yaw rate (a negative transfer function) shows what happens when the gain is increased in this case. With respect to Figure 12.34, any bit of gain will drive the system unstable.

The two transfer functions for ϕ lived up to expectations. Equation 12.38 shows that of the aileron δ_a, and Equation 12.39 that of the rudder δ_r.

$$\frac{\phi(s)}{\delta_a(s)} = \frac{13s^2 + 10.74s - 137.5}{s^4 + 2.097s^3 + 9.745s^2 + 2.681s + 7.472} \tag{12.38}$$

$$\frac{\phi(s)}{\delta_r(s)} = \frac{9s^2 + 1.338s - 515.5}{s^4 + 2.097s^3 + 9.745s^2 + 2.681s + 7.472} \tag{12.39}$$

Here is an opportunity to use one of those often overlooked but very useful facts from polynomial theory: by inspection we know that each of these transfer functions has exactly one zero in the left half and one in the right half of the complex plane.

It is *Descartes' rule of signs* and it goes like this: with the polynomial arranged in descending powers of the variable (s in this case), the number of positive roots of the polynomial is either equal to the number of sign differences between consecutive nonzero coefficients, or is less than it by a multiple of 2. By substituting $-s$ for s, the number of negative roots is the number of sign changes, or fewer than it by a multiple of 2.

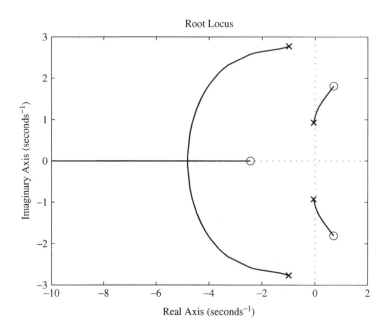

Figure 12.34 Root locus, $-r(s)/\delta_r(s)$.

Each of the numerator polynomials has exactly one sign change with $+s$ so there is one positive root, and since there are only two roots the other must be negative. Negating s yields $13s^2 - 10.74s - 137.5$ for the first and $9s^2 - 1.338s - 515.5$ and again, only one sign change, so our previous conclusion is confirmed, one negative and one positive root.

A root-locus plot of the transfer function $\phi(s)/\delta_a(s)$ (Equation 12.38) is shown in Figure 12.35 to illustrate the points. A root locus of $\phi(s)/\delta_r(s)$ is similar.

Both the roll-rate transfer functions had third-order numerators with sign changes, and have zeros in the left-half plane, at the origin, and in the right-half plane. Very small amounts of gain drove these transfer functions unstable.

The two sideslip transfer functions had no right-half plane zeros. The root locus from aileron to sideslip is shown in Figure 12.36, and of rudder to sideslip in Figure 12.37.

The aileron is clearly not going to help the coupled roll–spiral mode back to uncoupled modes, as that root locus is intercepted by two zeros before reaching the real axis.

Only the rudder to sideslip transfer function appears to offer hope. A gain of 18.8 yielded a roll mode at -0.488 s^{-1}, spiral mode of -0.254 s^{-1}, and Dutch roll roots at -0.859 ± 8.32 s^{-1}. The Dutch roll damping is 0.103 which is marginal for any class airplane.

Of greater concern is the value of the gain, which has units of degrees of rudder per degree of sideslip. Thus, one degree of sideslip would result in 18.8 deg of rudder deflection. One degree of sideslip is not very much, and more would be easily encountered in light turbulence. This would place a very high demand on the rudder system, likely driving it from stop-to-stop.

Thus, we reject the idea of any conventional feedback scheme to deal with the coupled roll–spiral mode.

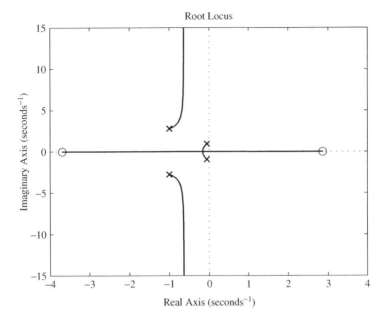

Figure 12.35 Root locus, $\phi(s)/\delta_a(s)$.

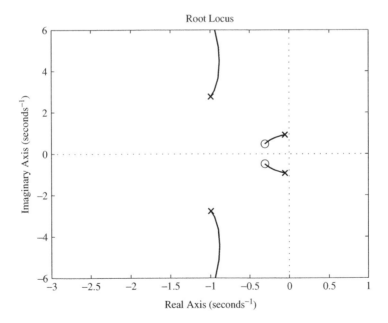

Figure 12.36 Root locus, $\beta(s)/\delta_a(s)$.

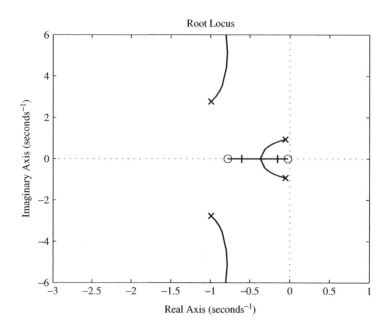

Figure 12.37 Root locus, $\beta(s)/\delta_r(s)$.

The aileron–rudder interconnect

The actual M2-F2 did have a roll stability-augmentation system but it was not based on the control effectiveness matrix in Equation 12.35. The aircraft had an *aileron–rudder interconnect* (ARI) such that aileron deflections automatically deflected the rudder through a mechanical linkage (sometimes called *ganging*). The objective of the ARI was to cancel out the *adverse yaw* caused by aileron deflections. (It is called 'adverse' yaw because positive aileron yields negative roll acceleration but positive yaw acceleration.) The use of the ARI is ubiquitous in airplane design; the F-18 *Hornet* called it the rolling-surface rudder interconnect, or RSRI.

For the M2-F2, the ratio could be made by algebraically cancelling the contribution to $\dot{r}$ from the aileron (the 3, 1 element of B_{LD}) with the appropriate amount of rudder (the 3, 2 element of B_{LD}). Thus, the M2-F2 engineers might have chosen

$$\delta_r = -\frac{2.17}{5.13}\delta_a = -0.423\delta_a$$

With this gearing, assuming the pilot did not attempt to deflect the rudders in addition to the deflection from the ARI, the control effectiveness matrix, with the lateral stick the sole controller, would be

$$B_{LD} = \begin{bmatrix} 8.416 \times 10^{-3} \\ 9.19 \\ 0 \\ 0 \end{bmatrix}$$

The new control, δ_{ARI}, will presumably deal with yaw rate automatically, and so we look at the transfer function to roll rate, shown in Equation 12.40.

$$\frac{p(s)}{\delta_{ARI}(s)} = \frac{9.193s^3 + 10.17s^2 + 80.51s}{s^4 + 2.097s^3 + 9.745s^2 + 2.681s + 7.472} \quad (12.40)$$

This resulted in two zeros near the Dutch roll roots that attracted those poles, while a zero at the origin and another at $-\infty$ attracted the coupled roll–spiral roots to the real axis, there to split into conventional roll and spiral modes. The root locus plot used to select the feedback gain is at Figure 12.38.

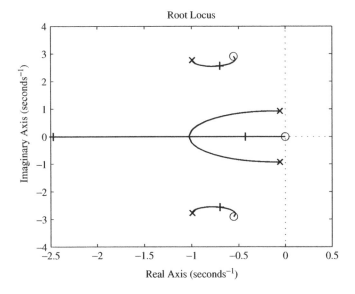

Figure 12.38 M2-F2 roll-rate feedback to the ARI.

The selected gain was $k_{ARI} = 0.240$. The spiral mode was $\lambda_S = -0.428$, the roll mode $\lambda_R = -2.47$, and the Dutch roll $\lambda_{DR} = -0.7 \pm j2.56$, rad/s. The gain, with units of degrees of ARI per degree per second of roll rate, appears reasonable – it is not to be imagined that M2-F2 pilots were performing air show maneuvers requiring snappy rolls. As to the eigenvalues, there are airplanes flying today with worse lateral–directional characteristics than those achieved here.

Assuming that the designers of the M2-F2 roll stability augmentation system used an approach similar to that shown here, and even though the lateral–directional dynamics now appear to have acceptable flying qualities, the aircraft was still susceptible to PIO. In fact, the pilot on the last flight had experienced a lateral PIO shortly before (attempting) landing.

As an interesting aside, the ARI gearing on the M2-F2 could be adjusted in flight by the pilot, as would be required for different flight conditions. Since the gearing would influence the roll stability-augmentation system, then presumably the pilot could also vary the feedback gain.

Problems

1. The airplane in this and problem 2 resulted from relaxing its static stability (Stevens and Lewis 1992). The short period and phugoid roots have combined to create two real roots and one complex conjugate pair. Consider the system:

$$A_{Long} = \begin{bmatrix} -1.93 \times 10^{-2} & 8.82 & -0.575 & -32.2 \\ -2.54 \times 10^{-4} & -1.02 & 0.905 & 0 \\ 2.95 \times 10^{-12} & 0.822 & -1.08 & 0 \\ 0 & 0 & 1 & 0 \end{bmatrix}$$

$$B_{Long} = \begin{bmatrix} 0.174 \\ -2.15 \times 10^{-3} \\ -0.176 \\ 0 \end{bmatrix}$$

The states are V, α, q, and θ; the single control is elevator, δ_e. Velocity is in ft/s, state angle measures are in radians, and elevator in degrees.

(a) What flying qualities specifications are relevant to this configuration?
(b) Use α and q feedback to place the short-period roots in the vicinity of $\lambda_{SP} = -2 \pm j2$. Show all steps, including root-locus plots.

2. Consider the system:

$$A_{LD} = \begin{bmatrix} 0.322 & 0.0364 & -0.992 & 0.064 \\ -30.6 & -3.68 & 0.665 & 0 \\ 8.54 & -0.0254 & -0.476 & 0 \\ 0 & 1 & 0 & 0 \end{bmatrix}$$

$$B_{LD} = \begin{bmatrix} 0.000295 & 0.000806 \\ 0.733 & 0.132 \\ 0.0319 & -0.062 \\ 0 & 0 \end{bmatrix}$$

The states are β, p, r, and ϕ; the controls are aileron δ_a and rudder δ_r. All state angle measures are in radians, controls are in degrees.

(a) What flying qualities specifications are relevant to this configuration?
(b) Use appropriate state feedback to either or both of the controls to place the Dutch roll roots so that Dutch roll damping $\zeta_{DR} \approx 0.7$. Show all steps, including root-locus plots.
(c) Examine the design when using an aileron–rudder interconnect as was done for the M2-F2.

3. The design of the flight-path controller mentioned two proposed configurations in the patent. The other proposal is shown in Figure 12.39.

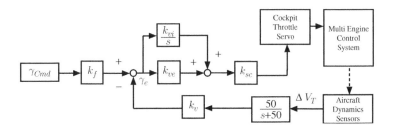

Figure 12.39 Alternate design, 'Emergency Control Aircraft System Using Thrust Modulation'.

Note that this design was proposed as a stand-alone solution with no flight-path feedback. Consider the application of this design to the example aircraft.
(a) The patent refers to γ_e as the 'phugoid damped flightpath angle error signal'. Is γ_e actually an error signal? Under what steady conditions will γ_e be zero?
(b) Consider the presence of an integrator downstream of the 'phugoid damped flight-path angle error signal': if γ_e doesn't vanish or at least change sign, What can you expect the throttle to (attempt to) do?
(c) Review Section 10.1.5, especially as regards the response of the airplane to a step input in throttle. Can this design work? If not, why not?

References

Burken, J.J., Maine, T.A., Burcham, F.W., Jr.,, and Kahler, J.A. (1996) *Longitudinal Emergency Control System Using Thrust Modulation Demonstration on an MD-11 Airplane*, NASA Technical Memorandum NASA TM-104318, July 1996.

Burken, J.J. *et al.* (2000), *Thrust-Control System for Emergency Control of an Airplane*. U.S. Patent 6,041,273.

Chalk, C.R. *et al.* (1969) *Background Information and User Guide for MIL-F-8785B(ASG), "Military Specification–Flying Qualities of Piloted Airplanes"*, Technical Report AFFDL-TR-69-72, Air Force Flight Dynamics Laboratory, Wright-Patterson Air Force Base, Ohio, August 1969.

D'Azzo, J.J. and Houpis, C.H., (1988) *Linear Control System Analysis and Design: Conventional and Modern*, McGraw-Hill, New York.

Langton, R. (2006) *Stability and Control of Aircraft Systems*, John Wiley & Sons, Ltd, Chichester.

Pyle, J.S. and Swanson, R.H. (1967) *Lift and Drag Characteristics of the M2-F2 Lifting Body During Subsonic Gliding Flight*, NASA Technical Memorandum TM-X1431, August 1967.

Stevens, B.L. and Lewis, F.L. (1992) *Aircraft Control and Simulation*, 1st edn, John Wiley & Sons, Inc., New York, pp. 255–259.

U.S. Naval Test Pilot School Flight Test Manual USNTPS-FTM-No. 103 (1997) *Fixed Wing Stability And Control–Theory and Flight Test Techniques*, figure 5.86.

13

Trends in Automatic Flight Control[1]

13.1 Overview

Many modern aircraft utilize a very different form of automatic flight control called *dynamic inversion*. It was demonstrated in the F-15 ACTIVE (Advanced Control Technology for Integrated Vehicles) program and was the basis for the flight control system design of the X-35 (Walker and Allen, 2002), which began in the Joint Strike Fighter program and became the F-35 fighter.

Dynamic inversion is conceptually simple. It is sometimes spoken of as subtracting the old dynamics and adding the desired ones. Its success depends on having on-board the aircraft a fairly extensive set of aerodynamic data (representing the old dynamics) and fast computational capabilities to access the data and put it in a form that dynamic inversion can use.

The output of the dynamic inversion process may be loosely viewed as vectors of desired moments for the control effectors to generate. Allocating the three moments to three control effectors is a simple process. However, modern aircraft may have many more than three independent control effectors. The F-15 ACTIVE (Figure 13.1) is a rather extreme example of this. All the following controls were capable of independent actuation: two rudders, left and right horizontal tail, left and right canards, and left and right axisymmetric thrust vectoring.

While no flight control is redundant in the sense of not being needed on the airplane, they may be redundant in their effects. Rolling moments may be generated by ailerons, diffential horizontal tail, or some combination of both. When the number of independent control effectors is greater than the effects they are to generate, the problem of how best to mix them becomes the *control allocation* problem.

The control allocation problem is compounded by the fact that the control effectors have their own position limits that must be considered. Optimal control allocation is not a simple problem to visualize, and some of the mathematics involved can be challenging. However, some of the methods of generating approximately optimal solutions are accessible and

---

[1] Parts of the section of this chapter that deal with control allocation are based on the author's previous journal articles on that subject. The principle source is Durham, W. (1994) Attainable moments for the constrained control allocation problem. *Journal of Guidance, Control, and Dynamics*, **17**(6) 1371–1373.

Figure 13.1 F-15 ACTIVE. Source: NASA.

will be described. In particular we will examine a method due to Bordignon and Bessolo (2002) which was successfully used in the design of the X-35 flight control system.

13.2 Dynamic Inversion

The basic idea behind dynamic inversion is quite simple. Consider a scalar equation

$$\dot{x} = ax(t) + bu(t) \tag{13.1}$$

Now say we want to find a control law (as we did for feedback control) that calculates control inputs that will generate some desired (subscript d) dynamic response $\dot{x}_d(t)$. Assume that we know a and b and can measure $x(t)$ and consider the following control law:

$$u(t) = \frac{1}{b}(\dot{x}_d - ax(t)) \tag{13.2}$$

It is a trivial exercise to substitute the $u(t)$ from Equation 13.2 into Equation 13.1 to find that

$$\dot{x} = \dot{x}_d \tag{13.3}$$

Note that we didn't specify where $\dot{x}_d$ came from. It may help to think of this process in a flight control setting. The pilot's inputs are fed to a computer that applies them to a model of an airplane with good flying qualities and which calculates the resulting time histories of $\dot{x}_d$. The computer applies $\dot{x}_d$ to a model of the inverted dynamics of the actual aircraft (as in Equation 13.2) to calculate the control effector deflections to make the airplane respond in the desired manner (as in Equation 13.3).

Also note that while we specified a linear equation in Equation 13.1, that was not necessary. All that is needed is that we can 'solve' the equation for u and that we can calculate the dynamics that remain.

One of the biggest problems with extending this scalar example to our aircraft equations of motion is the $1/b$ in Equation 13.2. Since we have many equations but only a few controls, any matrix inversion is likely to be problematic.

One of the earlier treatments of dynamic inversion in flight control applications is due to Elgersma (1988), and the following description follows his notation. We will present the theory in general form, then we will use the numerical example of the M2-F2 as provided in Equation 12.35 to illustrate the points.

To begin, consider that the equations of motion of an airplane are of two distinct types: the *kinematic equations* used to describe the orientation of the aircraft, and the force and moment equations.

Of the force and moment equations there is another division possible: those that are directly controlled by the control effectors, and those that are not.

In conventional aircraft the primary moment generators are the ailerons, elevator, and rudder, or their equivalents. These control effectors are used primarily to shape the response of the aircraft in roll, pitch, and yaw. Thus we speak of the equations of motion for roll, pitch, and yaw as *controlled equations*.

The only other controller we have heretofore considered is the engine thrust used primarily to generate longitudinal force. However, in none of the flying qualities requirements we have discussed is the control of airspeed a major concern. Moreover, there is an instinctive appreciation that control of roll, pitch, and yaw is a more intensive problem than control of airspeed which is performed on a completely slower time scale. For purposes of this discussion then we do not consider the equation for $\dot{u}$ to be a controlled equation. We will either leave the throttle alone, or address it separately.

The force equations are therefore relegated to a separate status called the *complementary equations*. While the ailerons, elevator, and rudder do appear in the complementary equations they are not used to directly control the forces. Thus the rudder does generate a side force, but it is not used intentionally to control $\dot{v}$. It is only through the yawing moment which side force generates that we use the rudder.

We therefore write the vector equations with descriptive superscripts to reflect these three types of equations.

$$\dot{\mathbf{x}}^{Kine} = \mathbf{f}^{Kine}(\mathbf{x})$$
$$\dot{\mathbf{x}}^{Comp} = \mathbf{f}^{Comp}(\mathbf{x}, \eta) \qquad (13.4)$$
$$\dot{\mathbf{x}}^{Cont} = \mathbf{f}^{Cont}(\mathbf{x}, \mathbf{u})$$

In Equation 13.4 the dependency of the kinematic equations is on $\mathbf{x}$ only since there are no control effects. We will think of these equations as those for $\dot{\phi}$, $\dot{\theta}$, and $\dot{\psi}$. The dependency of the controlled equations consists of the states and the controls **u**, consisting of ailerons, elevator, and rudder. We think of these equations as being those for $\dot{p}$, $\dot{q}$, and $\dot{r}$. The use of η in the complementary equations is to suggest some subset of **u** but with a different notation to remind us that the controls are there only in that they create some unintended (and maybe undesirable) effect.

13.2.1 The controlled equations

General

Dynamic inversion relies on there being some way to invert the controllable states and effectively 'solve' for the control vector **u**. We represent this inversion as some functional $\mathbf{g}_1$ such that

$$\mathbf{u}^* = \mathbf{g}_1(\dot{\mathbf{x}}^{Cont}, \mathbf{x}) \tag{13.5}$$

From Equation 13.5 the control laws are obtained by replacing $\dot{\mathbf{x}}^{Cont}$ with $\dot{\mathbf{x}}_d^{Cont}$, the desired dynamics.

Because the control effectors enter into the dynamics through the forces and moments and their effects are generally available only as the results of wind-tunnel and flight-test data, we normally rely on the linearized dynamics to represent those effects. Thus, **g** will be the inverse of a matrix of control derivatives that resulted from the linearization process, that is, elements of the B matrix.

In the simplest form of dynamic inversion the controlled states are given desired dynamics $\dot{\mathbf{x}}_d^{Cont}$ in response to commanded values of the states $\mathbf{x}_c^{Cont}$ according to the control law

$$\dot{\mathbf{x}}_d^{Cont} = \mathbf{\Omega}_1 (\mathbf{x}_c^{Cont} - \mathbf{x}^{Cont}) \tag{13.6}$$

In Equation 13.6 the matrix $\mathbf{\Omega}_1$ is normally diagonal, and its diagonal entries determine the dynamics of the controlled states in response to commanded values. In flight dynamics the controllable states are generally taken to be roll rate p, pitch rate q, and yaw rate r, so Equation 13.6 becomes

$$\begin{Bmatrix} \dot{p}_d \\ \dot{q}_d \\ \dot{r}_d \end{Bmatrix} = \mathbf{\Omega}_1 \begin{Bmatrix} p_c - p \\ q_c - q \\ r_c - r \end{Bmatrix} \tag{13.7}$$

The control law for the controllable dynamics is therefore

$$\mathbf{u}^* = \mathbf{g}_1(\dot{\mathbf{x}}_d^{Cont}, \mathbf{x}) \tag{13.8}$$

It is difficult to generalize the form of $\mathbf{g}_1$. If the controllable equations are linearized then we have matrices and matrix inverses. If a particular problem has nonlinear control effectiveness that can somehow be inverted, then that may be used. Otherwise it is not required to linearize any portion of the equations other than the control effectiveness, in which case $\mathbf{g}_1$ becomes a matrix inverse and a vector-valued function. For our purposes here we will deal only with the linearized dynamics.

The M2-F2

The model for the M2-F2 was introduced in Chapter 12 and is repeated here. All angular measures are in degrees.

$$A_{LD} = \begin{bmatrix} -0.4183 & 0 & -1.0 & 0.0919 \\ -115.0 & -0.885 & 1.18 & 0 \\ 8.50 & 0.136 & -0.794 & 0 \\ 0 & 1.0 & 0 & 0 \end{bmatrix} \quad B_{LD} = \begin{bmatrix} 0.0214 & 0.0306 \\ 13.0 & 9.0 \\ -2.17 & -5.13 \\ 0 & 0 \end{bmatrix} \quad (13.9)$$

With respect to our M2-F2 example the kinematic equation is that for $\dot{\phi}$, the complementary is that for $\dot{\beta}$, and the controlled equations are those for $\dot{p}$ and $\dot{r}$. The states and controls are

$$\mathbf{x} = \begin{Bmatrix} \beta \\ p \\ r \\ \phi \end{Bmatrix}, \mathbf{u} = \begin{Bmatrix} \delta_a \\ \delta_r \end{Bmatrix}$$

For the controlled equations $\dot{\mathbf{x}}^{Cont} = \mathbf{f}^{Cont}(\mathbf{x}, \mathbf{u})$ the corresponding parts of the M2-F2 dynamics are:

$$\begin{Bmatrix} \dot{p} \\ \dot{r} \end{Bmatrix} = \begin{bmatrix} -115.0 & -0.885 & 1.18 & 0 \\ 8.50 & 0.136 & -0.794 & 0 \end{bmatrix} \mathbf{x} + \begin{bmatrix} 13.0 & 9.0 \\ -2.17 & -5.13 \end{bmatrix} \mathbf{u} \quad (13.10)$$

The first step in dynamic inversion is to find an expression for the controls. From Equation 13.10 we simply invert the 2 × 2 control effectiveness matrix,

$$\mathbf{u} = \begin{bmatrix} 0.109 & 0.191 \\ -0.0460 & -0.276 \end{bmatrix} \begin{Bmatrix} \dot{p} \\ \dot{r} \end{Bmatrix}$$

$$- \begin{bmatrix} -10.89 & -0.0703 & -0.0203 & 0 \\ 2.95 & 3.23 \times 10^{-3} & 0.0165 & 0 \end{bmatrix} \mathbf{x}$$

We now need to assign dynamics via our choice of Ω_1. Equation 13.7 applied to the M2-F2 in scalar form is (with p_c corresponding to lateral stick and r_c to rudder pedal inputs)

$$\dot{p}_d = \Omega_{1_{11}}(p_c - p)$$
$$\dot{r}_d = \Omega_{1_{22}}(r_c - r) \quad (13.11)$$

Thus, when the control law is applied with these desired dynamics, $\Omega_{1_{11}}$ will become the reciprocal of the roll-mode time-constant, $1/\tau_R$. The other element, $\Omega_{1_{22}}$, will play a corresponding role in controlling the yaw rate.

Because Ω_1 is diagonal the roll commands and yaw commands of the aircraft will be completely *uncoupled*: lateral stick will result in nothing but roll and pedal input will result in nothing but yaw. Either or both the roll and yaw rate will in general influence the other dynamics, β and ϕ.

For simplicity let us take Ω_1 to be the identity matrix. The control law becomes

$$\mathbf{u}^* = \begin{bmatrix} 0.109 & 0.191 \\ -0.0460 & -0.276 \end{bmatrix} \begin{Bmatrix} p_c - p \\ r_c - r \end{Bmatrix}$$
$$- \begin{bmatrix} -10.89 & -0.0703 & -0.0203 & 0 \\ 2.95 & 3.23 \times 10^{-3} & 0.0165 & 0 \end{bmatrix} \mathbf{x} \quad (13.12)$$

When Equation 13.12 is substituted into Equation 13.10 the result is

$$\dot{p} = -p + p_c$$
$$\dot{r} = -r + r_c$$

To see how this control law affects the other lateral–directional dynamics, partition A_{LD} and B_{LD} into three sub-matrices

$$A_{LD} = \begin{bmatrix} A_1^{1 \times 4} \\ A_2^{2 \times 4} \\ A_3^{1 \times 4} \end{bmatrix} \quad B_{LD} = \begin{bmatrix} B_1^{1 \times 2} \\ B_2^{2 \times 2} \\ 0^{1 \times 2} \end{bmatrix}$$

In terms of these partitions our control law is

$$\mathbf{u}^* = \begin{bmatrix} 0^{2 \times 1} & B_2^{-1} & 0^{2 \times 1} \end{bmatrix} \begin{Bmatrix} \beta \\ p_c - p \\ r_c - r \\ \phi \end{Bmatrix} - B_2^{-1} A_2 \mathbf{x} \quad (13.13)$$

The term $B_{LD}\mathbf{u}^*$ becomes

$$B_{LD}\mathbf{u}^* = \begin{bmatrix} 0 & B_1 B_2^{-1} & 0 \\ 0 & I & 0 \\ 0 & 0 & 0 \end{bmatrix} \begin{Bmatrix} \beta \\ p_c - p \\ r_c - r \\ \phi \end{Bmatrix} - \begin{bmatrix} B_1 B_2^{-1} A_2 \\ A_2 \\ 0 \end{bmatrix} \mathbf{x}$$

$$= \begin{bmatrix} 0 & -B_1 B_2^{-1} & 0 \\ 0 & -I & 0 \\ 0 & 0 & 0 \end{bmatrix} \mathbf{x} - \begin{bmatrix} B_1 B_2^{-1} A_2 \\ A_2 \\ 0 \end{bmatrix} \mathbf{x} + \begin{bmatrix} B_1 B_2^{-1} \\ I \\ 0 \end{bmatrix} \begin{Bmatrix} p_c \\ r_c \end{Bmatrix} \quad (13.14)$$

To the old A_{LD} we add the parts of $B_{LD}\mathbf{u}^*$ that multiply $\mathbf{x}$.
The system dynamics (Equation 13.9) become

$$A_{p_c r_c} = \begin{bmatrix} -0.276 & 4.88 \times 10^{-4} & -1.00 & 0.0919 \\ 0 & -1.0 & 0 & 0 \\ 0 & 0 & -1.0 & 0 \\ 0 & 1.0 & 0 & 0 \end{bmatrix} \quad (13.15)$$

$$B_{p_c r_c} = \begin{bmatrix} 9.15 \times 10^{-4} & -4.36 \times 10^{-3} \\ 1.0 & 0 \\ 0 & 1.0 \\ 0 & 0 \end{bmatrix} \quad (13.16)$$

The new system matrix has two roots at $\lambda = -1\,\text{s}^{-1}$ associated with $\dot{p}$ and $\dot{r}$, as well as a root at $\lambda = 0$ associated with the $\dot{\phi}$ dynamics. These three states are completely independent of the $\dot{\beta}$ dynamics, which contributes a root at $\lambda = -0.276\,\text{s}^{-1}$.

The block diagram for this control law is shown in Figure 13.2.

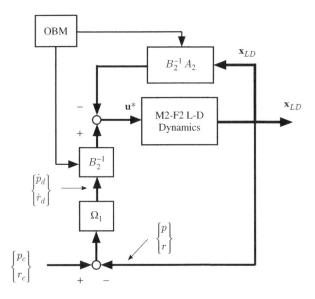

Figure 13.2 Dynamic inversion: controlled lateral–directional states.

In Figure 13.2 the thicker lines carry more than one signal, as noted. The block Ω_1 has been made the identity matrix in the current example. The two blocks B_2^{-1} and $B_2^{-1}A_2$ have elements provided by an on-board model, 'OBM'. The flight control computer accesses the model and recalcualtes all the matrix operations at a rate necessary to accomodate the changing flight conditions.

The multiplications by B_{LD} and B_1 in Equation 13.14 take place inside the block labeled 'M2-F2 L-D Dynamics'.

Note that everything in the block 'M2-F2 L-D Dynamics' represents the actual dynamics; the on-board model blocks represent what we think we know about what is in that block. If the on-board model is not sufficiently 'close' to the actual dynamics then the control law will exhibit degraded performance. The determination of what 'close' means in this context is a separate but very important problem, but beyond the scope of this book.

13.2.2 The kinematic equations

In dynamic inversion, control of the kinematic equations is accomplished by treating the controllable states as controls. Their dynamics are assigned as described previously, and then the kinematic equations are viewed as

$$\dot{\mathbf{x}}^{Kine} = \mathbf{f}^{Kine}\left(\mathbf{x}^{Other}, \mathbf{x}^{Cont}\right)$$

All this notation means is that we have viewed **x** as being made up of the controllable states, $\mathbf{x}^{Cont}$, and all the others, $\mathbf{x}^{Other}$. Now the kinematic equations are in a form similar to the controllable equations, but with $\mathbf{x}^{Other}$ instead of **x** and $\mathbf{x}^{Cont}$ instead of **u**. The rest follows the treatment of the controllable equations.

First, the inversion:

$$\mathbf{x}_c^{Cont} = \mathbf{g}_2\left(\dot{\mathbf{x}}_d^{Kine}, \mathbf{x}^{Other}\right) \tag{13.17}$$

Last is the assignment of dynamics:

$$\dot{\mathbf{x}}_d^{Kine} = \boldsymbol{\Omega}_2\left(\mathbf{x}_c^{Kine} - \mathbf{x}^{Kine}\right) \tag{13.18}$$

Commanded kinematic states, $\mathbf{x}_c^{Kine}$, are compared with their actual values, $\mathbf{x}^{Kine}$, to determine the desired kinematic dynamics, $\dot{\mathbf{x}}_d^{Kine}$, as specified in Equation 13.18. Those calculated dynamics are used in Equation 13.17 to determine the commanded controllable states, $\mathbf{x}_c^{Cont}$.

The commanded controllable states, $\mathbf{x}_c^{Cont}$ are then used to specify the desired controllable state rates, $\dot{\mathbf{x}}_d^{Cont}$, in Equation 13.11.

Flight dynamics applications

Control of the kinematic equations in aircraft applications is fairly intuitive once one appreciates how airplanes are controlled by pilots. We briefly discussed such a method earlier when examining times required to change bank angle (Section 12.2.1). Indeed, the simple equation $\dot{\phi} = p(t)$ suggests that there is only one way to control this particular kinematic equation: by treating $p(t)$ as a control.

The first step in this process is to 'solve' the kinematic equations for the controllable states. In flight dynamics our controllable states have been taken to be roll rate p, pitch rate q, and yaw rate r; and the kinematic states (using Euler angles) as bank angle ϕ, pitch angle θ, and heading angle ψ.

We already have an expression for p, q, and r in terms of $\dot{\phi}$, $\dot{\theta}$, and $\dot{\psi}$, back in Equation 4.5. With the substitutions for the customary names of the angles and angle rates used in flight dynamics,

$$\begin{Bmatrix} p \\ q \\ r \end{Bmatrix} = \begin{bmatrix} 1 & 0 & -\sin\theta \\ 0 & \cos\phi & \sin\phi\cos\theta \\ 0 & -\sin\phi & \cos\phi\cos\theta \end{bmatrix} \begin{Bmatrix} \dot{\phi} \\ \dot{\theta} \\ \dot{\psi} \end{Bmatrix} \tag{13.19}$$

Following Etkin's notation

$$\begin{Bmatrix} p \\ q \\ r \end{Bmatrix} = \mathbf{R}(\phi, \theta, \psi) \begin{Bmatrix} \dot{\phi} \\ \dot{\theta} \\ \dot{\psi} \end{Bmatrix}$$

This becomes Equation 13.17,

$$\begin{Bmatrix} p_c \\ q_c \\ r_c \end{Bmatrix} = \mathbf{R} \begin{Bmatrix} \dot{\phi}_d \\ \dot{\theta}_d \\ \dot{\psi}_d \end{Bmatrix} \tag{13.20}$$

Trends in Automatic Flight Control

Equation 13.18 becomes

$$\begin{Bmatrix} \dot{\phi}_d \\ \dot{\theta}_d \\ \dot{\psi}_d \end{Bmatrix} = \boldsymbol{\Omega}_2 \begin{Bmatrix} \phi_c - \phi \\ \theta_c - \theta \\ \psi_c - \psi \end{Bmatrix} \tag{13.21}$$

Equations 13.8, 13.11, 13.20 and 13.21 are sufficient to implement the control law, but to combine the elements let us back-substitute to get a closed-form equation for $\dot{\mathbf{x}}_d^{Cont}$.
First, Equation 13.21 into 13.20:

$$\begin{Bmatrix} p_c \\ q_c \\ r_c \end{Bmatrix} = \mathbf{R}\boldsymbol{\Omega}_2 \begin{Bmatrix} \phi_c - \phi \\ \theta_c - \theta \\ \psi_c - \psi \end{Bmatrix} \tag{13.22}$$

Then, Equation 13.22 into 13.11:

$$\begin{Bmatrix} \dot{p}_d \\ \dot{q}_d \\ \dot{r}_d \end{Bmatrix} = \boldsymbol{\Omega}_1 \mathbf{R} \boldsymbol{\Omega}_2 \begin{Bmatrix} \phi_c - \phi \\ \theta_c - \theta \\ \psi_c - \psi \end{Bmatrix} - \boldsymbol{\Omega}_1 \begin{Bmatrix} p \\ q \\ r \end{Bmatrix} \tag{13.23}$$

Then, once the inverse function $\mathbf{g}_1$ is determined, the desired rates are substituted into Equation 13.8 to make the control law:

$$\mathbf{u}^* = \mathbf{g}_1\left(\dot{\mathbf{x}}_d^{Cont}, \mathbf{x}\right)$$

Having described the dynamic inversion method for controlling the kinematic states, it should be noted that the method doesn't find much utility in actual aircraft control.

Most emphatically the heading angle ψ is not controlled by commanding specific body-axis angle rates as implied here. To change the heading angle the aircraft is banked and sufficient back-stick employed to generate enough extra lift so the vertical component equals the weight of the aircraft. The horizontal component of lift accelerates the aircraft in a curved flight path. It is true that during a turn the body-axis angular rates are present in a certain relationship to each other, but that is a consequence of the manner of controlling heading angle, not the method of doing it.

With respect to the pitch angle θ, that angle is not of as much concern as the flight-path angle γ, and γ depends on the angle-of-attack α, which is a complementary equation.

The roll angle ϕ is an angle that pilots care about, but a pilot would find a control system wherein lateral stick commanded a bank angle to be very strange indeed: he expects lateral stick to cause a roll rate, as described above.

However, a system to command and hold a bank angle is quite common as a function of autopilots, but the angle is not commanded by the lateral stick. See, for example, the autopilot control panel of the F-8 *Crusader* in Figure 13.3. Turning the bank control knob to the left or right would result in a constant bank angle, around 30° at full deflection. If the pilot manually moved the lateral stick, force sensors in the control stick would detect the torque and disengage the bank-angle hold.

If the altitude hold were also engaged, then the pilot could dial in a bank angle and fly in a circle without intervening. If he, as the author once did, dozed off, and the prevailing winds blew his circling aircraft over Hainan Island, he might be rudely awakened by warnings of surface-to-air missiles on their way. As the author once was.

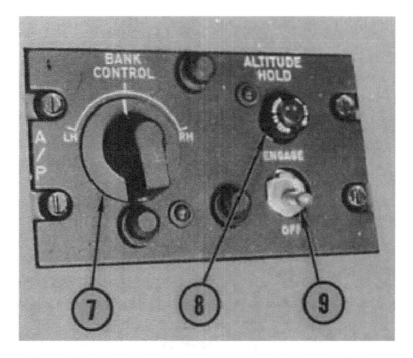

Figure 13.3 Autopilot control panel, F-8 *Crusader*. Source: U.S. Navy

The M2-F2

We begin with the M2-F2 where we left off. As discussed previously, we do not desire direct control of the yaw rate, so we limit the current design to that of a bank-angle command system to be commanded by some auxilliary cockpit device.

From Equations 13.15 and 13.16 the only dynamics that are of interest are in the scalar equation for $\dot{\phi}$. The commanded controllable state is p_c, and the inverse function is 1 times $\dot{\phi}$. Then, the commanded roll rate in response to desired bank angle dynamics becomes

$$p_c = \dot{\phi}_d$$

The implementation of Equation 13.18 is

$$\dot{\phi}_d = \Omega_2(\phi_c - \phi)$$

Substituting into Equation 13.11,

$$\dot{p}_d = \Omega_2(\phi_c - \phi) - p$$
$$= -p - \Omega_2 \phi + \Omega_2 \phi_c \qquad (13.24)$$

The dynamics of p and ϕ are independent of those for r and β, so they constitute their own system:

$$\begin{Bmatrix} \dot{p} \\ \dot{\phi} \end{Bmatrix} = \begin{bmatrix} -1 & -\Omega_2 \\ 0 & -\Omega_2 \end{bmatrix} \begin{Bmatrix} p \\ \phi \end{Bmatrix} + \begin{bmatrix} \Omega_2 \\ \Omega_2 \end{bmatrix} \phi_c$$

The transfer functions for this system are easily determined to be

$$\frac{p(s)}{\phi_c(s)} = \frac{\Omega_2 s}{(s+1)(s+\Omega_2)}$$

$$\frac{\phi(s)}{\phi_c(s)} = \frac{\Omega_2}{(s+\Omega_2)}$$

The second transfer function, $\phi(s)/\phi_c(s)$, is what we specified, and we may expect the bank angle to rise to its commanded value with a time constant of $1/\Omega_2$. The transfer function for the roll rate, however, has a zero at the origin, $s = 0$.

To examine the response of $p(s)/\phi_c(s)$ in the time domain we use partial fraction expansion from which, if $\Omega_2 \neq 1$,

$$\frac{p(s)}{\phi_c(s)} = \frac{\Omega_2}{1-\Omega_2}\left[\frac{1}{s+1} - \frac{\Omega_2}{s+\Omega_2}\right]$$

This formulation tells us that the step response will be the sum of the two parts, and that one seeks a negative steady state and the other a positive steady state. It also shows that if $\Omega_2 < 1$ (slower response in ϕ) then the first term will be the positive contributor, and the greater magnitude of the two terms, and vice versa if $\Omega_2 > 1$.

Figures 13.4 and 13.5 show the responses of the individual terms and of the overall response (the sum of the individual terms) for $\Omega_2 = 2.0\,\mathrm{s}^{-1}$ and $\Omega_2 = 0.5\,\mathrm{s}^{-1}$. All this is interesting, but offers no insight into how to choose Ω_2.

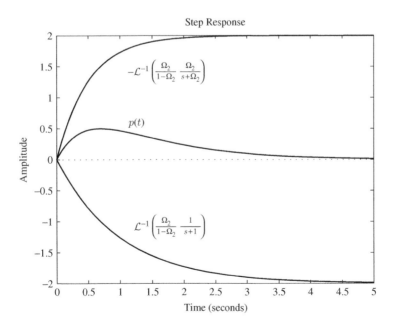

Figure 13.4 $\Omega_2 = 2.0\,\mathrm{s}^{-1}$.

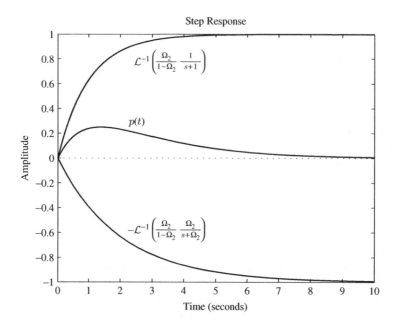

Figure 13.5 $\Omega_2 = 0.5\,\text{s}^{-1}$.

Let us now consider the demands being made on the control effectors (aileron and rudder). We know that $\dot{p}$ is directly proportional to the moment being generated by the controls. It is also pretty clear that the greatest acceleration in roll will be at $t = 0$, when the step is applied. We may determine the initial acceleration using the *initial value theorem*. This theorem is very much like the final value theorem except the limit is taken as $s \to \infty$. The transfer function of $\dot{p}$ is just s times the transfer function of $p(t)$. We want the initial value in response to a unit step whose LaPlace transform is $1/s$. The derivative s cancels the step input s and we are left with the transfer function $p(s)/\phi_c(s)$ to which we apply the limit. Forgetting about entropy and denoting $\dot{p}(0)$ as the limit of $\dot{p}(t)$ as $0 \leftarrow t$,

$$\dot{p}(0) = \lim_{s \to \infty} \frac{p(s)}{\phi_c(s)} = \lim_{s \to \infty} \frac{\Omega_2 s}{(s+1)(s+\Omega_2)}$$

We divide the numerator and denominator by s and then take the limit:

$$\dot{p}(0) = \lim_{s \to \infty} \frac{\Omega_2}{(1+1/s)(1+\Omega_2/s)} = \Omega_2$$

Thus, the larger is Ω_2, the greater is the initial roll acceleration, and the greater the effort required of the control effectors. This tells us what was probably obvious all along, that if we want to get to the desired bank angle faster it will take more effort from the control effectors.

For subsequent design purposes, we will take a slow response to ϕ_c:

$$\Omega_2 = 0.5\,\text{s}^{-1}$$

Trends in Automatic Flight Control

The system matrices at this point are

$$A_{\phi_c r_c} = \begin{bmatrix} -0.276 & 4.8597 \times 10^{-4} & -1.00 & .0919 \\ 0 & -1.0 & 0 & -0.5 \\ 0 & 0 & -1.0 & 0 \\ 0 & 0 & 0 & -0.5 \end{bmatrix} \quad (13.25)$$

$$B_{\phi_c r_c} = \begin{bmatrix} 9.1985 \times 10^{-4} & -4.3511 \times 10^{-3} \\ 0.5 & 0 \\ 0 & 1.0 \\ 0.5 & 0 \end{bmatrix} \quad (13.26)$$

There are two eigenvalues at -1 s^{-1} and one at -0.5 s^{-1} as we designed. The fourth eigenvalue is at -0.276 s^{-1} and it is associated with the dynamics of sideslip, β. While this root is stable, we can improve on the dynamics by using r_c.

The block diagram for the system after this step is shown in Figure 13.6. The only part of Figure 13.2 that is changed is the meaning of p_c, so we show just that portion.

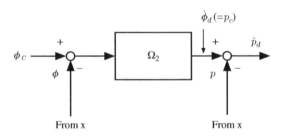

Figure 13.6 Kinematic control.

13.2.3 The complementary equations

It is possible to treat the controllable states as controls in the complementary equations, but the result is not nearly so nice as it was for the kinematic equations. We have used our control of the M2-F2 roll rate for a bank-angle command system, so here we turn to a use of our ability to command yaw rate. The only thing left to control in the M2-F2 dynamics is the sideslip angle, a complementary equation, so that is the aim.

First we examine whether or not we could or should attempt to invert the complementary equation to solve for the controlled variable r. Consider the expression for $\dot{v}$

$$\dot{v} = \frac{Y}{m} + g \sin\phi \cos\theta + pw - ru$$

It appears that we can pretty easily solve for r, since the velocity component u isn't likely to be zero in normal flight. Even taking into account that the side force Y may have some r dependency, we could still probably come up with an expression such as

$$r_c = \frac{m}{mu - Y_r}\left(\frac{Y'}{m} + g \sin\phi \cos\theta + pw - \dot{v}_d\right)$$

The approach wherein the complementary equations are inverted using this technique or some other has been performed, at least on a theoretical level. Control of complementary variables may be mixed in the inversion process permitting various combinations of interpretations of the commanded variables. Using the three moment equations (and assuming independent control of velocity) Azam and Singh (1994) developed controllers for ϕ, α, and β; and another for p, α, and β. Snell et al. (1992) developed a controller for α, β, and ϕ_w. (In Snell et al., the assignment of desired dynamics for wind-axis bank angle does not exactly follow the development described previously in that it incorporates filtering to reduce sensitivity to pilot inputs.)

Modern industry practice appears to follow a somewhat different path for the complementary equations. The Lockheed-Martin X-35 decouples the controllable states and then uses them in a rather conventional manner as controllers for the other aircraft modes. Some detail is available in Walker and Allen (2002).

We will therefore deal with the remaining complementary equation in a conventional fashion in which we consider the commanded yaw rate the control, and proceed with a root locus design. The system of interest is in Equations 13.25 and 13.26:

$$A_{\phi_c r_c} \begin{bmatrix} -0.276 & 4.8597 \times 10^{-4} & -1.00 & .0919 \\ 0 & -1.0 & 0 & -0.5 \\ 0 & 0 & -1.0 & 0 \\ 0 & 0 & 0 & -0.5 \end{bmatrix}$$

$$B_{\phi_c r_c} = \begin{bmatrix} 9.1985 \times 10^{-4} & -4.3511 \times 10^{-3} \\ 0.5 & 0 \\ 0 & 1.0 \\ 0.5 & 0 \end{bmatrix}$$

Our control vector now consists of ϕ_c and r_c, and the dynamics of interest are in the first row, $\dot{\beta}$. Using MATLAB® we find the transfer function $x_1(s)/u_2(s)$ which is $\beta(s)/r_c(s)$,

$$\frac{\beta(s)}{r_c(s)} = \frac{-0.004351s - 1.005}{s^2 + 1.276s + 0.2755} \tag{13.27}$$

The second-order characteristic polynomial results from pole-zero cancellations for the rolling modes.

Note that this transfer function is negative, and since MATLAB® assumes negative feedback for its root locus plots, we will need to supply $-\beta(s)/r_c(s)$ as the argument. When we do, the root locus in Figure 13.7 is obtained.

Obviously we need to change the scale, which we do, and Figure 13.8 results. The roots at -1 and -0.276 move toward each other, meet and split into complex roots (that continue in the large arcs in Figure 13.7).

The MATLAB® command rlocfind is next applied to select the roots. A feedback gain $k_\beta = 0.520$ is chosen (and we make a mental note to change its sign before applying it to the system, since we changed the sign of the transfer function). The roots that accompany this selection are $\lambda = -0.639 \pm j0.624$. This choice gives us a nice rise time with one small overshoot. The natural frequency could have been increased but with lower damping ratio (and greater control effort).

Trends in Automatic Flight Control

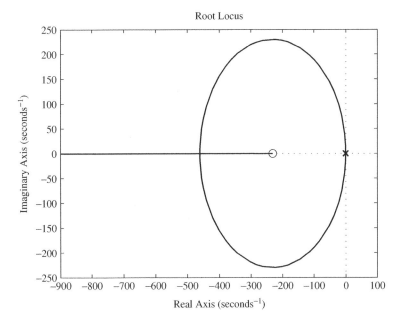

Figure 13.7 Root locus, β feedback.

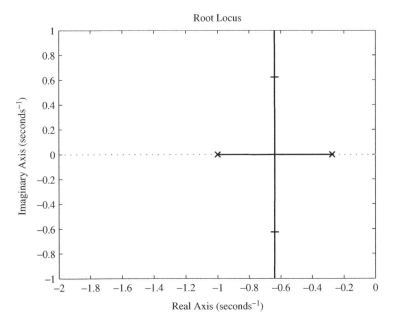

Figure 13.8 Close up of Figure 13.7.

The final system is

$$A_{\phi_c \beta_r} \begin{bmatrix} -0.278 & 4.8597 \times 10^{-4} & -1.00 & .0919 \\ 0 & -1.0 & 0 & -0.5 \\ 0.520 & 0 & -1.0 & 0 \\ 0 & 0 & 0 & -0.5 \end{bmatrix} \quad (13.28)$$

$$B_{\phi_c \beta_r} = \begin{bmatrix} 9.1985 \times 10^{-4} & -4.3511 \times 10^{-3} \\ 0.5 & 0 \\ 0 & 1.0 \\ 0.5 & 0 \end{bmatrix} \quad (13.29)$$

The controls are the commanded bank angle ϕ_c and the reference sideslip β_r. There may be no need to provide an explicit value for the reference sideslip, letting it default to $\beta_r = 0$ so that as the M2-F2 is flown about, that part of the control system works continuously to regulate the sideslip to zero.

The block diagram for the system with all loops closed is shown in Figure 13.9. The results of the preceding steps are lumped together into a block representing the system dynamics in Equations 13.25 and 13.26.

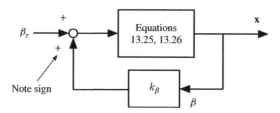

Figure 13.9 Complementary control. Conventional root locus to regulate sideslip.

13.3 Control Allocation

13.3.1 Background

Classically, airplane flight control effectors are designed with the idea of a single moment-generating controller for each rotational degree-of-freedom. Modern tactical aircraft have more than the classical three moment generators. In the High Angle-of-Attack Research Vehicle (HARV) (Davidson et al., 1992), one has potentially 13 or more independent moment controllers: horizontal tail, aileron, leading edge flap, trailing edge flap, and rudder, each left and right; and thrust-vectoring moment generators. To this add spoilers on the leading edge extensions, vortical lift and side-force generators, and the number of control effectors nears 20. These control effectors are all constrained to certain limits, determined by the physical geometry of the control actuators, or in some cases by aerodynamic considerations. The allocation, or blending, of these several control effectors to achieve specific objectives is the control allocation problem.

The control allocation problem is a very important concern in flight control system design. Any flight control law that does not utilize all available flight control effectors in the most efficient manner is carrying around extra weight and system complexity. That is, if a given suite of control effectors are used inefficiently then they could be replaced by smaller or fewer control effectors that would achieve the same results if used efficiently.

In the context of dynamic inversion, we are looking to find $\mathbf{g}_1$ in Equation 13.5. We generally look to solve this problem for the case in which the control effectiveness has been linearized, and may be expressed as a matrix times the control vector, $B\mathbf{u}$. The difficulty arises now because the matrix B is no longer square: there is no B^{-1} in the usual sense of the expression. For example, say the engineers added differential horizontal tail motion to the M2-F2, so that the matrix multiplying the controls in Equation 13.10 is no longer 2×2 but 2×3. What is the next step to be?

Let us first formally state the problem. Then we will examine a few decidedly suboptimal methods of solving it, and then discuss better solutions.

13.3.2 Problem statement

The three-moment problem

Suppose we have a vector of three desired moments $\mathbf{m}_d$ for our control effectors to generate, perhaps related to the desired accelerations in the dynamic inversion of the controllable equations. Suppose too that we have m control effectors, so the vector of control effectors previously named $\mathbf{u}$ is of dimension $m \times 1$. We also have the matrix of control effectiveness from the linearization process B which is now of dimension $3 \times m$, where $m > 3$. We wish to find a vector $\mathbf{u}$ that satisfies the equation

$$B\mathbf{u} = \mathbf{m}_d \tag{13.30}$$

We cannot solve this with B^{-1}, so let us denote the solution as $B^\dagger$. Note that $B^\dagger$ may represent a matrix, or it could just be a place marker for a procedure to be defined.

$$\mathbf{u} = B^\dagger \mathbf{m}_d \tag{13.31}$$

Moreover, we know that each of the u_i in $\mathbf{u}$ has certain limits, minimum and maximum values that are admissible. We combine all the lower limits into a vector $\mathbf{u}_{Min}$ and all the upper limits into $\mathbf{u}_{Max}$ so that

$$\mathbf{u}_{Min} \leq \mathbf{u} \leq \mathbf{u}_{Max} \tag{13.32}$$

By Equation 13.32 we mean, of course, that for each u_i, $i = 1 \ldots m$, $u_{i_{Min}} \leq u_i \leq u_{i_{Max}}$.

There is a mathematical notation that is often used to describe this problem. One sees expressions like $\mathbf{x} \in \Re^n$ which means that the elements of the vector $\mathbf{x}$ are real numbers and that $\mathbf{x}$ is of dimension n. Likewise, $B \in \Re^{m \times n}$ means B is an $m \times n$ matrix of real numbers. Thus,

$$\mathbf{u} \in \Re^m, \quad \mathbf{m}_d \in \Re^3, \quad B \in \Re^{3 \times m} \tag{13.33}$$

The notion that each control effector is constrained to certain limits give rise to defining a *set* of admissible controls called Ω. The boundary of Ω represents combinations of control effectors with one or more at their upper or lower limits, said to be *saturated*. In defining the set the u_i that satisfy the expression in {} are its members, the vertical bar | means 'such that', and the set Ω is itself a subset of the real numbers. Thus,

$$\Omega = \{u_i \mid u_{i_{Min}} \leq u_i \leq u_{i_{Max}}\} \subset \Re \tag{13.34}$$

We denote the boundary of this set as $\partial(\Omega)$, combinations of controls for which none can be displaced further. If a control vector lies on that boundary it is called $\mathbf{u}^*$:

$$\mathbf{u}^* = \mathbf{u} \subset \partial(\Omega) \tag{13.35}$$

Our control effectors generate moments according to the numbers in the B matrix. The matrix B is said to *map* vectors in $\Re^m$ (the control effectors) into $\Re^3$ (the moments).

$$B : \Re^m \to \Re^3 \tag{13.36}$$

In other words, $\mathbf{m} = B\mathbf{u}$. Inasmuch as $\mathbf{u}$ is constrained to Ω, so too must $\mathbf{m}$ be limited to a finite set of what is attainable, the Attainable Moment Subset (AMS). We call this set Φ and define it as

$$\Phi = \{\mathbf{m} \mid \mathbf{m} = B\mathbf{u}, \ \mathbf{u} \subset \Omega\} \tag{13.37}$$

It may require proof to some, but we will just state what seems obvious: the boundary of Φ (or $\partial(\Phi)$) is the mapping to the moments of the boundary of Ω. In other words, the maximum moments attainable are achieved by controls that lie at the limits of the set of admissible controls.

$$\partial(\Phi) = \{\mathbf{m} \mid \mathbf{m} = B\mathbf{u}^*\} \tag{13.38}$$

The admissible controls

Ω is easy to visualize for two control effectors: it is a rectangle. Figure 13.10 shows a possibility for the M2-F2 with its original two controls. The aileron can range from minimum to maximum at the same time the rudder ranges from minimum to maximum. The upper and lower bounds of the rectangle represent the aileron movement while the rudder is saturated at its upper or lower limit, and the left and right bounds say the same

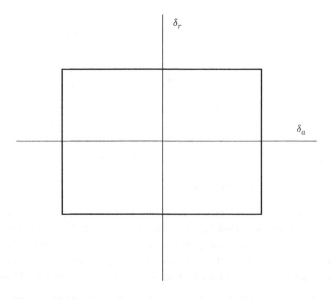

Figure 13.10 Two-dimensional set of admissible controls, Ω.

thing for the rudder while the aileron is saturated. The four vertices of this Ω are the four conditions where both effectors are saturated: min-min, min-max, max-min, and max-max.

If our M2-F2 had the additional differential horizontal tail then its Ω would be a rectangular prism, like a matchbox. Each of the edges of this prism would represent one control varying while the other two were fixed at some combination of their limits, and the faces, or facets, that bound this prism would represent two controls varying while the third is saturated. The vertices would represent the eight combinations of all three effectors being saturated. Figure 13.11 is intended to show such an Ω, but naturally can't be right because the page of this book is only two dimensional. What is drawn in Figure 13.11 is actually the projection of a rectangular prism onto the plane of the page, and we rely on our imaginations to add the extra depth.

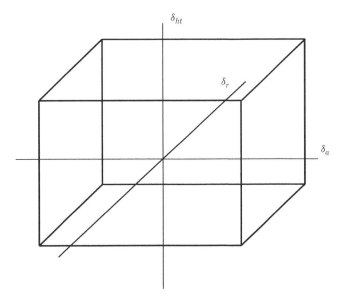

Figure 13.11 Three-dimensional set of admissible controls, Ω.

Higher dimensional examples of Ω could be drawn, but nothing new would be learned. One of the biggest problems when drawing higher dimensional rectangular prisms is the notion that the things on the boundary of the prism are of dimension one less than the dimension of the prism. Thus, a four-dimensional prism is bounded by three-dimensional prisms, and every point within one of those three-dimensional objects is outside the four-dimensional object. Five-dimensional prisms of admissible controls are bounded by four-dimensional objects, and so on.

We will stop trying to visualize such objects and trust the mathematics not to lead us astray.

The attainable moments

Now, given the admissible controls, we may characterize the attainble moments, or Φ. Φ may be visualized by taking all the key points in Ω–vertices and edges–and multiplying them by B to be plotted on axes representing the moments.

As an aid to this process, we will use *singular value decomposition* (SVD, and in MATLAB®, svd, q.v). The singular value decomposition takes a given matrix and makes it the product of three matrices of special properties. Thus, our matrix B of dimension $n \times m$ may be decomposed as

$$B = USV^T \qquad (13.39)$$

U is an $n \times n$ unitary matrix, S is an $n \times m$ rectangular diagonal matrix with non-negative numbers on the diagonal, and V^T (the transpose of V) is an $m \times m$ unitary matrix. The diagonal entries of S are known as the singular values of B. The n columns of U and the m columns of V are called the left-singular vectors and right-singular vectors of B, respectively. In the general case this definition holds for complex numbers, in which case V^T is replaced by V^* to indicate the conjugate transpose.

A unitary matrix for our purposes is a rotation matrix, just like all the transformation matrices we derived in Chapter 3 and with the same properties. They perform rotations on the vectors they operate on, with V^T rotating m-dimensional vectors (vectors of controls).

The matrix S is not square, and the diagonal elements (for $m > n$) are in the first n columns of S that make up a square partition of the matrix. The rest of the entries are zero. This matrix operates on the vectors that were rotated by V^T by stretching (or contracting) each vector by the singular value, and projecting that result into an n-dimensional space. The projection comes from the fact that there is nothing to multiply the last $m - n$ components of each m-dimensional control vector.

Let us see an example. Using the MATLAB® function rand to generate a 3×5 matrix of random numbers,

$$B = \begin{bmatrix} 0.6294 & 0.8268 & -0.4430 & 0.9298 & 0.9143 \\ 0.8116 & 0.2647 & 0.0938 & -0.6848 & -0.0292 \\ -0.7460 & -0.8049 & 0.9150 & 0.9412 & 0.6006 \end{bmatrix} \qquad (13.40)$$

For convenience we will take our control limits to be

$$-1 \le u_i \le 1, \; i = 1 \ldots 5 \qquad (13.41)$$

The MATLAB® command [U,S,V]=svd(B) yields up

$$U = \begin{bmatrix} -0.1115 & 0.9937 & 0.0122 \\ -0.4447 & -0.0608 & 0.8936 \\ 0.8887 & 0.0942 & 0.4487 \end{bmatrix}$$

$$S = \begin{bmatrix} 2.0002 & 0 & 0 & 0 & 0 \\ 0 & 1.7216 & 0 & 0 & 0 \\ 0 & 0 & 0.7123 & 0 & 0 \end{bmatrix}$$

$$V = \begin{bmatrix} -0.5470 & 0.2938 & 0.5590 & 0.5125 & -0.1984 \\ -0.4626 & 0.4238 & -0.1608 & -0.6880 & -0.3275 \\ 0.4104 & -0.2089 & 0.6865 & -0.3588 & -0.4335 \\ 0.5186 & 0.6124 & -0.2503 & 0.3013 & -0.4501 \\ 0.2224 & 0.5616 & 0.3573 & -0.2110 & 0.6804 \end{bmatrix}$$

To make the problem easier to deal with let the control vector **u** be

$$\mathbf{u} = \begin{Bmatrix} 1 \\ 0 \\ 0 \\ 0 \\ 0 \end{Bmatrix}$$

Then $B\mathbf{u} = USV\mathbf{u}$. The first multiplication $V\mathbf{u}$ is just the first column of V. The next is S times this vector,

$$SV\mathbf{u} = \begin{bmatrix} 2.0002 & 0 & 0 & 0 & 0 \\ 0 & 1.7216 & 0 & 0 & 0 \\ 0 & 0 & 0.7123 & 0 & 0 \end{bmatrix} \begin{Bmatrix} -0.5470 \\ -0.4626 \\ 0.4104 \\ 0.5186 \\ 0.2224 \end{Bmatrix} = \begin{Bmatrix} -1.0941 \\ -0.7963 \\ 0.2923 \end{Bmatrix}$$

The last two elements of the result of the first multiplication, $V\mathbf{u}$, are effectively discarded by the operation of S. This is the projection from five dimensions to three.

We have, in fact, already seen an illustration of this process, albeit in lower dimensions. Consider Figure 13.11 for what it really is: the projection of a three-dimensional figure into the two-dimensional plane of the page. The missing dimension is perpendicular to the page. Components of vectors in that direction in the three-dimensional figure were simply removed.

In general, then, the m-dimensional set of admissible controls represented by the rectangular prism Ω is rotated, projected into three-dimensions, and then rotated again, to get the attainable moments, Φ. We cannot in general draw pictures of either one because we are limited to two dimensions. Software is available to draw representations of Φ, a very complicated version of Figure 13.11, but there is little to be learned from them.

We may, nonetheless, characterize the structure of the Attainable Moment Subset Φ in terms of the size of the objects that bound it.

- The boundary of the Φ for the two-moment problem consists of one-dimensional edges that represent B times the corresponding edges of the bounding objects of Ω. Along these edges all controls save one are saturated.
- The boundary of the Φ for the three-moment problem consists of two-dimensional facets that represent B times the corresponding facets of the bounding objects of Ω. On these facets all controls save two are saturated.

Control failure reconfiguration

The first desired characteristic of any solution to the control allocation problem is that we be able to calculate the solutions in real-time in a flight control computer. There are generally two ways of calculating the solution: from aerodynamic data carried on board, or as pre-calculated solutions that are retrieved in flight according to the flight condition.

Pre-calculated solutions are less desirable for one major reason: they cannot easily be reconfigured to compensate for changes in control effectiveness through control failure or other degradation. If the solution is being solved on the fly, then a failed control may be accomodated by simply removing that control from consideration, that is, delete its column from the B matrix and continue.

The two-moment problem

Throughout this section we will have need of examples that make the point of the material without being too challenging to the author's ability to draw the figures. Thus we will from time-to-time take examples from the two-moment problem. Its problem statement is identical to the earlier one, with the number three replaced by the number two.

The two-moment problem has practical applications. Our whole treatment of the M2-F2's lateral–directional dynamics involved just rolling and yawing moments. In fact, the author has seen flight control system design teams at very high levels of military procurement in which there were two teams: one for the longitudinal parts of the control law, and another for the lateral–directional parts.

We note that Figure 13.11 could be representative of a simple control allocation problem with three controls and two moments. If we view the figure as a three-dimensional object it is Ω, and if we view it as a two-dimensional object, and relabel the axes, it is Φ, and could appear as in Figure 13.12.

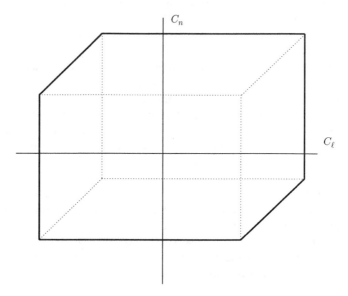

Figure 13.12 Possible two-dimensional set of attainable moments, Φ, corresponding to the admissible controls in Figure 13.11.

There are a few interesting points to be made from Figure 13.12. First, we have left traces of the six edges and two vertices that were projected to the interior of Φ. From this information we can immediately conclude that it is not a sufficient condition that all controls be at their limits for the maximum attainable moments to be attained. Clearly there are points interior to Φ at which all the controls are saturated, but the resulting moment is not on the boundary.

So far as necessary conditions for a set of controls to generate a moment on the boundary of Φ, for the two-moment problem, the vertices and edges of the two-dimensional Φ are mappings of the vertices and edges of Ω. The vertices of an m-control Ω are points where all m-controls are at a limit, and on the edges $m - 1$ controls are saturated.

The necessary conditions for optimality of a solution to the three-moment problem are similar, except that now some solutions can lie on the two-dimensional facets of Φ, wherein all controls but two are saturated. It bears repeating that this condition is only necessary: satisfying it does not assure optimality.

While we are contemplating Figure 13.12, think of a line piercing the page at right angles through the origin of the moments. Now picture this line back in the three-dimensional Figure 13.11, where it has some meaning. Every point on that line is some combination of the three control effectors, but in Figure 13.12 they project to a point at the origin. That is to say, there are combinations of the three controls that cancel each other's effects so that the net moment produced is zero. That direction lies within the *null space* of the matrix B and it can be calculated by the MATLAB® command null.

13.3.3 Optimality

We will define an optimal solution to the control allocation problem as one that can yield every attainable solution over the whole of those generated by the admissible set of controls. That is, for all the moments in the attainable moment subset Φ, the solution method will find controls that generate those moments. For moments in the interior of Φ the solutions are not unique (consider the null space of B for example). However, for moments on the boundary of Φ they are unique, and they correspond in one-to-one fashion with combinations of controls on the boundary of the admissible controls Ω.

As a foreshadowing of our treatment of optimal solutions, the reader is invited to consider the following question: given a B matrix and specified control limits, what control deflections yield the maximum rolling moment this system can attain (by the first row of B)? Answer this for the matrix and limits given in Equations 13.40 and 13.41. Note that nothing is said about the pitching and yawing moments, they may fall where they may.

This question can be answered with *no* calculations, by inspection of the B matrix and consideration of the control limits alone. If you don't see it, consider that the answer is obtained the same way for *any* shape of B, including one of three moments and one control.

When you have answered that question then you will see that you can determine, by inspection of the B matrix and consideration of the control limits, the maximum and minimum rolling moments, maximum and minimum pitching moments, and maximum and minimum yawing moments that are attainable.

Now, change the problem so that the first row of the B matrix has two entries that are not just small, but identically zero, say u_1 and u_2:

$$B = \begin{bmatrix} 0 & 0 & -0.4430 & 0.9298 & 0.9143 \\ 0.8116 & 0.2647 & 0.0938 & -0.6848 & -0.0292 \\ -0.7460 & -0.8049 & 0.9150 & 0.9412 & 0.6006 \end{bmatrix} \quad (13.42)$$

What now characterizes the maximum attainable rolling moment with regard to the facets that bound Φ and their counterparts in Ω?

13.3.4 Sub-optimal solutions

Generalized inverses

We begin with the assumption that our control effectiveness matrix B is of dimension $n \times m$, where $n = 3$ for the threee-moment problem and $n = 2$ for the two-moment problem, that $m > n$, and that B is of full rank robustly, which means that any square matrix formed by taking three columns from B is non-singular and may be inverted. A generalized inverse of such a matrix is any matrix $B^\dagger$ such that

$$BB^\dagger = I \qquad (13.43)$$

The pseudo-inverse

A particular generalized inverse is one that will minimize the *2-norm* of our control vector **u** when solving the control allocation problem. The 2-norm of a vector is just its length, the square-root of the sum of the squares of the individual controls. We call this generalized inverse the *pseudo-inverse*, sometimes without the hyphen, and often the 'Moore–Penrose pseudoinverse', and we name it P.

With any background in optimization it isn't hard to derive P. The optimization to be performed is to minimize $\mathbf{u}^T\mathbf{u}$ (the sum of the squares of the controls) subject to the constraint that $B\mathbf{u} = \mathbf{m}$ for arbitrary **m**. We don't require the square root of $\mathbf{u}^T\mathbf{u}$ since if $\mathbf{u}^T\mathbf{u}$ is minimum, so is its square root.

Using LaGrange multipliers we define the scalar function

$$\mathcal{H}(\mathbf{u}, \lambda) = \frac{1}{2}\mathbf{u}^T\mathbf{u} + \lambda^T(\mathbf{m} - B\mathbf{u})$$

Here λ is an n-vector of LaGrange multipliers. The factor of $1/2$ anticipates that there will be a 2 to cancel. $\mathcal{H}$ will be a minimum (or maximum) when

$$\frac{\partial \mathcal{H}}{\partial \mathbf{u}} = 0, \quad \frac{\partial \mathcal{H}}{\partial \lambda} = 0$$

Performing the operations yields

$$\frac{\partial \mathcal{H}}{\partial \mathbf{u}} = \mathbf{u}^T - \lambda^T B = 0$$

Hence we require that $\mathbf{u}^T = \lambda^T B$, or $\mathbf{u} = B^T \lambda$.

$$\frac{\partial \mathcal{H}}{\partial \lambda} = \mathbf{m} - B\mathbf{u} = 0$$

So that $\mathbf{m} = B\mathbf{u}$. Now combining the two results,

$$\mathbf{m} = B\mathbf{u} = BB^T \lambda$$

Since B is full rank, BB^T is too, and since BB^T is square it is invertible. Thus

$$\lambda = [BB^T]^{-1}\mathbf{m}$$

Since $\mathbf{u} = B^T\lambda$ we have $\mathbf{u} = B^T[BB^T]^{-1}\mathbf{m}$ and

$$P = B^T[BB^T]^{-1} \tag{13.44}$$

This looks like a good solution to the control allocation problem. The most computationally intensive tasks our flight control computer will be called on to perform are matrix multiplications and calculating the inverse of a 3×3 matrix.

The reader is probably wondering at this point where the control limits were considered in reaching this solution. The answer is that they weren't, and that fact results in this solution being non-optimal. If the control limits were not part of the problem solution, then there is no reason to think that the solution will respect them. The non-optimality of the pseudo-inverse may be proven but for now let us demonstrate the fact. We will make up a problem with random entries in B and assume control limits of ± 1. If we can find a point in Φ for which the pseudo-inverse solution fails to yield admissible controls, then the point is made.

We will use the 3×5 matrix of random numbers from earlier,

$$B = \begin{bmatrix} 0.6294 & 0.8268 & -0.4430 & 0.9298 & 0.9143 \\ 0.8116 & 0.2647 & 0.0938 & -0.6848 & -0.0292 \\ -0.7460 & -0.8049 & 0.9150 & 0.9412 & 0.6006 \end{bmatrix}$$

The MATLAB® function pinv generates the pseudo-inverse,

$$P = \begin{bmatrix} 0.2096 & 0.8126 & 0.1252 \\ 0.2676 & -0.1139 & -0.2837 \\ -0.1317 & 0.7775 & 0.6034 \\ 0.3203 & -0.4510 & 0.1062 \\ 0.3179 & 0.3790 & 0.3546 \end{bmatrix}$$

Now we take a control vector that is certainly admissible,

$$\mathbf{u} = \begin{Bmatrix} 1 \\ 1 \\ 1 \\ 1 \\ 1 \end{Bmatrix}$$

The moment generated by this combination of controls is

$$\mathbf{m} = B\mathbf{u} = \begin{Bmatrix} 2.8573 \\ 0.4560 \\ 0.9058 \end{Bmatrix}$$

However, if we were given this moment as $\mathbf{m}_d$ without knowing how it was created, and used the pseudo-inverse to solve for the controls, we would find that

$$\mathbf{u} = P\mathbf{m}_d = \begin{Bmatrix} 1.0829 \\ 0.4558 \\ 0.5247 \\ 0.8057 \\ 1.4023 \end{Bmatrix}$$

The first and fifth controls are past their limits, by quite a bit in the case of the fifth. Since the controls have physical limits, the airplane's response to the control system's solution would be to saturate the first and fifth controls so that

$$\mathbf{u}_{Sat} = \begin{Bmatrix} 1.0 \\ 0.4558 \\ 0.5247 \\ 0.8057 \\ 1.0 \end{Bmatrix}$$

The actual moment generated by this combination of controls is

$$\mathbf{m}_{Sat} = B\mathbf{u}_{Sat} = \begin{Bmatrix} 2.4372 \\ 0.4005 \\ 0.7261 \end{Bmatrix}$$

This result is different from $\mathbf{m}_d$ in both magnitude and direction. Although it is somewhat similar, one should not be designing control laws that give unpredictable results.

Other closed-form generalized inverses

The pseudo-inverse is just one of an infinity of closed-form inverses. An entire family of these solutions can be obtained from an optimization problem that aims to minimize other norms of $\mathbf{u}$, or a weighted 2-norm. The weighted 2-norm is seen frequently, and the problem is to minimize $\mathbf{u}^T W \mathbf{u}$, where W is a diagonal matrix. The diagonal terms are positive, and represent the aim to spare activity in some controls at the expense of more demands on the others.

The optimization proceeds as before, and results in

$$P_W = W^{-1} B^T [B W^{-1} B^T]^{-1} \tag{13.45}$$

Note, however, that Equation 13.45 is valid for any choice of W for which W is non-singular. Although there are infinitely many ways to select W, there doesn't appear to be much physical significance to the off-diagonal terms, or to a diagonal matrix with negative entries.

Moreover, it is possible to 'tailor' a pseudo-inverse so that it will provide optimal solutions in a small number of directions in Φ. The calculation of tailored generalized inverses requires a detailed knowledge of the AMS Φ, which we will see is pretty difficult to calculate. For that reason, the tailored generalized inverses would probably have to be calculated off-line and stored in the flight control computer for retrieval and use. The pseudo-inverse can be calculated 'on the fly' and can easily adapt to changing flight conditions.

For a proof that no single generalized inverse can, for arbitrary moment demands, yield solutions that attain the maximum available moment without violating some control constraint, the interested reader is referred to Bordignon (1996).

Other methods

Aside from the widespread application of generalized inverse solutions to the problem other methods of interest include techniques called 'pseudo controls' (Lallman, 1985), and 'daisy chaining'. The use of pseudo controls is an application of generalized inverses, imbedded in a dynamic inversion control law.

Daisy chaining was introduced at the High-Angle-of-Attack Projects and Technology Conference held at the NASA Dryden Flight Research Facility in April 1992. The main idea is to use conventional controls until one or more is commanded past its limits, and then to bring other controls into the solution. Daisy chaining likewise cannot everywhere yield maximum available moments, and additionally tends to demand higher control actuator rates than other methods.

These other methods of solving the control allocation problem are interesting in themselves, but do not seem to have found much application in current flight control system design.

13.3.5 Optimal solutions

We presented our definition of optimality in Section 13.3.3. We submit without proof that there is no closed-form solution that is optimal. There are various procedures that one may use to find optimal solutions to the control allocation problem, but to-date none has been found that is attractive in the setting of real-time flight control. The optimal solution is important however, since knowing it will allow us to analyze other methods to determine what has been given up by their use.

The method of determining optimal solutions to be described in this section consists of two parts. First we determine the entire set of attainable moments Φ. Second we consider arbitrary desired moments $\mathbf{m}_d$ and determine the largest vector of moments attainable in the direction of the desired moment. That solution is found on the boundary of Φ. If the maximum is greater than that desired, then the maximum is uniformly scaled to produce that desired.

Determination of Φ

Earlier in Section 13.3.3 we asked a few questions regarding the extreme points of rolling, pitching, and yawing moments. The answer is that the sign of the entries in the B matrix determine the controls at those extreme moments. Thus the first row of B multiplies the control vector $\mathbf{u}$ to generate the rolling moment. If we want the maximum rolling moment then we want the sum of entries times control deflections to be the greatest. If an entry is positive, then the corresponding control should be placed at its upper limit; if the entry is negative then the control should be placed at its lower limit. Therefore with

$$B = \begin{bmatrix} 0.6294 & 0.8268 & -0.4430 & 0.9298 & 0.9143 \\ 0.8116 & 0.2647 & 0.0938 & -0.6848 & -0.0292 \\ -0.7460 & -0.8049 & 0.9150 & 0.9412 & 0.6006 \end{bmatrix}$$

$$\mathbf{u}_{C_\ell Max} = \begin{Bmatrix} 1 \\ 1 \\ -1 \\ 1 \\ 1 \end{Bmatrix}$$

Then we changed the problem so that the first row of the B matrix has two entries that are identically zero. All the other controls should be set to their upper or lower limits, but the presence of zeros means that the corresponding controls have *no* influence on the rolling moment and they are free to vary. This condition, that all controls saturated save two, describes a two-dimensional facet of Φ, and since it lies at a maximum distance from the origin, it is on the boundary of Φ.

Normally we don't find B matrices with conveniently placed zeros, but it is easy to rotate the axes so that there are the two desired zeros. In short, we want to find a rotation matrix T_f such that the transformation $T_f B$ results in exactly two zeros in the first row.

We have selected the first row arbitrarily; it could have been any one of the three. Also note that we don't really care what the transformation does to the other two rows of B since they will contain no useful information. Therefore we only need the first row of T_f.

When we have an algorithm for finding such matrices T_f, we then proceed methodically through every permutation of controls taken two at a time, perform the transformation, determine the signs of the non-zero entries and hence determine the saturated controls that define the facet at the maximum in this new direction. Moreover, we can simultaneously determine the facet at the minimum along the new axis by setting the saturated control to its lower limit for positive entries and at its upper limit for negative entries.

Denote the first row of T_f as $\mathbf{t} = [t_{11}\ t_{12}\ t_{13}]$. Suppose we are just starting through the n choices for the first control and the $n-1$ choices for the second so the two entries to be zeroed correspond to u_1 and u_2. The problem to be solved is, given B, find t_{11}, t_{12}, and t_{13} such that

$$\begin{bmatrix} t_{11} & t_{12} & t_{13} \end{bmatrix} \begin{bmatrix} b_{11} & b_{12} \\ b_{21} & b_{22} \\ b_{31} & b_{32} \end{bmatrix} = \begin{bmatrix} 0 & 0 \end{bmatrix} \quad (13.46)$$

This problem is overdetermined, so we should be able to assign one of the elements in $\mathbf{t}$ arbitrarily. We select t_{13} and solve for the other two elements in terms of it. First we transpose the matrices in Equation 13.46,

$$\begin{bmatrix} b_{11} & b_{21} & b_{31} \\ b_{12} & b_{22} & b_{32} \end{bmatrix} \begin{bmatrix} t_{11} \\ t_{12} \\ t_{13} \end{bmatrix} = \begin{bmatrix} 0 \\ 0 \end{bmatrix} \quad (13.47)$$

Now note that Equation 13.47 may be written as

$$\begin{bmatrix} b_{11} & b_{21} \\ b_{12} & b_{22} \end{bmatrix} \begin{bmatrix} t_{11} \\ t_{12} \end{bmatrix} + t_{13} \begin{bmatrix} b_{31} \\ b_{32} \end{bmatrix} = \begin{bmatrix} 0 \\ 0 \end{bmatrix} \quad (13.48)$$

Now, if the square matrix in Equation 13.48 is singular this just means that $t_{13} = 0$, so we must pick another of the elements of $\mathbf{t}$. We are guaranteed by our assumption that B is robustly full rank that there will be some solution to Equation 13.46. Without loss of generality assume the matrix is non-singular so that

$$\begin{bmatrix} t_{11} \\ t_{12} \end{bmatrix} = -t_{13} \begin{bmatrix} b_{11} & b_{21} \\ b_{12} & b_{22} \end{bmatrix}^{-1} \begin{bmatrix} b_{31} \\ b_{32} \end{bmatrix} \quad (13.49)$$

Trends in Automatic Flight Control

In Equation 13.49 t_{13} is arbitrary, so we set $t_{13} = 1$ and solve for t_{11} and t_{12}. We could at this point normalize **t** so that its length is one, as it would be in a rotation matrix, but this will not affect the signs of the entries in the rest of the first row of B, just their magnitudes.

Then the row vector $\mathbf{t} = [t_{11} \ t_{12} \ 1]$ times B will yield a row vector with zeros in the u_1 and u_2 positions. The signs of the remaining terms define the facet required.

Let us apply this to our randomly generated sample B matrix, Equation 13.40.

$$B = \begin{bmatrix} 0.6294 & 0.8268 & -0.4430 & 0.9298 & 0.9143 \\ 0.8116 & 0.2647 & 0.0938 & -0.6848 & -0.0292 \\ -0.7460 & -0.8049 & 0.9150 & 0.9412 & 0.6006 \end{bmatrix} \quad (13.50)$$

For the facet of Φ corresponding to u_1 and u_2 (as in the example above)

$$\begin{bmatrix} t_{11} \\ t_{12} \end{bmatrix} = -\begin{bmatrix} 0.6294 & 0.8116 \\ 0.8268 & 0.2647 \end{bmatrix}^{-1} \begin{bmatrix} -0.7460 \\ -0.8049 \end{bmatrix} = \begin{bmatrix} 0.9036 \\ 0.2184 \end{bmatrix} \quad (13.51)$$

From this we obtain $\mathbf{t} = [0.9036 \ 0.2184 \ 1]$. Then,

$$\mathbf{t}B = \begin{bmatrix} 0 & 0 & 0.5352 & 1.6318 & 1.4204 \end{bmatrix} \quad (13.52)$$

The only information we need from Equation 13.52 is

$$\mathbf{t}B = \begin{bmatrix} 0 & 0 & + & + & + \end{bmatrix}$$

From this we conclude that one of the facets of the subset of attainable moments Φ for the sample allocation problem (Equations 13.40 and 13.41) is that where u_3, u_4, and u_5 are at their upper limits while u_1 and u_2 are free to vary.

Given B, $\mathbf{u}_{Min}$, and $\mathbf{u}_{Max}$, the steps to determine the facets that define Φ are:
For $i = 1 \ldots n-1$ and $j = i+1 \ldots n$:

1. Solve

$$\begin{bmatrix} t_{11} \\ t_{12} \end{bmatrix} = -\begin{bmatrix} B(1,i) & B(2,i) \\ B(1,j) & B(2,j) \end{bmatrix}^{-1} \begin{bmatrix} B(3,i) \\ B(3,j) \end{bmatrix}$$

2. Form the row vector $\mathbf{t} = [t_{11} \ t_{12} \ 1]$.
3. Evaluate $\mathbf{r} = \mathbf{t}B$.
4. There should be zeros in $\mathbf{r}$ at the ith and jth positions. The signs of the remaining entries define which of the remaining controls should be placed at their upper or lower limits to define the facet.

One facet is determined by setting each of the controls corresponding to entries in $\mathbf{r}$ to its upper limit if the sign of the entry is positive, and to its lower limit if the sign of the entry is negative. Call this vector of controls $\mathbf{u}^*$. Now create four control vectors from $\mathbf{u}^*$ by setting the ith and jth controls to their four limiting combinations of lower and upper limits. Here, the first subscript refers to u_i and the second to u_j, and a 0 means the lower limit and 1 the upper limit.

$$\mathbf{u}^*_{00} \quad \mathbf{u}^*_{01} \quad \mathbf{u}^*_{10} \quad \mathbf{u}^*_{11}$$

Then the four vertices of the facet thus identified are

$$\mathbf{m}^*_{00} = B\mathbf{u}^*_{00}$$
$$\mathbf{m}^*_{01} = B\mathbf{u}^*_{01}$$
$$\mathbf{m}^*_{10} = B\mathbf{u}^*_{10}$$
$$\mathbf{m}^*_{11} = B\mathbf{u}^*_{11}$$

5. Another facet is determined by setting each of the controls corresponding to entries in **r** to its lower limit if the sign of the entry is positive, and to its upper limit if the sign of the entry is negative. Determine the moments at the vertices of this facet following a procedure similar to that in step 4.

Before leaving this section, we note that each facet may be considered as the base of a trapezoidal pyramid with the origin of Φ ($\mathbf{m} = \mathbf{0}$) as its vertex. Thus, given all the coordinates of the facets of Φ the volume of each pyramid may be calculated. The sum of all these volumes is the volume of Φ, and is a measure of the maximum capabilities of this particular control allocation problem, against which other methods of solution may be compared.

Finally, there are $2n(n-1)$ facets for a control allocation problem of three moments and n controls. This means the number of calculations needed to determine all of Φ is proportional to n^2.

Determination of optimal solutions

Given that we may generate the entire ensemble of facets that define the attainable moment set, we wish to find solutions $\mathbf{u} = B\mathbf{m}_d$ such that if $\mathbf{m}_d$ is in Φ then an admissible set of controls is found. Solutions for moments in the interior of Φ are not unique, but those on the boundary of Φ are. If we can find the unique admissible solutions to the boundary moments, then those in the interior should be easy.

We will therefore treat the desired moments $\mathbf{m}_d$ as directions in the three-dimensional space of moments and find the intersection of a line in that direction with the boundary of Φ. If the moment thus generated at the boundary is greater than $\mathbf{m}_d$ then we will uniformly scale it back until it is the same. This will ensure that as greater demands are made for desired moments that the solutions will proceed continuously up to their maximum with no changes in the method of solution along the way.

We do not address what to do if the desired moment $\mathbf{m}_d$ is greater than the maximum attainable in that direction. There are a few ways of dealing with this case, but the situation should never occur: a well designed control law does not ask more of the control effectors than they can deliver. Once the desired moments are unattainable every assumption under which the control law was designed are invalid.

Let us begin by assuming we know which facet of the boundary of Φ contains the intersection of a line in the direction of the desired moment with that facet, and ask how we are to evaluate the controls that obtain at that point. Figure 13.13 shows the geometry of that point on the facet.

The named features of Figure 13.13 are

- The origin of Φ, $\mathbf{m} = \mathbf{0}$.
- $\mathbf{m}_d$, the desired moment.

Trends in Automatic Flight Control

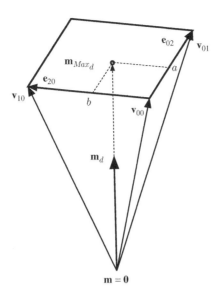

Figure 13.13 A facet of Φ.

- A vector to the point of intersection with the facet, named $\mathbf{m}_{Max_d}$, the maximum moment in the desired direction. It is shown here as a dashed extension to $\mathbf{m}_d$. This vector is some scalar multiple of $\mathbf{m}_d$, say

$$\mathbf{m}_{Max_d} = c\mathbf{m}_d$$

  Note that for $\mathbf{m}_d$ to be attainable on this facet, $c \geq 1$.
- $\mathbf{v}_{00}$, $\mathbf{v}_{01}$, and $\mathbf{v}_{10}$, vectors from the origin to three of the vertices of the facet. The 0 and 1 in the subscripts are shorthand for denoting which of the two controls that define that facet is at its lower limit (0) or its upper limit (1), the same as was done previously.
- $\mathbf{e}_{20}$ and $\mathbf{e}_{02}$, vectors (edges) from $\mathbf{v}_{00}$ to $\mathbf{v}_{10}$ and $\mathbf{v}_{01}$, respectively. The subscripts 0 and 1 mean the same as above, while the additional subscript 2 means that the control is varying between its minimum and maximum, so that it starts out as a 0, becomes a 2 in transit, and then is a 1 (or vice versa).
- Two scalar numbers a and b which denote distances along each of the two edges $\mathbf{e}_{20}$ and $\mathbf{e}_{02}$.

Using simple vector addition,

$$\begin{aligned} \mathbf{e}_{20} &= \mathbf{v}_{10} - \mathbf{v}_{00} \\ \mathbf{e}_{02} &= \mathbf{v}_{01} - \mathbf{v}_{00} \end{aligned} \quad (13.53)$$

Some of the advantages of the bookkeeping subscripts become apparent when comparing the left and right side subscripts of these equations.

The most important piece of vector addition is

$$\mathbf{m}_{Max_d} = \mathbf{v}_{00} + a\mathbf{e}_{02} + b\mathbf{e}_{20} = c\mathbf{m}_d \quad (13.54)$$

Equation 13.54 offers a way to test a facet to see if it contains the extended $\mathbf{m}_d$. We form the system of equations

$$-a\mathbf{e}_{02} - b\mathbf{e}_{20} + c\mathbf{m}_d = \begin{bmatrix} -\mathbf{e}_{02} & -\mathbf{e}_{20} & \mathbf{m}_d \end{bmatrix} \begin{Bmatrix} a \\ b \\ c \end{Bmatrix} = \mathbf{v}_{00} \qquad (13.55)$$

Equation 13.55 is easily solved for a, b, and c:

$$\begin{Bmatrix} a \\ b \\ c \end{Bmatrix} = \begin{bmatrix} -\mathbf{e}_{02} & -\mathbf{e}_{20} & \mathbf{m}_d \end{bmatrix}^{-1} \mathbf{v}_{00} \qquad (13.56)$$

If the particular facet under consideration contains the intersection, then it is necessary and sufficient that:
$$\begin{array}{c} c \geq 1 \\ 0 \leq a \leq 1 \\ 0 \leq b \leq 1 \end{array} \qquad (13.57)$$

- If the matrix inversion fails due to singularity then the desired moment is parallel to the plane of the facet under consideration, and there is no intersection.
- If either a or b or both are not in the range (0, 1) then the desired moment $\mathbf{m}_d$ does not point toward this facet but toward another. The value of c has no significance in this case.
- If $0 \leq a \leq 1$ and $0 \leq b \leq 1$ but $c \leq -1$ then the correct facet is this facet's opposite (see step 5 in facet determination).
- If $0 \leq a \leq 1$ and $0 \leq b \leq 1$ but $|c| < 1$ then the desired moment is not attainable on this or any other facet.
- If $0 \leq a \leq 1$ and $0 \leq b \leq 1$ and $c \geq 1$ then this facet is the correct one. The controls that generated the moments used in the determination of a, b, and c are unique and correspond to those moments in a one-to-one relationship. Denote the controls that generate the three vertices in Figure 13.13 as $\mathbf{u}_{00}$, $\mathbf{u}_{01}$, and $\mathbf{u}_{10}$ with the same meaning as was used to identify this facet earlier.
  The controls that generate $\mathbf{m}_d$ are therefore

$$\mathbf{u} = \frac{1}{c}[\mathbf{u}_{00} + a(\mathbf{u}_{01} - \mathbf{u}_{00}) + b(\mathbf{u}_{10} - \mathbf{u}_{00})] \qquad (13.58)$$

Note that this solution is the solution at the boundary scaled by the factor $1/c$.

Direct method of solution

The combination of the algorithms in the two previous sections is referred to as the *direct method* of solution of the control allocation problem. At any instant the flight control computer performs dynamic inversion up to the point where a desired moment $\mathbf{m}_d$ is generated. The facets of the attainable moment set are examined one at a time until the proper facet is found.

An advantage of the direct method is that it will always find a solution to the control allocation problem if it exists. The big disadvantage is obviously the requirement that all the facets be searched. Even if the correct facet is found on the first one examined, there is no guarantee of that and the worst case must be assumed.

The direct method has been presented here because of the insights it gives into the geometry of the control allocation problem. Those insights should help greatly in evaluating other methods of solution, and in determining what went wrong when something does.

13.3.6 Near-optimal solutions

The direct method of solving the control allocation problem is optimal. As suggested previously, in the process of identifying the facets of the attainable moments set there is sufficient information to determine the volume of Φ for comparison with other methods. Finding the volume associated with non-optimal methods ranges from analytical solutions (generalized inverses) to the requirement to generate as many valid solutions in as many different directions and magnitudes of $\mathbf{m}_d$ as possible, and then fitting a mathematical wrapper around those points.

One non-direct control allocation method consistently performs well compared with direct allocation, and it is relatively fast and easy to implement. The method is called *cascaded generalized inverses* (CGI) and it is the method used in the design of the X-35 control laws (Bordignon and Bessolo, 2002).

In its simplest form the CGI method uses a single generalized inverse – we will assume the unweighted pseudo-inverse – exclusively until one or more of the control effectors becomes saturated. The saturated controls are left at whatever limit they were at when saturated and removed from the problem: their columns are removed from the B matrix. What happens next depends on which of three cases obtain with respect to the remaining control effectors.

1. The remaining system is over-determined. This is the case we have been discussing throughout the section on control allocation: more controls than moments. If this is the case, then the pseudo-inverse of this system is calculated and applied to whatever remains of the desired moments with consideration of the moments generated by the saturated controls.
2. The remaining system is square, with three controls for the three moments. Here the matrix inverse is calculated and applied to the remaining moments.
3. The remaining system is under-determined, with fewer controls than moments. The solution method of choice is another kind of pseudo-inverse that minimizes the square of the error between what is desired and what may be attained. The solution is similar to the pseudo-inverse for the over-determined case. Call the under-determined B matrix B_u and the pseudo-inverse P_u:

$$P_u = [B_u^T B_u]^{-1} B_u^T \tag{13.59}$$

If use of the original pseudo-inverse does not saturate any controls then nothing extra is required. Once any controls are saturated, the saturated controls are removed and the applicable case in the preceding list is applied to the remaining problem. If that step creates more saturated controls then those controls are removed and the procedure is repeated, until the desired moment is attained or the method fails.

We have the beginnings of a cascaded generalized inverse problem above, wherein we were demonstrating that the single application of a pseudo-inverse was not an optimal solution. We have:

$$B = \begin{bmatrix} 0.6294 & 0.8268 & -0.4430 & 0.9298 & 0.9143 \\ 0.8116 & 0.2647 & 0.0938 & -0.6848 & -0.0292 \\ -0.7460 & -0.8049 & 0.9150 & 0.9412 & 0.6006 \end{bmatrix}$$

$$-1 \leq u_i \leq 1, \; i = 1 \ldots 5 \tag{13.60}$$

The desired moment is

$$\mathbf{m}_d = \begin{Bmatrix} 2.8573 \\ 0.4560 \\ 0.9058 \end{Bmatrix}$$

The pseudo-inverse was applied to this problem and yielded

$$\mathbf{u} = P\mathbf{m} = \begin{Bmatrix} 1.0829 \\ 0.4558 \\ 0.5247 \\ 0.8057 \\ 1.4023 \end{Bmatrix}$$

We now remove the two saturated controls from the problem by removing the columns of B that correspond to them (the first and the fifth). The remaining B, here called B_1, is

$$B_1 = \begin{bmatrix} 0.8268 & -0.4430 & 0.9298 \\ 0.2647 & 0.0938 & -0.6848 \\ -0.8049 & 0.9150 & 0.9412 \end{bmatrix}$$

The moment that is generated by the two saturated controls, $\mathbf{m}'$, is found from the two removed columns of B times the saturated values of those controls,

$$\mathbf{m}' = \begin{bmatrix} 0.6294 & 0.9143 \\ 0.8116 & -0.0292 \\ -0.7460 & 0.6006 \end{bmatrix} \begin{Bmatrix} 1 \\ 1 \end{Bmatrix} = \begin{Bmatrix} 1.5437 \\ 0.7824 \\ -0.1454 \end{Bmatrix}$$

The remaining moment is the original desired moment minus $\mathbf{m}'$

$$\mathbf{m}_{d_1} = \begin{Bmatrix} 2.8573 \\ 0.4560 \\ 0.9058 \end{Bmatrix} - \begin{Bmatrix} 1.5437 \\ 0.7824 \\ -0.1454 \end{Bmatrix} = \begin{Bmatrix} 1.3136 \\ -0.3263 \\ 1.0513 \end{Bmatrix}$$

Now we just invert B_1 and multiply $\mathbf{m}_{d_1}$, with the result

$$\mathbf{u}_1 = B_1^{-1}\mathbf{m}_{d_1} = \begin{Bmatrix} 1 \\ 1 \\ 1 \end{Bmatrix}$$

When this result is recombined with the two saturated controls the result is

$$\mathbf{u} = \begin{Bmatrix} 1 \\ 1 \\ 1 \\ 1 \\ 1 \end{Bmatrix}$$

We know, because the demonstration was thus contrived, that this is the control vector used to generate the given desired moment. While this desired moment is not obviously on the boundary of the attainable moment set it was confirmed that it indeed is. Thus, in this case at least, the cascaded generalized inverse method yields an optimal solution.

In the course of extensive testing with randomly generated problems and with actual airplane problems, the cascaded generalized inverse method results seldom failed to yield optimal or very nearly optimal solutions.

Problems

1. When MATLAB® is used to create a matrix of transfer functions for the dynamic inversion example in Equations 13.15 and 13.16 (with $C_{LD} = I$ and $D_{LD} = 0$) the denominators of all the transfer functions are second-order polynomials, while the system is fourth order. Explain.
2. What would be the physical significance if the control effectiveness matrix in Equation 13.10 were singular?
3. Aircraft carrier approaches are flown at constant angle of attack, normally maintained by judicious changes in throttle and elevator, all the while controlling the rate of descent. This design problem will simplify one part of the task by making the pilot's stick input command angle of attack, leaving him to control his flight path by use of throttle.
We revisit our relaxed static-stability fighter (Stevens and Lewis, 1992) from problem 12.3-1. Consider the system:

$$A = \begin{bmatrix} -1.93 \times 10^{-2} & 8.82 & -0.575 & -32.2 \\ -2.54 \times 10^{-4} & -1.02 & 0.905 & 0 \\ 2.95 \times 10^{-12} & 0.822 & -1.08 & 0 \\ 0 & 0 & 1 & 0 \end{bmatrix}$$

$$B = \begin{bmatrix} 0.174 \\ -2.15 \times 10^{-3} \\ -0.176 \\ 0 \end{bmatrix}$$

The states are V, α, q, and θ, and the control is δ_e. Velocity is in ft/s and angle measures are in radians except elevator is in degrees.

Design a dynamic-inversion control law such that the pilot's stick commands angle of attack. Place the short-period roots in the vicinity of $\lambda_{SP} = -2 \pm j2$. Show all steps, including root-locus plots. How does this problem relate to that in problem 12.3-1? How do your results compare with the results in that problem?

4. Augment the M2-F2 equations by adding the dynamics of the heading angle ψ and consider a dynamic inversion control law.

   (a) Can the controls be used to change the aircraft heading without changing bank angle?
   (b) If so, what happens to the other states while this is going on? If not, why not?

5. A velocity-vector roll is one in which the roll is about the wind-axis, performed at constant angle-of-attack with no sideslip.

   (a) Identify the appropriate equations for designing a dynamic-inversion control law such that lateral stick performs the desired maneuver.
   (b) Identify the kinematic, controlled, and complementary equations in this set of equations, and state any assumptions necessary.
   (c) Show the steps you would take in the design and identify the parameters that would have to be chosen.

6. Show the necessary steps to prove Equation 13.45.

7. Assume that Figures 13.11 and 13.12 are representative of some actual control allocation problem. Of the six numbers in the B matrix, how many are zero, and where are they? Explain.

8. Repeat the determination of optimal solutions to the control allocation problem with application to the 2−moment problem.

9. Apply the CGI method to the example problem in Equations 13.40 and 13.41 using the following desired moment, and determine whether or not the moment is optimal (on the boundary of attainable moments):

$$\mathbf{m}_d = \left\{ \begin{array}{c} -0.6559 \\ 1.2963 \\ 0.6333 \end{array} \right\}$$

You should need the original over-determined pseudo-inverse, one more over-determined pseudo-inverse, and one under-determined pseudo-inverse.

References

Azam, M. and Singh, Sahjendra S.N. (1994) Invertibility and trajectory control for nonlinear maneuvers of aircraft. *Journal of Guidance, Control, and Dynamics*, **17**(1), 192–200.

Bordignon, K. and Bessolo, J. (2002) Control Allocation for the X-35B, Proceedings of the 2002 Biennial International Powered Lift Conference and Exhibit 5-7 November 2002, Paper number 2002-6020.

Bordignon, K.A. (1996) *Constrained control allocation for systems with redundant control effectors*, Ph.D. dissertation URN etd-08082007-161936, Virginia Polytechnic Institute & State University.

Davidson, J.B., et al. (1992) Development of a Control Law Design Process Utilizing Advanced Synthesis Methods With Application to the NASA F-18 HARV, High-Angle-of-Attack Projects and Technology Conference, NASA Dryden Flight Research Facility, 21-23 April 1992.

Durham, W. (1994) Attainable moments for the constrained control allocation problem. *Journal of Guidance, Control, and Dynamics*, **17**(6), 1371–1373.

Elgersma, M.R. (1988) *Control of nonlinear systems using partial dynamic inversion*, Ph.D. Thesis, Control Sciences Department, University of Minnesota.

Lallman, F.J. (1985) Relative Control Effectiveness Technique With Application to Airplane Control Coordination, NASA Technical Paper 2416, April, 1985.

Snell, S.A., Enns, D.F., and Garrard, W.L., Jr., (1992) Nonlinear inversion flight control for a supermaneuverable aircraft. *Journal of Guidance, Control, and Dynamics,* **15**(4, July–August), 976984.

Stevens, B.L. and Lewis, F.L. (1992) *Aircraft Control and Simulation*, 1st edn, John Wiley & Sons, Inc., pp. 255–259.

Walker, G.P., and Allen, D.A. (2002) X-35B Stovl Flight Control Law Design and Flying Qualities, Proceedings of the 2002 Biennial International Powered Lift Conference and Exhibit 5-7, November 2002, Paper number AIAA 2002-6018.

Appendix A

Example Aircraft

The example is based on data taken from Nelson (1998) and other data needed for this example. See Tables A.1 and A.2. The data for sea-level flight at Mach 0.4 were utilized. In order to produce the four matrices used in the analyses a MATLAB® M-File was prepared. The function *Lin* just evaluates the quantities determined in the previous chapters. Using MATLAB® for this purpose is a bit of an overkill; however, it is a substantial savings over using a slide rule. Also, invoking the function from MATLAB® leaves the matrices in the workspace for future manipulation.

Table A.1 Aerodynamic data for the A-4 *Skyhawk*.

Longitudinal		Lat–Dir	
$C_{L_{Ref}}$	0.28	C_{y_β}	−0.98
$C_{D_{Ref}}$	0.03	C_{ℓ_β}	−0.12
C_{L_α}	3.45	C_{n_β}	0.25
C_{D_α}	0.30	C_{ℓ_p}	−0.26
C_{m_α}	−0.38	C_{n_p}	0.022
$C_{L_{\dot\alpha}}$	0.72	C_{ℓ_r}	0.14
$C_{m_{\dot\alpha}}$	−1.1	C_{n_r}	−0.35
C_{L_q}	0.0	$C_{\ell_{\delta_a}}$	0.08
C_{m_q}	−3.60	$C_{n_{\delta_a}}$	0.06
C_{L_M}	0.0	$C_{y_{\delta_r}}$	0.17
C_{D_M}	0.0	$C_{\ell_{\delta_r}}$	−0.105
C_{m_M}	0.0	$C_{n_{\delta_r}}$	0.032
$C_{L_{\delta_e}}$	0.36		
$C_{m_{\delta_e}}$	−0.50		
C_{T_V}	−0.06		
$C_{T_{\delta_T}}$	See notes		

Aircraft Flight Dynamics and Control, First Edition. Wayne Durham.
© 2013 John Wiley & Sons, Ltd. Published 2013 by John Wiley & Sons, Ltd.

Table A.2 Physical data for the A-4 *Skyhawk*.

Property	Value
Weight	17,578 lb
I_{xx}	8,090 slug·ft^2
I_{yy}	25,900 slug·ft^2
I_{zz}	29,200 slug·ft^2
I_{xz}	1,300 slug·ft^2
S	260 ft^2
b	27.5 ft
$\bar{c}$	10.8 ft

The calling and returned arguments are as follows:

```
function [aLong,bLong,aLD,bLD] = Lin(Ref,Phys,D)
% function [aLong,bLong,aLD,bLD] = Lin(Ref,Phys,D)
%
% Lin Takes reference flight conditions,
% aircraft physical characteristics,
% nondimensional derivatives; and returns
% dimensional a and b matrices.
%
% Inputs:
% Ref is a 6-vector of reference flight conditions,
%   Ref(1)=density (slugs/ft^3)
%   Ref(2)=TAS (ft/s)
%   Ref(3)=Mach
%   Ref(4)=CL (trim)
%   Ref(5)=CD (trim)
%   Ref(6)=Gamma, flight path angle (radians)
% Phys is an 9-vector of physical characteristics,
%   Phys(1) = Weight (pounds)
%   Phys(2) = Ixx (slug-ft^2)
%   Phys(3) = Iyy (slug-ft^2)
%   Phys(4) = Izz (slug-ft^2)
%   Phys(5) = Ixz (slug-ft^2)
%   Phys(6) = Area (ft^2)
%   Phys(7) = Span (ft)
%   Phys(8) = Chord (ft)
%   Phys(9) = Thrust angle (radians)
% D is a 29-vector of nondimensional derivatives
%   D(1)=CLAlpha
%   D(2)=CDAlpha
%   D(3)=CmAlpha
%   D(4)=CLAlphaDot
```

```
%   D(5)=CmAlphaDot
%   D(6)=CLq
%   D(7)=Cmq
%   D(8)=CLM
%   D(9)=CDM
%   D(10)=CmM
%   D(11)=CLDeltaM
%   D(12)=CDDeltaM
%   D(13)=CMDeltaM
%   D(14)=CTV
%   D(15)=CTDeltaT
%   D(16)=CyBeta
%   D(17)=ClBeta
%   D(18)=CnBeta
%   D(19)=Clp
%   D(20)=Cnp
%   D(21)=Cyp
%   D(22)=Clr
%   D(23)=Cnr
%   D(24)=Cyr
%   D(25)=ClDeltaL
%   D(26)=CnDeltaL
%   D(27)=ClDeltaN
%   D(28)=CnDeltaN
%   D(29)=CyDeltaN
%
% Outputs
% aLong and aLD are 4x4 longitudinal
% and lateral/directional system matrices
% bLong and bLD are 4x2 longitudinal
% and lateral/directional control matrices
```

The rest of the code simply multiplies and adds terms, taking sines and cosines where necessary:

```
gee=32.174;
mass=Phys(1)/gee;
q=0.5*Ref(1)*Ref(2)^2;
qS=q*Phys(6);
qSb=qS*Phys(7);
qSc=qS*Phys(8);
V=Ref(2);
mV=mass*V;
qSoV=qS/V;
qSoM=qS/mass;
CW=Phys(1)/qS;
```

```
    CosEps=cos(Phys(9));
    SinEps=sin(Phys(9));
    CosGam=cos(Ref(6));
    SinGam=sin(Ref(6));
    CT=CW*sin(Ref(6)-Phys(9))+Ref(5)*CosGam+Ref(4)*SinGam;

aLong(4,1)=0;
aLong(4,2)=0;
aLong(4,3)=1;
aLong(4,4)=0;

Xu=qSoV*(2*CW*SinGam-2*CT*CosEps-Ref(3)*D(9));
Tu=qSoV*(2*CT+D(14));
Tu=0;
aLong(1,1)=Xu+Tu*CosEps;
Xw=qSoV*(Ref(4)-D(2));
aLong(1,2)=Xw;
aLong(1,3)=0;
aLong(1,4)=-Phys(1)*CosGam;
aLong(1,:)=aLong(1,:)/mass;

Zu=-qSoV*(2*Ref(4)+Ref(3)*D(8));
aLong(2,1)=Zu+Tu*SinEps;
Zw=-qSoV*(Ref(5)+D(1));
aLong(2,2)=Zw;
Zq=-qSc*D(6)/(2*V);
aLong(2,3)=Zq+mass*V;
aLong(2,4)=-Phys(1)*SinGam;
m_ZwDot=mass+qSc*D(4)/(2*V^2);
aLong(2,:)=aLong(2,:)/m_ZwDot;

MwDot=qSc*Phys(8)*D(5)/(2*V^2);
Mu=Ref(3)*qSc*D(10)/V;
aLong(3,1)=Mu+MwDot*aLong(2,1);
Mw=qSc*D(3)/V;
aLong(3,2)=Mw+MwDot*aLong(2,2);
Mq=qSc*Phys(8)*D(7)/(2*V);
aLong(3,3)=Mq+MwDot*aLong(2,3);
aLong(3,4)=MwDot*aLong(2,4);
aLong(3,:)=aLong(3,:)/Phys(3);

bLong(1,1)=qSoM*D(15)*CosEps;
bLong(1,2)=-qSoM*D(12);
bLong(2,1)=qS*D(15)*SinEps/m_ZwDot;
bLong(2,2)=-qSoM*D(11)/m_ZwDot;
bLong(3,1)=0;
```

```
MDeltaM=qSc*D(13);
bLong(3,2)=(MDeltaM+MwDot*bLong(2,2))/Phys(3);
bLong(4,1)=0;
bLong(4,2)=0;

qSboV=qSoV*Phys(7);

Yv=qSoV*D(16);
Yp=qSboV*D(21)/2;
Yr=qSboV*D(24)/2;
aLD(1,1)=Yv;
aLD(1,2)=Yp;
aLD(1,3)=Yr-mass*Ref(2);
aLD(1,4)=Phys(1)*CosGam;
aLD(1,:)=aLD(1,:)/mass;

Lv=qSboV*D(17);
Lp=qSboV*Phys(7)*D(19)/2;
Lr=qSboV*Phys(7)*D(22)/2;
Nv=qSboV*D(18);
Np=qSboV*Phys(7)*D(20)/2;
Nr=qSboV*Phys(7)*D(23)/2;
aLD(2,1)=Phys(4)*Lv+Phys(5)*Nv;
aLD(2,2)=Phys(4)*Lp+Phys(5)*Np;
aLD(2,3)=Phys(4)*Lr+Phys(5)*Nr;
aLD(2,4)=0;
aLD(3,1)=Phys(5)*Lv+Phys(2)*Nv;
aLD(3,2)=Phys(5)*Lp+Phys(2)*Np;
aLD(3,3)=Phys(5)*Lr+Phys(2)*Nr;
aLD(3,4)=0;
aLD(2:3,:)=aLD(2:3,:)/(Phys(2)*Phys(4)-Phys(5)^2);

aLD(4,1)=0;
aLD(4,2)=1;
aLD(4,3)=SinGam/CosGam;
aLD(4,4)=0;

bLD(1,1)=0;
bLD(1,2)=qSoM*D(29);
LdL=qSb*D(25);
NdL=qSb*D(26);
LdN=qSb*D(27);
NdN=qSb*D(28);
bLD(2,1)=Phys(4)*LdL+Phys(5)*NdL;
bLD(2,2)=Phys(4)*LdN+Phys(5)*NdN;
bLD(3,1)=Phys(5)*LdL+Phys(2)*NdL;
```

```
bLD(3,2)=Phys(5)*LdN+Phys(2)*NdN;
bLD(2:3,1:2)=bLD(2:3,1:2)/(Phys(2)*Phys(4)-Phys(5)^2);
bLD(4,1)=0;
bLD(4,2)=0;
```

The state and control vectors are:

$$\mathbf{x}_{Long} = \begin{Bmatrix} u \\ w \\ q \\ \theta \end{Bmatrix} \quad \mathbf{u}_{Long} = \begin{Bmatrix} \delta_T \\ \delta_e \end{Bmatrix}$$

$$\mathbf{x}_{LD} = \begin{Bmatrix} v \\ p \\ r \\ \phi \end{Bmatrix} \quad \mathbf{u}_{LD} = \begin{Bmatrix} \delta_a \\ \delta_r \end{Bmatrix}$$

Six of the parameters needed by the function are not given in the data:

```
Ref(1)=density (slugs/ft 3)
Ref(2)=TAS (ft/s)
```

A table of properties of the standard atmosphere was used to obtain the density ρ and speed of sound a at sea-level. True air speed is the Mach number times speed of sound.

```
Ref(6)=Gamma, flight path angle (radians)
```

Not specified, and assumed to be zero.

```
Phys(9) = Thrust angle (radians)
```

Not specified, and assumed to be zero.

```
D(14)=CTV
```

Results for constant thrust (jet engine) were used, $C_{T_V} = -2C_{T_{Ref}}$. Since we have assumed level flight and zero thrust angle, $C_{T_V} = -2C_{D_{Ref}}$.

```
D(15)=CTDeltaT
```

A made-up value for *CTDeltaT* was used. The A-4D utilized a J-52 engine that produced 11,200 lb of thrust at sea-level. The throttle was modeled as a control with range of value from zero to one, with a one-to-one correspondence between zero and full thrust. Therefore,

$$C_{T_{\delta_T}} = \frac{\partial C_T}{\partial \delta_T} = \frac{\partial T / \partial \delta_T}{\bar{q}S} = \frac{11,200}{\bar{q}S}$$

Example Aircraft

The output from the function *Lin* is based on the states v and w. These states were transformed to β and α for the analysis. For small disturbances the expressions for β and α are linearized by $\beta = v/V_{Ref}$ and $\alpha = w/V_{Ref}$, whence the transformations:

$$\begin{Bmatrix} u \\ \alpha \\ q \\ \theta \end{Bmatrix} = \begin{bmatrix} 1 & 0 & 0 & 0 \\ 0 & 1/V_{Ref} & 0 & 0 \\ 0 & 0 & 1 & 0 \\ 0 & 0 & 0 & 1 \end{bmatrix} \begin{Bmatrix} u \\ w \\ q \\ \theta \end{Bmatrix}$$

$$\begin{Bmatrix} \beta \\ p \\ r \\ \phi \end{Bmatrix} = \begin{bmatrix} 1/V_{Ref} & 0 & 0 & 0 \\ 0 & 1 & 0 & 0 \\ 0 & 0 & 1 & 0 \\ 0 & 0 & 0 & 1 \end{bmatrix} \begin{Bmatrix} v \\ p \\ r \\ \phi \end{Bmatrix}$$

Section 3.3 pertains.

Reference

Nelson, R.C. (1998) *Flight Stability and Automatic Control*, 2nd edn, WCB/McGraw-Hill.

Appendix B

Linearization

All results are valid for straight, symmetric flight using stability axes.

B.1 Derivation of Frequently Used Derivatives

$$\left.\frac{\partial V}{\partial u}\right|_{Ref} = \left.\frac{\partial (u^2+v^2+w^2)^{1/2}}{\partial u}\right|_{Ref} = \left.\frac{u}{(u^2+v^2+w^2)^{1/2}}\right|_{Ref} = \frac{u_{Ref}}{V_{Ref}} = 1$$

$$\left.\frac{\partial V}{\partial v}\right|_{Ref} = \frac{v_{Ref}}{V_{Ref}} = 0, \quad \left.\frac{\partial V}{\partial w}\right|_{Ref} = \frac{w_{Ref}}{V_{Ref}} = 0$$

$$\left.\frac{\partial \overline{q}}{\partial u}\right|_{Ref} = \left.\frac{\rho}{2}\frac{\partial (u^2+v^2+w^2)}{\partial u}\right|_{Ref} = \rho_{Ref} V_{Ref}$$

$$\left.\frac{\partial \overline{q}}{\partial v}\right|_{Ref} = \rho_{Ref} v_{Ref} = 0, \quad \left.\frac{\partial \overline{q}}{\partial w}\right|_{Ref} = \rho_{Ref} w_{Ref} = 0$$

$$\left.\frac{\partial \hat{p}}{\partial u}\right|_{Ref} = p_{Ref} \left.\frac{\partial (b/2V)}{\partial u}\right|_{Ref} = 0$$

The latter derivative vanishes because $p_{Ref} = 0$. Similarly,

$$\left.\frac{\partial \hat{p}}{\partial v}\right|_{Ref} = \left.\frac{\partial \hat{p}}{\partial w}\right|_{Ref} = \left.\frac{\partial \hat{q}}{\partial u}\right|_{Ref} = \left.\frac{\partial \hat{q}}{\partial v}\right|_{Ref} = \left.\frac{\partial \hat{q}}{\partial w}\right|_{Ref}$$
$$= \left.\frac{\partial \hat{r}}{\partial u}\right|_{Ref} = \left.\frac{\partial \hat{r}}{\partial v}\right|_{Ref} = \left.\frac{\partial \hat{r}}{\partial w}\right|_{Ref} = 0$$

$$\left.\frac{\partial \hat{p}}{\partial p}\right|_{Ref} = \left.\frac{\partial pb/2V}{\partial p}\right|_{Ref} = \left.\frac{b}{2V}\right|_{Ref} = \frac{b}{2V_{Ref}}$$

$$\left.\frac{\partial \hat{q}}{\partial q}\right|_{Ref} = \frac{\overline{c}}{2V_{Ref}}, \quad \left.\frac{\partial \hat{r}}{\partial r}\right|_{Ref} = \frac{b}{2V_{Ref}}$$

Aircraft Flight Dynamics and Control, First Edition. Wayne Durham.
© 2013 John Wiley & Sons, Ltd. Published 2013 by John Wiley & Sons, Ltd.

$$\left.\frac{\partial\alpha}{\partial u}\right|_{Ref} = \left.\frac{\partial\tan^{-1}(w/u)}{\partial u}\right|_{Ref} = \left.\frac{-w}{u^2\sec^2\alpha}\right|_{Ref} = 0$$

$$\left.\frac{\partial\alpha}{\partial v}\right|_{Ref} = \left.\frac{\partial\tan^{-1}(w/u)}{\partial v}\right|_{Ref} = 0$$

$$\left.\frac{\partial\alpha}{\partial w}\right|_{Ref} = \left.\frac{\partial\tan^{-1}(w/u)}{\partial w}\right|_{Ref} = \left.\frac{1}{u\sec^2\alpha}\right|_{Ref} = \frac{1}{V_{Ref}}$$

$$\left.\frac{\partial\beta}{\partial v}\right|_{Ref} = \left.\frac{\partial\sin^{-1}(v/V)}{\partial v}\right|_{Ref} = \left.\frac{V - v(\partial V/\partial v)}{V^2\cos\beta}\right|_{Ref} = \frac{1}{V_{Ref}}$$

$$\left.\frac{\partial\beta}{\partial u}\right|_{Ref} = \left.\frac{\partial\beta}{\partial w}\right|_{Ref} = 0$$

The thrust derivative, $C_T(\hat{V}, \delta_T) = T/\bar{q}S$:

$$\frac{\partial C_T}{\partial \hat{V}} \equiv C_{T_V} = \left.\frac{1}{\bar{q}S}\frac{\partial T}{\partial V}\frac{\partial V}{\partial \hat{V}} + \frac{T}{S}\frac{\partial(1/\bar{q})}{\partial V}\frac{\partial V}{\partial \hat{V}}\right|_{Ref}$$

$$= \left.\frac{V}{\bar{q}S}\frac{\partial T}{\partial V} - \frac{\rho V T}{2\bar{q}^2 S}\frac{\partial(V^2)}{\partial \hat{V}}\right|_{Ref}$$

$$= \left.\frac{V}{\bar{q}S}\frac{\partial T}{\partial V} - 2T\frac{\rho V^2/2}{\bar{q}^2 S}\right|_{Ref}$$

$$= \frac{1}{\bar{q}_{Ref}S}\left(V_{Ref}\left.\frac{\partial T}{\partial V}\right|_{Ref} - 2T_{Ref}\right)$$

In the case of a rocket, thrust does not vary with velocity, and the same is very nearly true in the case of a jet engine. This leads to

$$C_{T_V} = -2T_{Ref}/\bar{q}_{Ref}S = -2C_{T_{Ref}} \text{ (Constant thrust)}$$

For propeller-driven aircraft with constant speed propellers, constant thrust-horsepower is usually assumed ($TV = \text{constant} = T_{Ref}V_{Ref}$) so

$$T = T_{Ref}V_{Ref}/V$$

$$\left.\frac{\partial T}{\partial V}\right|_{Ref} = -\frac{T_{Ref}}{V_{Ref}}$$

$$C_{T_V} = -3C_{T_{Ref}} \text{ (Constant thrust-horsepower)}$$

Linearization

B.2 Non-dimensionalization of the Rolling Moment Equation

Dimensional form:

$$\dot{p} = \frac{1}{I_D}[(I_{zz}L_v + I_{xz}N_v)v + (I_{zz}L_p + I_{xz}N_p)p$$
$$+ (I_{zz}L_r + I_{xz}N_r)r + (I_{zz}L_{\delta_\ell} + I_{xz}N_{\delta_\ell})\Delta\delta_\ell$$
$$+ (I_{zz}L_{\delta_n} + I_{xz}N_{\delta_n})\Delta\delta_n]$$

Substitute:

$\frac{\partial \vec{}}{\partial \downarrow}\bigg|_{Ref}$	L	N
v	$\left(\frac{\bar{q}_{Ref}Sb}{V_{Ref}}\right)C_{\ell_\beta}$	$\left(\frac{\bar{q}_{Ref}Sb}{V_{Ref}}\right)C_{n_\beta}$
p	$\left(\frac{\bar{q}_{Ref}Sb^2}{2V_{Ref}}\right)C_{\ell_p}$	$\left(\frac{\bar{q}_{Ref}Sb^2}{2V_{Ref}}\right)C_{n_p}$
r	$\left(\frac{\bar{q}_{Ref}Sb^2}{2V_{Ref}}\right)C_{\ell_r}$	$\left(\frac{\bar{q}_{Ref}Sb^2}{2V_{Ref}}\right)C_{n_r}$
δ_ℓ	$(\bar{q}_{Ref}Sb)C_{\ell_{\delta_\ell}}$	$(\bar{q}_{Ref}Sb)C_{n_{\delta_\ell}}$
δ_n	$(\bar{q}_{Ref}Sb)C_{\ell_{\delta_n}}$	$(\bar{q}_{Ref}Sb)C_{n_{\delta_n}}$

$$\dot{p} = \frac{\bar{q}_{Ref}Sb}{I_D}\left[\left(I_{zz}C_{\ell_\beta} + I_{xz}C_{n_\beta}\right)\frac{v}{V_{Ref}}\right.$$
$$+ (I_{zz}C_{\ell_p} + I_{xz}C_{n_p})\frac{pb}{2V_{Ref}}$$
$$+ (I_{zz}C_{\ell_r} + I_{xz}C_{n_r})\frac{rb}{2V_{Ref}}$$
$$+ (I_{zz}C_{\ell_{\delta_\ell}} + I_{xz}C_{n_{\delta_\ell}})\Delta\delta_\ell$$
$$\left.+ \left(I_{zz}C_{\ell_{\delta_n}} + I_{xz}C_{n_{\delta_n}}\right)\Delta\delta_n\right]$$

For small changes in v, $v/V_{Ref} \approx \beta$. Also, $pb/2V_{Ref} = \hat{p}$ and $rb/2V_{Ref} = \hat{r}$. Make substitutions $I_{xx} = \hat{I}_{xx}\rho S(b/2)^3$, $I_{zz} = \hat{I}_{zz}\rho S(b/2)^3$, and $I_{xz} = \hat{I}_{xz}\rho S(b/2)^3$, so that

$$I_D = I_{xx}I_{zz} - I_{xz}^2 = [\rho S(b/2)^3]^2(\hat{I}_{xx}\hat{I}_{zz} - \hat{I}_{xz}^2) \equiv [\rho S(b/2)^3]^2 \hat{I}_D$$

Now evaluate the $\dot{p}$ equation with appropriate substitutions:

$$\dot{p} = \frac{(\bar{q}_{Ref}Sb)}{\hat{I}_D[\rho S(b/2)^3]}\left[\left(\hat{I}_{zz}C_{\ell_\beta} + \hat{I}_{xz}C_{n_\beta}\right)\beta\right.$$
$$+ (\hat{I}_{zz}C_{\ell_p} + \hat{I}_{xz}C_{n_p})\hat{p}$$
$$+ (\hat{I}_{zz}C_{\ell_r} + \hat{I}_{xz}C_{n_r})\hat{r}$$
$$+ (\hat{I}_{zz}C_{\ell_{\delta_\ell}} + \hat{I}_{xz}C_{n_{\delta_\ell}})\Delta\delta_\ell$$
$$\left.+ \left(\hat{I}_{zz}C_{\ell_{\delta_n}} + \hat{I}_{xz}C_{n_{\delta_n}}\right)\Delta\delta_n\right]$$

The factor on the right side simplifies:
$$\frac{\bar{q}_{Ref} S b}{\rho S (b/2)^3} = \frac{V_{Ref}^2}{(b/2)^2}$$

When this factor is inverted and multiplied, the left-hand side becomes:
$$\frac{(b/2)^2}{V_{Ref}^2}\dot{p} = \left(\frac{b}{\bar{c}}\right)\left(\frac{\bar{c}}{2V_{Ref}}\frac{d}{dt}\right)\left(\frac{pb}{2V_{Ref}}\right) = AD\hat{p}$$

Therefore the completely non-dimensional rolling moment equation is
$$D\hat{p} = \frac{1}{A\hat{I}_D}\left[\left(\hat{I}_{zz}C_{\ell_\beta} + \hat{I}_{xz}C_{n_\beta}\right)\beta\right.$$
$$+ \left(\hat{I}_{zz}C_{\ell_p} + \hat{I}_{xz}C_{n_p}\right)\hat{p}$$
$$+ \left(\hat{I}_{zz}C_{\ell_r} + \hat{I}_{xz}C_{n_r}\right)\hat{r}$$
$$+ \left(\hat{I}_{zz}C_{\ell_{\delta_\ell}} + \hat{I}_{xz}C_{n_{\delta_\ell}}\right)\Delta\delta_\ell$$
$$\left. + \left(\hat{I}_{zz}C_{\ell_{\delta_n}} + \hat{I}_{xz}C_{n_{\delta_n}}\right)\Delta\delta_n\right]$$

B.3 Body Axis Z-Force and Thrust Derivatives

The required dependencies for Z come from $Z = C_Z \bar{q} S$. Using Equation 6.1 we have for C_Z:
$$C_Z = -C_D(M, \alpha, \delta_m)\sin\alpha\sec\beta$$
$$-C_Y(\beta, \hat{p}, \hat{r}, \delta_n)\sin\alpha\tan\beta$$
$$-C_L(M, \alpha, \hat{\dot{\alpha}}, \hat{q}, \delta_m)\cos\alpha$$

Therefore,
$$Z_u = \left.\frac{\partial\left(\bar{q}SC_Z\right)}{\partial u}\right|_{Ref}$$
$$= SC_{Z_{Ref}}\left.\frac{\partial\bar{q}}{\partial u}\right|_{Ref} + S\bar{q}_{Ref}\left.\frac{\partial C_Z}{\partial u}\right|_{Ref}$$
$$= \rho_{Ref} V_{Ref} SC_{Z_{Ref}} + S\bar{q}_{Ref}\left.\frac{\partial C_Z}{\partial u}\right|_{Ref}$$

In the first term,
$$C_{Z_{Ref}} = -C_{D_{Ref}}\sin\alpha_{Ref}\sec\beta_{Ref} - C_{Y_{Ref}}\sin\alpha_{Ref}\tan\beta_{Ref} - C_{L_{Ref}}\cos\alpha_{Ref}$$
$$C_{Z_{Ref}} = -C_{L_{Ref}}$$

Therefore,
$$\rho_{Ref} V_{Ref} SC_{Z_{Ref}} = -2\left(\frac{\rho_{Ref} V_{Ref}^2 S}{2V_{Ref}}\right) C_{L_{Ref}} = -\left(\frac{\bar{q}S}{V_{Ref}}\right)(2C_{L_{Ref}})$$

In order to evaluate $\partial C_Z/\partial u|_{Ref}$ consider each term in
$$C_Z = -C_D \sin\alpha \sec\beta - C_Y \sin\alpha \tan\beta - C_L \cos\alpha$$

None of the derivatives in the first two terms will survive the application of reference conditions because $\alpha_{Ref} = 0$ or because (as we have shown in B.1) both $\partial\alpha/\partial u|_{Ref}$ and $\partial\beta/\partial u|_{Ref}$ are zero. For similar reasons we need not bother taking derivatives with respect to α in the last term. We are therefore left with

$$\left.\frac{\partial C_Z}{\partial u}\right|_{Ref} = -\cos\alpha_{Ref}\left[\frac{\partial C_L}{\partial M}\frac{\partial M}{\partial V}\frac{\partial V}{\partial u} + \frac{\partial C_L}{\partial \hat{\dot{\alpha}}}\frac{\partial \hat{\dot{\alpha}}}{\partial \dot{\alpha}}\frac{\partial \dot{\alpha}}{\partial u}\right]_{Ref}$$

$$= -\frac{1}{a_{Ref}}C_{LM}$$

$$= -\frac{M_{Ref}}{V_{Ref}}C_{LM}$$

In this expression we have used $\partial\dot{\alpha}/\partial u|_{Ref} = 0$, and a_{Ref} is the speed of sound at V_{Ref} at the specified altitude. As a result,

$$Z_u = -\frac{\bar{q}_{Ref} S}{V_{Ref}}(2C_{L_{Ref}} + M_{Ref}C_{LM})$$

Derivation of Z_v and Z_w are left as exercises. Arguments similar to those used in deriving Z_u may be used to show that

$$Z_v = 0$$

$$Z_w = -\frac{\bar{q}_{Ref} S}{V_{Ref}}(C_{D_{Ref}} + C_{L_\alpha})$$

The thrust derivatives can be a little tricky to get into the desired form. For the u derivative, we have

$$\left.\frac{\partial T}{\partial u}\right|_{Ref} = \left.\frac{\partial(\bar{q}SC_T)}{\partial u}\right|_{Ref}$$

$$= SC_{T_{Ref}}\left.\frac{\partial\bar{q}}{\partial u}\right|_{Ref} + S\bar{q}_{Ref}\left.\frac{\partial C_T}{\partial u}\right|_{Ref}$$

$$= \rho_{Ref} V_{Ref} SC_{T_{Ref}} + \bar{q}_{Ref} S\left.\frac{\partial C_T}{\partial \hat{V}}\frac{\partial \hat{V}}{\partial V}\frac{\partial V}{\partial u}\right|_{Ref}$$

$$= \frac{\bar{q}_{Ref} S}{V_{Ref}}(2C_{T_{Ref}} + C_{T_V})$$

This is a perfectly correct result, except that we normally will be given C_W, $C_{L_{Ref}}$, and $C_{D_{Ref}}$, but not $C_{T_{Ref}}$. Also, the thrust term appears twice in the body-axis force equations, as $X + T\cos\epsilon_T$ and $Z + T\sin\epsilon_T$. The needed relationships come from the wind-axis force equations. Since we are working on the Z-force equation, we want the wind-axis force equation that contains the lift L. For steady, straight flight this requires

$$L_{Ref} - T_{Ref}\sin(\epsilon_T - \alpha_{Ref}) = mg\cos\mu_{Ref}\cos\gamma_{Ref}$$

$$L_{Ref} - T_{Ref}\sin\epsilon_T = mg\cos\gamma_{Ref}$$

Dividing by $\bar{q}S$ and rearranging yields

$$C_{T_{Ref}}\sin\epsilon_T = C_{L_{Ref}} - C_W\cos\gamma_{Ref}$$

Now, using the results for Z_u and T_u to evaluate $Z_u + T_u\sin\epsilon_T$,

$$Z_u + T_u\sin\epsilon_T = \frac{\bar{q}_{Ref}S}{V_{Ref}}\left[\left(-2C_{L_{Ref}} - M_{Ref}C_{L_M}\right) + \left(2C_{T_{Ref}} + C_{T_V}\right)\sin\epsilon_T\right]$$

$$= \frac{\bar{q}_{Ref}S}{V_{Ref}}\left(-M_{Ref}C_{L_M} - 2C_W\cos\gamma_{Ref} + C_{T_V}\sin\epsilon_T\right) \quad (B.1)$$

Of course, if the engine is a rocket or jet, $T_u = 0$ and the previous result for Z_u is used. The last three thrust derivatives are straightforward,

$$T_v = 0$$
$$T_w = 0$$
$$T_{\delta_T} = \bar{q}_{Ref}SC_{T_{\delta_T}}$$

B.4 Non-dimensionalization of the Z-Force Equation

Dimensional form:

$$(m - Z_{\dot{w}})\dot{w} = (Z_u + T_u\sin\epsilon_T)\Delta u + Z_w w + (Z_q + mV_{Ref})q$$
$$-mg\sin\gamma_{Ref}\Delta\theta + T_{\delta_T}\sin\epsilon_T\Delta\delta_T + Z_{\delta_m}\Delta\delta_m$$

Substitute (and using Equation B.1):

$\left.\frac{\partial\vec{\ }}{\partial\downarrow}\right|_{Ref}$	Z	T
$\dot{w}$	$-\left(\frac{\bar{q}_{Ref}S\bar{c}}{2V_{Ref}^2}\right)C_{L_{\dot{\alpha}}}$	0
u	$\left(\frac{\bar{q}_{Ref}S}{V_{Ref}}\right)(-2C_{L_{Ref}} - M_{Ref}C_{L_M})$	$\left(\frac{\bar{q}_{Ref}S}{V_{Ref}}\right)(2C_{T_{Ref}} + C_{T_V})$
w	$-\left(\frac{\bar{q}_{Ref}S}{V_{Ref}}\right)(C_{D_{Ref}} + C_{L_\alpha})$	0
q	$-\left(\frac{\bar{q}_{Ref}S\bar{c}}{2V_{Ref}}\right)C_{L_q}$	0
δ_m	$-(\bar{q}_{Ref}S)C_{L_{\delta_m}}$	0
δ_T	0	$(\bar{q}_{Ref}S)C_{T_{\delta_T}}$

Linearization

$$\left[m + \left(\frac{\bar{q}_{Ref}S\bar{c}}{2V_{Ref}^2}\right)C_{L\dot{\alpha}}\right]\dot{w} = \frac{\bar{q}_{Ref}S}{V_{Ref}}(-M_{Ref}C_{L_M} - 2C_W\cos\gamma_{Ref} + C_{T_V}\sin\epsilon_T)\Delta u$$
$$-\left(\frac{\bar{q}_{Ref}S}{V_{Ref}}\right)(C_{D_{Ref}} + C_{L\alpha})w - \left[\left(\frac{\bar{q}_{Ref}S\bar{c}}{2V_{Ref}}\right)C_{L_q} - mV_{Ref}\right]q - mg\sin\gamma_{Ref}\Delta\theta$$
$$+ \sin\epsilon_T(\bar{q}_{Ref}S)C_{T_{\delta_T}}\Delta\delta_T - (\bar{q}_{Ref}S)C_{L_{\delta_m}}\Delta\delta_m$$

Factor out $\bar{q}_{Ref}S$ from both sides and rearrange:

$$\left[\frac{m}{\bar{q}_{Ref}S} + \left(\frac{\bar{c}}{2V_{Ref}^2}\right)C_{L\dot{\alpha}}\right]\dot{w} = (-M_{Ref}C_{L_M} - 2C_W\cos\gamma_{Ref} + C_{T_V}\sin\epsilon_T)\frac{\Delta u}{V_{Ref}}$$
$$-(C_{D_{Ref}} + C_{L\alpha})\frac{w}{V_{Ref}} - \left[C_{L_q} - \frac{2mV_{Ref}^2}{\bar{q}S\bar{c}}\right]\frac{\bar{q}\bar{c}}{2V_{Ref}} - \frac{mg}{\bar{q}_{Ref}S}\sin\gamma_{Ref}\Delta\theta$$
$$+ \sin\epsilon_T C_{T_{\delta_T}}\Delta\delta_T - C_{L_{\delta_m}}\Delta\delta_m$$

On the left-hand side:

$$\left[\frac{m}{\bar{q}_{Ref}S} + \left(\frac{\bar{c}}{2V_{Ref}^2}\right)C_{L\dot{\alpha}}\right]\dot{w} = \left(\frac{2V_{Ref}^2 m}{\bar{q}_{Ref}S\bar{c}} + C_{L\dot{\alpha}}\right)\frac{\bar{c}\dot{w}}{2V_{Ref}^2}$$
$$= \left(\frac{4m}{\rho_{Ref}S\bar{c}} + C_{L\dot{\alpha}}\right)\left(\frac{\bar{c}}{2V_{Ref}}\frac{d}{dt}\right)\frac{w}{V_{Ref}}$$
$$= (2\hat{m} + C_{L\dot{\alpha}})D\alpha$$

The mV_{Ref} term on the right-hand side has become

$$\frac{2mV_{Ref}^2}{\bar{q}S\bar{c}} = 2\hat{m}$$

$$(2\hat{m} + C_{L\dot{\alpha}})D\alpha = (-M_{Ref}C_{L_M} - 2C_W\cos\gamma_{Ref} + C_{T_V}\sin\epsilon_T)\Delta\hat{V}$$
$$-(C_{D_{Ref}} + C_{L\alpha})\alpha + (2\hat{m} - C_{L_q})\hat{q} - C_W\sin\gamma_{Ref}\Delta\theta$$
$$+ \sin\epsilon_T C_{T_{\delta_T}}\Delta\delta_T - C_{L_{\delta_m}}\Delta\delta_m$$

Appendix C

Derivation of Euler Parameters[1]

We begin the derivation of Euler parameters with a discussion of further properties of the direction cosine matrix, its eigenvalues and eigenvectors. For now we drop the subscripts and take a generic matrix T that transforms representations of vectors in one orthogonal system to their representations in another. The thought of actually expanding the determinant $|\lambda I - T|$ using the direction cosine or Euler angle representations, and then solving $|\lambda I - T| = 0$ for $\lambda_i, i = 1 \ldots 3$, is daunting. Instead we will take an indirect approach. First, a few facts from linear algebra:

1. The determinant of the product of two matrices is the product of their determinants, or $|AB| = |A||B|$.
2. The determinant of the transpose of a matrix is the same as the determinant of the matrix, or $|A^T| = |A|$.
3. The determinant of the negative of a matrix is either the same as the determinant of the matrix if it is of even order, or the negative of the determinant of the matrix if it is of odd order, or $|-A| = (-1)^n |A|$ where A is $n x n$.
4. The sum of the transpose of two matrices is the transpose of their sum, or $A^T + B^T = [A + B]^T$.
5. The determinant of a matrix is equal to the product of its eigenvalues. For a $3x3$ matrix A, $|A| = \lambda_1 \lambda_2 \lambda_3$, where $\lambda_i, i = 1 \ldots 3$ satisfy $|\lambda_i I_3 - A| = 0$.
6. The trace of a matrix (sum of the terms on its principal diagonal) is equal to the sum of its eigenvalues. For the $3x3$ matrix A, $trace(A) = a_{11} + a_{22} + a_{33} = \lambda_1 + \lambda_2 + \lambda_3$.
7. Complex eigenvalues of real matrices occur only in complex conjugate pairs.

Consider now the expression $[T - I_3]T^T$. Expanding,

$$[T - I_3]T^T = [TT^T - T^T] = [I_3 - T^T]$$

Taking determinants of the left and right sides,

$$|T - I_3||T^T| = |I_3 - T^T|$$

[1] Adapted from the notes of Frederick H. Lutze.

We use the fact that $|T^T| = |T| = 1$ to arrive at

$$|T - I_3| = |I_3 - T^T| = |I_3^T - T^T| = |I_3 - T|^T = |I_3 - T|$$

With $|T - I_3| = |I_3 - T|$ we proceed as

$$\begin{aligned}|T - I_3| &= |I_3 - T| \\ &= |-(T - I_3)| \\ &= (-1)^n |T - I_3| \\ &= (-1)^3 |T - I_3| \\ &= -|T - I_3|\end{aligned}$$

The conclusion is that $|T - I_3| = -|T - I_3|$. The only number equal to its negative is zero, so we must have $|T - I_3| = |I_3 - T| = 0$. With $|I_3 - T| = 0$ we conclude that one of the eigenvalues of T is 1. Therefore let $\lambda_3 = 1$.

While we're at it let's learn about λ_1 and λ_2. We know that $\lambda_3 = 1$ and $|T| = \lambda_1 \lambda_2 \lambda_3$, so $\lambda_1 \lambda_2 = 1$. First, if λ_1 is complex then λ_2 is its complex conjugate, or $\lambda_2 = \lambda_1^*$ and $\lambda_1 \lambda_1^* = 1$. For complex numbers the product $\lambda_1 \lambda_1^*$ is the length of λ_1, so $\lambda_1 \lambda_1^* = |\lambda_1| = |\lambda_2| = 1$. We may therefore write $\lambda_1 = \cos \eta + j \sin \eta$, $\lambda_2 = \cos \eta - j \sin \eta$, or in polar form, $\lambda_1 = e^{j\eta}$ and $\lambda_2 = e^{-j\eta}$.

If λ_1 is real, then λ_2 is real as well. Using $trace(T) = \lambda_1 + \lambda_2 + \lambda_3$, we first note that none of the entries in T is greater than 1 or less than -1 (they are cosines of angles). We must have $-2 \leq \lambda_1 + \lambda_2 \leq 2$. Also $\lambda_1 \lambda_2 = 1$ so $\lambda_2 = 1/\lambda_1$. Therefore $-2 \leq \lambda_1 + 1/\lambda_1 \leq 2$. If λ_1 is positive then $-2\lambda_1 \leq \lambda_1^2 + 1 \leq 2\lambda_1$, and we must simultaneously satisfy $0 \leq \lambda_1^2 + 2\lambda_1 + 1$ and $\lambda_1^2 - 2\lambda_1 + 1 \geq 0$. It is easily shown that $0 \leq \lambda_1^2 + 2\lambda_1 + 1$ and $\lambda_1^2 - 2\lambda_1 + 1 \geq 0$ for any real λ_1, and that $\lambda_1^2 - 2\lambda_1 + 1 = 0$ only at $\lambda_1 = 1$ (and $\lambda_2 = 1$). If λ_1 is negative then it may be shown using similar arguments that $\lambda_1 = -1$ (and $\lambda_2 = -1$). These two cases are, of course, special instances of $\lambda_1 = \cos \eta - j \sin \eta$ with $\eta = 0$ and $\eta = 180 \deg$. In summary, we have shown that the eigenvalues of a direction cosine matrix are:

$$\begin{aligned}\lambda_1 &= \cos \eta + j \sin \eta \\ \lambda_2 &= \cos \eta - j \sin \eta \\ \lambda_3 &= 1\end{aligned}$$

Getting back to Euler parameters, we recall that the definition of an eigenvector $\mathbf{v}$ of T requires that $\lambda_i \mathbf{v}_{\lambda_i} = T \mathbf{v}_{\lambda_i}$, $\mathbf{v}_{\lambda_i} \neq \mathbf{0}$. Therefore, for $\lambda_3 = 1$ we have $\mathbf{v}_{\lambda_3} = T \mathbf{v}_{\lambda_3}$. This means there is a vector $\mathbf{v}_{\lambda_3}$ that is invariant under the transformation T. If the transformation T is $T_{2,1}$ from F_1 to F_2 then $\mathbf{v}_{\lambda_3}$ has exactly the same components in both F_1 and F_2. This is possible only if F_1 and F_2 differ by a rotation about a common axis in the same (or opposite) direction as $\mathbf{v}_{\lambda_3}$. Figure C.1 shows this rotation. The invariant vector defines the *eigenaxis* and a unit vector in the direction of $\mathbf{v}_{\lambda_3}$ is denoted $\mathbf{e}_\eta$. By now one is probably not surprised to learn that the angle through which one must rotate F_1 about $\mathbf{e}_\eta$ to align the axes with F_2 is the same angle η that define the other two eigenvalues λ_1 and λ_2.

We now proceed to find the matrix $T_{2,1}$ based on the axis $\mathbf{e}_\eta$ and angle η. The overall plan is to find out how an arbitrary vector that is fixed relative to $\mathbf{e}_\eta$ (and rotates through η) survives the rotation, and apply that result to each of $\mathbf{i}_1, \mathbf{j}_1$, and $\mathbf{k}_1$ to see how they become

Derivation of Euler Parameters

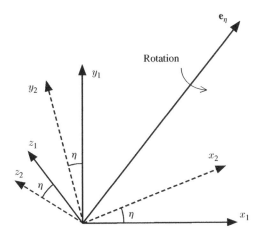

Figure C.1 Single rotation about $\mathbf{e}_\eta$ through angle η from F_1 to F_2.

$\mathbf{i}_2$, $\mathbf{j}_2$, and $\mathbf{k}_2$. The picture we have in mind is of a rotation of F_1 about $\mathbf{e}_\eta$ so that the final orientation is F_2.

In F_1 we may represent $\mathbf{e}_\eta$ by its direction cosines ξ, ζ, and χ as

$$\mathbf{e}_{\eta_1} = \xi \mathbf{i}_1 + \zeta \mathbf{j}_1 + \chi \mathbf{k}_1$$

In F_2 the direction cosines of $\mathbf{e}_\eta$ are the same, so

$$\mathbf{e}_{\eta_2} = \xi \mathbf{i}_2 + \zeta \mathbf{j}_2 + \chi \mathbf{k}_2$$

Because they are direction cosines, $\xi^2 + \zeta^2 + \chi^2 = 1$.

Now consider the effect of the rotation on an arbitrary vector fixed with respect to $\mathbf{e}_\eta$ at an angle ϕ (Figure C.2). We will denote this vector $\mathbf{u}$ before the rotation, and afterward, $\mathbf{v}$. The angle ϕ is of course unchanged as the tip of the aribitrary vector rotates in a circular arc through angle η about $\mathbf{e}_\eta$.

We now establish a local coordinate system that rotates with the vector (Figure C.3). In this coordinate system unit vector $\mathbf{e}_x$ is in the radial direction outwards, and $\mathbf{e}_y$ is tangential in the direction of positive rotation. The orthogonal coordinate system is completed by $\mathbf{e}_\eta$.

The coordinate system F_o is rotated into the coordinate system F_f in a manner completely analogous to our θ_z rotation in the discussion of Euler angles. Without further ado we may write

$$\{\mathbf{w}\}_f = \begin{bmatrix} \cos\eta & \sin\eta & 0 \\ -\sin\eta & \cos\eta & 0 \\ 0 & 0 & 1 \end{bmatrix} \{\mathbf{w}\}_o$$

or

$$\{\mathbf{w}\}_o = \begin{bmatrix} \cos\eta & -\sin\eta & 0 \\ \sin\eta & \cos\eta & 0 \\ 0 & 0 & 1 \end{bmatrix} \{\mathbf{w}\}_f$$

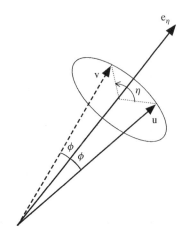

Figure C.2 Vector fixed relative to e_η.

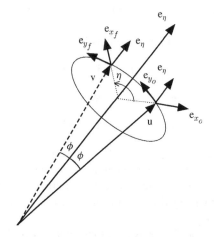

Figure C.3 Coordinate systems F_o and F_f.

Here we have used **w** to denote any vector, recalling that **u** and **v** are different vectors and that each may be represented in either F_o or F_f. For our purposes **w** may be either **u** or **v**. We wish to formulate $\mathbf{v}_f$ in terms of its length v and the angle ϕ, and then use the transformation to find $\mathbf{v}_o$. Examining the geometry of **v** in F_f (Figure C.4), we may write down $\mathbf{v}_f$:

$$\{\mathbf{v}\}_f = \begin{Bmatrix} v \sin\phi \\ 0 \\ v \cos\phi \end{Bmatrix}$$

Derivation of Euler Parameters

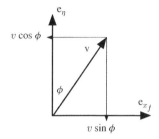

Figure C.4 Geometry of v in F_f.

Therefore,

$$\{v\}_o = \begin{bmatrix} \cos\eta & -\sin\eta & 0 \\ \sin\eta & \cos\eta & 0 \\ 0 & 0 & 1 \end{bmatrix} \{v\}_f$$

$$= \begin{bmatrix} \cos\eta & -\sin\eta & 0 \\ \sin\eta & \cos\eta & 0 \\ 0 & 0 & 1 \end{bmatrix} \begin{Bmatrix} v\sin\phi \\ 0 \\ v\cos\phi \end{Bmatrix}$$

$$= \begin{Bmatrix} v\sin\phi\cos\eta \\ v\sin\phi\sin\eta \\ v\cos\phi \end{Bmatrix}$$

In component form this is:

$$\{v\}_o = v\sin\phi\cos\eta\, \mathbf{e}_{x_o} + v\sin\phi\sin\eta\, \mathbf{e}_{y_o} + v\cos\phi\, \mathbf{e}_\eta$$

Since this vector we are rotating around is to be arbitrary, the result should not depend on the angle ϕ or the orientation of F_o. We can eliminate these dependencies by applying some vector algebra. First note that, with respect to the vector $\mathbf{u}$, the vector cross product

$$\mathbf{e}_\eta \times \mathbf{u} = v\sin\phi\, \mathbf{e}_{y_o}$$

So the second term in the expression for $\{v\}_o$ becomes

$$v\sin\phi\sin\eta\, \mathbf{e}_{y_o} = (\mathbf{e}_\eta \times \mathbf{u})\sin\eta$$

Also, $\mathbf{e}_{x_o} = \mathbf{e}_{y_o} \times \mathbf{e}_\eta$ and we can solve $\mathbf{e}_\eta \times \mathbf{u} = v\sin\phi\, \mathbf{e}_{y_o}$ for $\mathbf{e}_{y_o}$ and substitute:

$$\mathbf{e}_{x_o} = \frac{(\mathbf{e}_\eta \times \mathbf{u}) \times \mathbf{e}_\eta}{v\sin\phi}$$

Expanding the vector triple product yields

$$\mathbf{e}_{x_o} = \frac{\mathbf{u}(\mathbf{e}_\eta \cdot \mathbf{e}_\eta) - (\mathbf{e}_\eta \cdot \mathbf{u})\mathbf{e}_\eta}{v\sin\phi} = \frac{\mathbf{u} - (\mathbf{e}_\eta \cdot \mathbf{u})\mathbf{e}_\eta}{v\sin\phi}$$

Using this relationship the first term in the expression for $\{\mathbf{v}\}_o$ becomes

$$v \sin \phi \cos \eta \mathbf{e}_{x_o} = v \sin \phi \cos \eta \left(\frac{\mathbf{u} - (\mathbf{e}_\eta \cdot \mathbf{u}) \mathbf{e}_\eta}{v \sin \phi} \right)$$

$$= \cos \eta \mathbf{u} - \cos \eta (\mathbf{e}_\eta \cdot \mathbf{u}) \mathbf{e}_\eta$$

Collecting terms we may write the final result,

$$\mathbf{v} = \cos \eta \mathbf{u} + (\mathbf{e}_\eta \cdot \mathbf{u})(1 - \cos \eta) \mathbf{e}_\eta + (\mathbf{e}_\eta \times \mathbf{u}) \sin \eta$$

Now, given the eigenaxis $\mathbf{e}_\eta$, angle η, and any arbitrary vector $\mathbf{u}$ fixed with respect to $\mathbf{e}_\eta$, we may use this equation to find out what happens to the vector after the rotation. Let us apply this in turn to the vectors $\mathbf{i}_1$, $\mathbf{j}_1$, and $\mathbf{k}_1$. After these vectors are rotated they will be aligned with $\mathbf{i}_2$, $\mathbf{j}_2$, and $\mathbf{k}_2$. So first we take $\mathbf{u} = \mathbf{i}_1$ and $\mathbf{v} = \mathbf{i}_2$:

$$\mathbf{i}_2 = \cos \eta \mathbf{i}_1 + (\mathbf{e}_\eta \cdot \mathbf{i}_1)(1 - \cos \eta) \mathbf{e}_\eta + (\mathbf{e}_\eta \times \mathbf{i}_1) \sin \eta$$

Into this we insert $\mathbf{e}_\eta = \xi \mathbf{i}_1 + \zeta \mathbf{j}_1 + \chi \mathbf{k}_1$ so that $(\mathbf{e}_\eta \times \mathbf{i}_1) = \chi \mathbf{j}_1 + \zeta \mathbf{k}_1$ and $(\mathbf{e}_\eta \cdot \mathbf{i}_1) = \xi$. After collecting and arranging terms we arrive at

$$\mathbf{i}_2 = [\cos \eta + \xi^2 (1 - \cos \eta)] \mathbf{i}_1 \\ + [\xi \zeta (1 - \cos \eta) + \chi \sin \eta] \mathbf{j}_1 \\ + [\xi \chi (1 - \cos \eta) + \zeta \sin \eta] \mathbf{k}_1$$

This is the first row of our direction cosine matrix $T_{2,1}$. The second and third rows may be obtained in a similar fashion. A few trigonometric identities and definitions will clean up the expressions. We may express the trigonometric functions in terms of half-angles by using common trigonometric identities:

$$\cos \eta = 2 \cos^2(\eta/2) - 1$$

$$1 - \cos \eta = 2 \sin^2(\eta/2)$$

$$\sin \eta = 2 \sin(\eta/2) \cos(\eta/2)$$

We also define the four *Euler parameters*

$$q_0 \doteq \cos(\eta/2) \\ q_1 \doteq \xi \sin(\eta/2) \\ q_2 \doteq \zeta \sin(\eta/2) \\ q_3 \doteq \chi \sin(\eta/2)$$

Note that $q_0^2 + q_1^2 + q_2^2 + q_3^2 = 1$. We may now tidy up the expressions in $T_{2,1}$ considerably. For instance, the first term in the equation for $\mathbf{i}_2$ above (same as t_{11}) becomes:

$$\cos \eta + \xi^2(1 - \cos \eta) = [2\cos^2(\eta/2) - 1] + \xi^2[2\sin^2(\eta/2) - 1] \\ = 2q_0^2 - 1 + 2q_1^2 \\ = 2q_0^2 - (q_0^2 + q_1^2 + q_2^2 + q_3^2) + 2q_1^2 \\ = q_0^2 + q_1^2 - q_2^2 - q_3^2$$

The finished result is

$$T_{2,1} = \begin{bmatrix} (q_0^2 + q_1^2 - q_2^2 - q_3^2) & 2(q_1 q_2 + q_0 q_3) & 2(q_1 q_3 - q_0 q_2) \\ 2(q_1 q_2 - q_0 q_3) & (q_0^2 - q_1^2 + q_2^2 - q_3^2) & 2(q_2 q_3 + q_0 q_1) \\ 2(q_1 q_3 + q_0 q_2) & 2(q_2 q_3 - q_0 q_1) & (q_0^2 - q_1^2 - q_2^2 + q_3^2) \end{bmatrix}$$

Appendix D

Fedeeva's Algorithm

Fedeeva's algorithm is a recursive method for calculating $[sI - A]^{-1}$. Recall that

$$[sI - A]^{-1} = \frac{C(s)}{d(s)} = \frac{Adj[sI - A]}{d(s)}$$

The numerator $C(s)$ is the *adjoint* of $[sI - A]$, and is a matrix of polynomials in s with powers of s ranging from zero to $n - 1$, and may be expressed as the sum of matrices,

$$C(s) = C_1 s^{n-1} + C_2 s^{n-2} + C_3 s^{n-3} + \cdots + C_{n-2} s^2 + C_{n-1} s + C_n$$

The denominator is the characteristic polynomial,

$$d(s) = s^n + d_1 s^{n-1} + d_2 s^{n-2} + d_3 s^{n-3} + \cdots + d_{n-2} s^2 + d_{n-1} s + d_n$$

Note that

$$[sI - A]^{-1}[sI - A] = \frac{Adj[sI - A]}{|sI - A|}[sI - A] = I_n$$

$$Adj[sI - A] \cdot [sI - A] = |sI - A| I_n$$

We now expand both sides of this expression. The left-hand side is

$$[C_1 s^{n-1} + C_2 s^{n-2} + C_3 s^{n-3} + \cdots + C_{n-2} s^2 + C_{n-1} s + C_n]$$
$$= [C_1 s^n + C_2 s^{n-1} + C_3 s^{n-2} + \cdots + C_{n-2} s^3 + C_{n-1} s^2 + C_n s]$$
$$- [C_1 A s^{n-1} + C_2 A s^{n-2} + C_3 A s^{n-3} + \cdots + C_{n-2} A s^2 + C_{n-1} A s + C_n A]$$

The right-hand side,

$$|sI - A| I_n = I_n s^n + d_1 I_n s^{n-1} + d_2 I_n s^{n-2} + \cdots + d_{n-2} I_n s^2 + d_{n-1} I_n s + d_n I_n$$

Reassembled and grouped,

$$|sI - A| I_n = I_n s^n + d_1 I_n s^{n-1} + d_2 I_n s^{n-2} + \cdots + d_{n-2} I_n s^2 + d_{n-1} I_n s + d_n I_n$$
$$= C_1 s^n + [C_2 - C_1 A] s^{n-1} + [C_3 - C_2 A] s^{n-2} + \cdots$$
$$\cdots + [C_{n-1} - C_{n-2} A] s^2 + [C_n - C_{n-1} A] s^2 - C_n A$$

We now equate like powers of s:

$$C_1 s^n = I_n s^n \quad \Rightarrow C_1 = I_n$$
$$[C_2 - C_1 A]s^{n-1} = d_1 I_n s^{n-1} \quad \Rightarrow [C_2 - C_1 A] = d_1 I_n$$
$$[C_3 - C_2 A]s^{n-2} = d_2 I_n s^{n-2} \quad \Rightarrow [C_3 - C_2 A] = d_2 I_n$$
$$\vdots$$
$$[C_{n-1} - C_{n-2} A]s^2 = d_{n-2} I_n s^2 \Rightarrow [C_{n-1} - C_{n-2} A] = d_{n-2} I_n$$
$$[C_n - C_{n-1} A]s^2 = d_{n-1} I_n s \quad \Rightarrow [C_n - C_{n-1} A] = d_{n-1} I_n$$
$$[-C_n A] = d_n I_n$$

That is, if we know the coefficients $d_1 \ldots d_n$, we have the recursive formula:

$$C_1 = I_n$$
$$C_2 = C_1 A + d_1 I_n$$
$$C_3 = C_2 A + d_2 I_n$$
$$\vdots$$
$$C_{n-1} = C_{n-2} A + d_{n-1} I_n$$
$$C_n = C_{n-1} A + d_{n-1} I_n$$
$$0 = C_n A + d_n I_n$$

Note that this can also be written as

$$C_1 = I_n$$
$$C_2 = A + d_1 I_n$$
$$C_3 = A^2 + d_1 A + d_2 I_n$$
$$\vdots$$
$$C_k = A^{k-1} + d_1 A^{k-2} + \cdots + d_{k-2} A + d_{k-1} I_n$$
$$\vdots$$
$$C_{n-1} = A^{n-2} + d_1 A^{n-3} + \cdots + d_{n-3} A + d_{n-2} I_n$$
$$C_n = A^{n-1} + d_1 A^{n-2} + \cdots + d_{n-2} A + d_{n-1} I_n$$
$$0 = A^n + d_1 A^{n-1} + \cdots + d_{n-1} A + d_n I_n$$

The last equation in these expressions is the result of the well known *Cayley–Hamilton theorem*, that every square matrix satisfies its own characteristic equation. That is, given the characteristic equation in s,

$$s^n + d_1 s^{n-1} + d_2 s^{n-2} + d_3 s^{n-3} + \cdots + d_{n-2} s^2 + d_{n-1} s + d_n = 0$$

Replacing s with A (and d_n with $d_n I_n$) yields the characteristic equation in A:

$$A^n + d_1 A^{n-1} + d_2 A^{n-2} + d_3 A^{n-3} + \cdots + d_{n-2} A^2 + d_{n-1} A + d_n I_n = 0$$

We now borrow a result from the study of eigenvalues (Pettofrezzo, 1966). The sum of the diagonal elements of a matrix A is called the *trace* of A and is denoted by $tr(A)$. Let $t_1 = tr(A)$, $t_2 = tr(A^2)$, and so on. It can be shown that the coefficients of the characteristic equation are given by the equations:

$$d_1 = -t_1$$

$$d_2 = -\frac{1}{2}(d_1 t_1 + t_2)$$

$$d_3 = -\frac{1}{3}(d_2 t_1 + d_1 t_2 + t_3)$$

$$\vdots$$

$$d_k = -\frac{1}{k}(d_{k-1} t_1 + d_{k-2} t_2 + \cdots + d_1 t_{k-1} + t_k)$$

$$\vdots$$

$$d_{n-1} = -\frac{1}{n-1}(d_{n-2} t_1 + d_{n-3} t_2 + \cdots + d_1 t_{n-2} + t_{n-1})$$

$$d_n = -\frac{1}{n}(d_{n-1} t_1 + d_{n-2} t_2 + \cdots + d_1 t_{n-1} + t_n)$$

These results can be simplified by noting

$$\begin{aligned}
d_1 &= -t_1 \\
&= -tr(A) \\
&= -tr(C_1 A) \\
d_2 &= -\tfrac{1}{2}(d_1 t_1 + t_2) \\
&= -\tfrac{1}{2} tr(d_1 A + A^2) \\
&= -\tfrac{1}{2} tr[(d_1 I_n + A)A] \\
&= -\tfrac{1}{2} tr[(d_1 I_n + C_1 A)A] \\
&= -\tfrac{1}{2} tr(C_2 A)
\end{aligned}$$

Each step works the same, and in general

$$d_k = -\frac{1}{k} tr(C_{k-1} A)$$

The algorithm is now complete. In summary,

$$[sI - A]^{-1} = \frac{C(s)}{d(s)}$$

$C(s)$ and $d(s)$ are calculated by:

$$C(s) = C_1 s^{n-1} + C_2 s^{n-2} + C_3 s^{n-3} + \cdots + C_{n-2} s^2 + C_{n-1} s + C_n$$
$$d(s) = s^n + d_1 s^{n-1} + d_2 s^{n-2} + d_3 s^{n-3} + \cdots + d_{n-2} s^2 + d_{n-1} s + d_n$$

Where:

$$\begin{aligned}
C_1 &= I_n & d_1 &= -tr(C_1 A) \\
C_2 &= C_1 A + d_1 I_n & d_2 &= -\tfrac{1}{2} tr(C_2 A) \\
C_3 &= C_2 A + d_2 I_n & d_3 &= -\tfrac{1}{3} tr(C_3 A) \\
&\vdots & & \\
C_k &= C_{k-1} A + d_{k-1} I_n & d_k &= -\tfrac{1}{k} tr(C_k A) \\
&\vdots & & \\
C_n &= C_{n-1} A + d_{n-1} I_n & d_n &= -\tfrac{1}{n} tr(C_n A) \\
0 &= C_n A + d_n I_n & &
\end{aligned}$$

Reference

Pettofrezzo, A.J. (1966) *Matrices and Transformations,* Dover Publications, Inc., p. 84.

Appendix E

MATLAB® Commands Used in the Text

E.1 Using MATLAB®

Many flight dyanamics and control calculations become very tedious to solve by hand, and some relief is needed. There are many advantages to solving these problems by writing one's own code, at least for a few small problems. A primary advantage is that one develops an appreciation for the workings of specialized software packages that solve the problem as a mysterious 'black box'. A second reason is that when things go wrong, one knows better where to look to fix them.

Some of the manual methods of solving the problems can be quite arcane. For example, there are several rules for constructing root-locus plots by hand, and they require a pretty good understanding of complex variables. Thus a classically trained flight control engineer knows that as a feedback gain is increased, the locus of the roots will depart their original location in the complex plane at an angle he can calculate and arrive at a zero at another angle that he can also calculate.

Many of these manual methods have gone the way of the slide rule, and one no longer relies on them. We accept change and introduce MATLAB®, 'a high-level language and interactive environment for numerical computation, visualization, and programming'. There are other software packages, but at this writing MATLAB® appears to be ubiquitous in engineering schools.

Our treatment will be just an introduction to MATLAB® and will use no functions not found in the student edition. None of the highly specialized toolboxes will be used. This section assumes no previous knowledge of MATLAB® and is written to ensure that all the examples in the text that use MATLAB® may be repeated.

See the index for a listing of where MATLAB® is used in the text.

Aircraft Flight Dynamics and Control, First Edition. Wayne Durham.
© 2013 John Wiley & Sons, Ltd. Published 2013 by John Wiley & Sons, Ltd.

E.2 Eigenvalues and Eigenvectors

The basic command in MATLAB® to generate eigenvalues and eigenvectors is `eig`. `[v, e] = eig(A)` of an $n \times n$ matrix A produces a diagonal matrix e of eigenvalues and a matrix v whose columns are the corresponding eigenvectors.

If a vector of eigenvalues is desired then `diag` is used. The function `diag` returns a vector of the diagonal terms if its argument is a matrix, or a diagonal matrix of the entries of the argument if its argument is a vector. See Section E.6.2 for use of `diag`.

E.3 State-Space Representation

The general formulation of dynamical systems in MATLAB® has not just $\dot{x} = Ax(t) + Bu(t)$ that we previously developed, but an output equation of the form $y(t) = Cx(t) + Du(t)$ as well:

$$\dot{x} = Ax(t) + Bu(t)$$
$$y(t) = Cx(t) + Du(t) \tag{E.1}$$

MATLAB® also allows for the matrices A, B, C, and D to be time varying as well, but we won't have need of that. In the output equation the matrix C often represents the scaling and unit conversions involved in measuring the states, such as turning degrees to radians. The matrix D often is all zeros but could arise, for example, if the output being used is the load factor, g. Because g is proportional to lift, and because the elevator or horizontal tail deflection changes the lift, there will be a non-zero entry in D.

A state-space system in MATLAB® is defined using the command `ss(A,B,C,D)` where the matrices in the argument have been suitably defined. A matrix is defined by rows in square brackets, with each row terminated by a semicolon (;). Thus,

$$A = \begin{bmatrix} 1 & 2 \\ 3 & 4 \end{bmatrix} \quad B = \begin{bmatrix} 1 \\ 2 \end{bmatrix} \quad C = \begin{bmatrix} 1 & 0 \\ 0 & 1 \end{bmatrix} \quad D = \begin{bmatrix} 0 \\ 0 \end{bmatrix}$$

is generated by

```
A = [1 2; 3 4];
B = [1; 2];
C = [1 0;0 1];
D = [0; 0];
```

The semicolon at the end of the command suppresses the echoing of the result.

The C matrix could have been defined as `C = eye(2)` which creates a 2×2 identity matrix.

To create a system named `sysSS` that MATLAB® can manipulate,

```
sysSS = ss(A,B,C,D)
```

With no semicolon at the end of this line MATLAB® helpfully tells you what you've created, labeling the rows and columns of the matrices:

```
sysSS =

  a =
          x1  x2
      x1   1   2
      x2   3   4

  b =
           u1
      x1   1
      x2   2

  c =
          x1  x2
      y1   1   0
      y2   0   1

  d =
           u1
      y1   0
      y2   0

Continuous-time state-space model.
```

Note that a state-space model internally in MATLAB® is more than just four matrices. It is actually a model of the dynamical system specified that can be manipulated in a manner that simulates the real system.

E.4 Transfer Function Representation

We previously derived the transfer function matrices for systems of the form $\dot{x} = Ax(t) + Bu(t)$. We can include the output equation by redefining $G(s)$ to be the transfer function matrix from $\mathbf{u}(s)$ to $\mathbf{x}(s)$ to be that from $\mathbf{u}(s)$ to $\mathbf{y}(s)$. These transfer function matrices were:

$$\mathbf{x}(s) = [sI - A]^{-1} B\mathbf{u}(s)$$

From $\mathbf{y}(t) = C\mathbf{x}(t) + D\mathbf{u}(t)$ we assume the matrices C and D are not time-varying and have $\mathbf{y}(s) = C\mathbf{x}(s) + D\mathbf{u}(s)$. Then,

$$\begin{aligned} \mathbf{y}(s) &= C\mathbf{x}(s) + D\mathbf{u}(s) \\ &= C[sI - A]^{-1} B\mathbf{u}(s) + D\mathbf{u}(s) \\ &= \{C[sI - A]^{-1} B + D\}\mathbf{u}(s) \end{aligned}$$

Thus we redefine $G(s)$,

$$G(s) = \{C[sI - A]^{-1} B + D\}$$

$$\mathbf{y}(s) = G(s)\mathbf{u}(s) \qquad (\text{E.2})$$

MATLAB® represents polynomials as row vectors whose entries are the coefficients of the polynomial in descending powers of the independent variable. Thus $s^3 + 2s - 4$ is [1 0 2 -4]. If two such polynomials are defined, such as num = [2 1 3] and den = [1 0 2 -4] then the MATLAB® command

```
g = tf(num,den)
```

creates the transfer function

$$g = \frac{s^2 + s + 3}{s^3 + 2s - 4}$$

Once several transfer functions have been defined, they may be assembled into a transfer function matrix like $G(s)$ defined above. Given g11 and g21 then G=[g11; g21] yields

$$G(s) = \begin{bmatrix} g_{11}(s) \\ g_{21}(s) \end{bmatrix}$$

The way we have formulated the problem we will most often begin with the elements of a state-space model. MATLAB® can convert a state-space model directly to transfer functions (and vice-versa). Instead of the two vectors, the argument of the command tf can be the name of a defined state-space system. Thus, given the matrices defined above,

```
sysSS = ss(A,B,C,D);
sysTF = tf(sysSS);
```

If you won't be needing sysSS you can nest the commands into one:

```
sysTF = tf(ss(A,B,C,D));
```

The numerical example we have created can be used to highlight a few more points. First, typing the name of a defined system will echo the properties of that system.

```
sysTF

sysTF =

   From input to output...
           s - 6.106e-16
    1:   ---------
          s^2 - 5 s - 2

           2 s + 1
    2:   ---------
          s^2 - 5 s - 2
```

MATLAB® Commands Used in the Text   277

Second, the appearance of the number 6.106e-16 in the numerator results from the finite precision representation of numbers used internally to a computer by MATLAB®. It is easy to verify that the numerator of the transfer function $x_1(s)/u(s)$ should be simply s. The number produced in this position by MATLAB® may be different on different computers. The presence of the small but non-zero term may introduce problems in some applications. If necessary, the problem can be fixed in any of a number of ways. The most straightforward is to simply replace the transfer function with the correct one, that is,

```
sysTF(1)=tf([1 0], [1 -5 -2])
```

E.5 Root Locus

The root-locus function `rlocus` is called using a suitably defined system as its primary argument. The system may be a state-space or transfer-function representation. A second argument that defines a vector of feedback gains is often useful. Some points to keep in mind are:

- `rlocus` operates on a single-input single-output system. If one has defined a matrix of transfer functions then the proper one in that matrix must be specified. A few examples of this are shown in Section E.6.3.
- `rlocus` assumes negative feedback, which can sometimes be a problem if the transfer function of the system is negative. In the transfer functions created by MATLAB® the coefficient of the highest power of s in the numerator may be positive or negative, and if it is negative then positive feedback will be required. This is most easily accomplished by placing a minus sign in front of the argument, for example, `rlocus(-G)`. (There is nothing strange about negative transfer functions; in flight dynamics the sign can be changed by changing the sign convention of the control being used.)
- MATLAB® automatically scales the root locus plot, sometimes making it hard to see the area of interest. The command `axis` can change the appearance by specifying a vector with minimum x, maximum x, minimum y, and maximum y values: (`[xmin xmax ymin ymax]`). `axis` is used in Section E.6.3.

The companion to `rlocus` is `rlocfind` which permits the user to pick a point on the root locus plot and determine the gain that will result in that root, as well as the other roots that result. A root locus plot must have been previously created and scaled if necessary. Then the command `[k, p] = rlocfind(G)` will bring the root locus to the fore with a pair of cross-hairs to select the root. If the selection is made slightly off the locus MATLAB® will find the closest point. Note that if positive feedback has been used, the same must be assumed with `rlocfind`.

E.6 MATLAB® Functions (m-files)

Various MATLAB® tasks were repeated often and it was found convenient to place them in files (m-files) that listed the commands and could be called from the command line like a macro. This use of MATLAB® functions is very low level—they are usually used for more complex and sophisticated data calculations and manipulations.

E.6.1 Example aircraft

Appendix A contains a MATLAB® function, called an `m-file`, that performs basic mathematical functions to calculate the terms in the state-space representation of the equations of motion. It also assembles the terms into matrices, which are returned.

The function is evoked from the MATLAB® command line by repeating the first line of the function, replacing any named element except the function name with existing named elements. Those on the right side must be compatible with the function. Those on the left side are returned; they will be created if they don't exist, or overwritten if they do.

```
function [aLong,bLong,aLD,bLD] = Lin(Ref,Phys,D)
```

A plain text file with the commands is created and given a name with the suffix `.m` and then placed in MATLAB®'s *search path* (see the MATLAB® documentation).

E.6.2 Mode sensitivity matrix

This MATLAB® function (`m-file`) calculates the sensitivity matrix described in Section 9.4.1. It does no error checking, and does not even ensure that the matrix passed to it is square. If it is used additional code should be inserted to make it more robust.

```
function S=Sens(a)

[nrows,ncols]=size(a);
[v,e]=eig(a);
diag(e);
for i=1:nrows;v(:,i)=v(:,i)/norm(v(:,i));end;
m=inv(v);
S=[];

for i=1:nrows
S(i,:)=abs(v(i,:)*diag(m(:,i)));
end

for i=1:nrows
sum=0;for j=1:nrows;sum=sum+S(i,j);end;S(i,:)=S(i,:)/sum;
end
```

E.6.3 Cut-and-try root locus gains

When control system design requires repeated applications of `rlocus` and `rlocfind` using different values of parameters and gains (as will often occur when performing successive loop closures), an `m-file` can be created that repeats the actions.

Following is a script that was used in determining gains for the glideslope command example.

```
[aLong, bLong, aLD, bLD] = A4Low;
tLong = eye(4);tLong(2,2)=1/Ref(2);
aLong=tLong*aLong*inv(tLong);
bLong=tLong*bLong(:,1);
r2d = 180/pi;
cLong=r2d*eye(4);cLong(1,1)=1;cLong(4,2)=-r2d;
SSLong=ss(aLong, bLong, cLong, 0);
TFLong=tf(SSLong);
figure(2);
rlocus(TFLong(1));
axis([-.2 .05 -.2 .2])
[k_v, p_v] = rlocfind(TFLong(1));
LongInner = feedback(TFLong, k_v, 1, 1)
k_gamma = 1;
PIzero = input('Enter value for PI zero: ')
fz = tf([1/PIzero 1], [1 0]);
FL_k_vi = LongInner*k_gamma*fz;
figure(5);
rlocus(FL_k_vi(4));
axis([-.2 .05 -.2 .2])
[k_vi, p_vi] = rlocfind(FL_k_vi(4))
TheSys = feedback(fz*k_vi*LongInner, k_gamma, 1, 4);
figure(6);
step(TheSys, 60)
```

- The first line calls the function A4Low which is just a version of lin.m with all the input values hard-coded. The returned matrices are the longitudinal and lateral-directional system and control effectiveness matrices.
- The matrix tLong is the conversion from w to α; see the notes in Appendix A. Ref(2) is the true airspeed previously passed to lin.m in the vector Ref. The matrices are transformed, after which the radians-to-degrees conversion is created.
- The first column of bLong becomes bLong, discarding the elevator effectiveness as it is not needed. bLong(:,1) is all rows (:) and the first column (1) of bLong.
- The output C matrix cLong is then created. When finished it is

$$cLong = \begin{bmatrix} 1 & 0 & 0 & 0 \\ 0 & r2d & 0 & 0 \\ 0 & 0 & r2d & 0 \\ 0 & -r2d & 0 & rd2 \end{bmatrix}$$

In the output equation this matrix multiplies the state vector, which at this point, with angle measurements in radians, is

$$\mathbf{x}_{Long} = \begin{Bmatrix} u \\ \alpha \\ q \\ \theta \end{Bmatrix}$$

The output then becomes, with angle measurements in degrees,

$$\mathbf{x}_{Long} = \begin{Bmatrix} u \\ \alpha \\ q \\ \theta - \alpha \end{Bmatrix}$$

Where $\theta - \alpha = \gamma$.
- The state-space and transfer function representations of the system are created. Note that one can save a few seconds of work in the invoking of ss by using the number 0 instead of a matrix of zeros for the D matrix.
- A root locus is made for TFLong(1) which is $u(s)/\delta_T(s)$. The axes of the root locus plot are specified to be close to the origin, otherwise the short-period roots would cause the automatic scaling to render the area of interest too small. rlocfind is used to pick the desired roots. This root locus is shown in Figure 12.24.
- The inner loop is closed using feedback creating LongInner. The last two arguments specify the first output and the first input of the system are to be used.
- The γ feedback gain is made $k_y = 1$ as discussed in the text. The proportional-integral zero (PIzero) location is got through user input. The filter fz is defined and used to create a forward loop FL_k_vi shown in Figure 12.30. Root locus is performed on that system, and k_{vi} is selected.
- TheSys is the system with all the gains as determined. A 60 second time history in response to a step input is determined and becomes Figure 12.31.

E.7 Miscellaneous Applications and Notes

E.7.1 Matrices

Matrices may be manually entered by enclosing the entries in square brackets, with elements in a row separated by a white space (space or tab) and the start of a new row signified by a semicolon or return (or new-line).

```
A=[1 2 3;4 5 6
7 8 9]

A =

    1   2   3
    4   5   6
    7   8   9
```

Also helpful is the way MATLAB® builds matrices. One may assign a value to any element of a non-existent matrix without specifying undefined elements of lesser row or column size. Thus, `A(3,3)=1` creates a 3×3 matrix of zeros except for the (3, 3) element which is 1. Once defined one may add elements anywhere at all, so that after creating A we could, for example, append a vector to the right side of A and then add a row to the bottom of that result:

```
A=[A [1;2;3];[1 1 1 1]]

A =

    0    0    0    1
    0    0    0    2
    0    0    1    3
    1    1    1    1
```

Special matrices used in the preparation of examples in the text include `eye(3)` which returns a 3×3 identity matrix, and `rand(3,5)` which returns a 3×5 matrix of random numbers.

Matrices can be mixed up with scalars if the meaning is clear, so that in order to make the numbers in the random matrix uniformly distributed between -1 and 1 one specifies the lower limit (-1) and the range (2) in the command

```
B=-1+2*rand(3,5)

B =

    0.6294    0.8268   -0.4430    0.9298    0.9143
    0.8116    0.2647    0.0938   -0.6848   -0.0292
   -0.7460   -0.8049    0.9150    0.9412    0.6006
```

The -1 was internally converted to a 3×5 matrix of -1s.

Also of general use is the means of changing the displayed format of numbers. The command `format` is used followed by a keyword (see `help format` for a list).

```
format shorte; a = 2.5

a =

    2.5000e+00
```

E.7.2   Commands used to create Figures 10.2 and 10.3

```
t = 0:.01:3;
omega = 3.06;
```

```
sigma = -1.17;
d2r = pi/180;
phi_alpha = 32.9*d2r;
phi_theta = 17.4*d2r;
theta = .0897*cos(omega*t+phi_theta);
alpha = .094*cos(omega*t+phi_alpha);
figure(1);
plot(t,alpha,'-',t,theta,'--')
theta=.0897*cos(omega*t+phi_theta).*exp(sigma*t);
alpha=.094*cos(omega*t+phi_alpha).*exp(sigma*t);
figure(2);
plot(t,alpha,'-',t,theta,'--')
```

Notes:

- `t = 0:.01:3;` 101 points of time from 0 to 3.00 by 0.01 second.
- `omega = 3.06;` Damped frequency.
- `sigma = -1.17;` Damping term; real part of the eigenvalue.
- `d2r = pi/180;` Degrees to radians conversion.
- `phi_alpha = 32.9*d2r;` Angle-of-attack phase angle.
- `phi_theta = 17.4*d2r;` Pitch-attitude phase angle.
- `theta = .0897*` ... Time history of θ with no damping. Note that the vector t is multiplied by scalars to present a vector argument to the `cos` function, which then returns a vector of cosines.
- `alpha = .094*` ... Time history of α with no damping.
- `figure(1);` Create or use existing figure number 1.
- `plot(t,alpha,'-',t,theta,'--')` Plot alpha with solid line, theta with dashed line.
- `theta = .0897* ....*exp(sigma*t);` Time history of θ with damping. The operator `.*` performs mulitiplications of the two vectors on a term-by-term basis.
- `alpha = .094* ....*exp(sigma*t);` Time history of α with damping.
- `figure(2);` Create or use existing figure number 2.
- `plot(t,alpha,'-',t,theta,'--')` Plot alpha with solid line, theta with dashed line.

Index

A-4 *Skyhawk*, 73, 127, 140, 146, 148, 149, 166, 183, 185, 188, 189, 247
Adverse yaw, 204
Aerodynamic angles
   angle of attack, 58
   angle-of-attack, *see also* Stability and control derivatives, 15
   sideslip, 58
   sideslip, *see also* Stability and control derivatives, 15
Aileron–rudder interconnect (ARI), 204
Argand diagram, 117, 119, 142
   Dutch roll, 142
   phugoid, 132
   short period, 130
Attainable moments
   volume of, 238
Automatic flight control, 169
   autopilot, 169
   control augmentation system (CAS), 169
   coupled roll–spiral oscillation, 198
   flight-path control, 188
   fly by wire, 170
   stability augmentation system (SAS), 169
Autopilot, 169

Balanced flight, 84
Bermuda triangle, 60

Calspan, 200
Control allocation, 224
   admissible controls, 225, 233
   attainable moments, 226, 235
   cascaded generalized inverse, 241
   control failure reconfiguration, 229
   daisy chaining, 234
   direct method, 240
   generalized inverses, 232
   optimal solutions, 231, 235, 238
   pseudo controls, 234
   pseudo-inverse, 232–234
      weighted, 234
Control Augmentation System (CAS), 169
Coordinate system, *see* Reference frame, 7
Coordinate system transformations, 17
   direction cosines, 18
      properties, 20
   Euler angles, 21
   Euler parameters, 25
   quaternions, 26
   transformations of systems of equations, 26
Customs and conventions, 6, 14, 27, 38, 53, 72, 105
   added mass, 105
   aerodynamic angles, 15
   angular velocity components, 38, 54
   body axes, 14
   dimensional derivatives, 105
   Euler angles, 27
      principal values, 27
   lateral–directional groupings, 56
   latitude and longitude, 14
   linear velocity components, 53
   longitudinal groupings, 56
   stability and control derivatives, 72
   symmetric flight, 57

Descartes' rule of signs, 201
Direction cosines, 18, 263–265, 268
  properties, 20
  time varying, 33, 34
Dutch roll. *see* Lateral–directional modes
Dynamic inversion, 210
  complementary equations, 211, 221–222
  controlled equations, 211–215
  desired dynamics, 212
  kinematic equations, 211, 215–219

Equations of motion, 75
  body axes, 75
    force, 57, 75, 76
    kinematic, 77
    moment, 58, 76, 77
    navigation, 77
  wind axes
    force, 58, 78
    kinematic, 80
    moment, 58
    navigation, 81
Euler angles, 21, 27, 28, 265
  principal values, 27
  time varying, 34
Euler parameters, 25, 263, 264, 268
  time varying, 36

F-104 *Starfighter*, 198
F-14 *Tomcat*, 142
F-15 *Eagle*, 12, 53
F-15 ACTIVE, 61, 62, 209, 210
F-16XL, 15
F-18 *Hornet*, 12
F-4 *Phantom II*, 15
F-8 *Crusader*, 10, 29
Feedback control
  aircraft applications, 178
    phugoid mode, 188
    roll mode, 178
    short-period mode, 184
  first-order systems, 170
  mass–spring–damper system, 172, 178
    position feedback, 175
    rate feedback, 175
  rate feedback
    effect on steady-state, 182, 188
  second-order systems, 172
Final value theorem, 125
Flat-earth assumption, 47, 59, 77, 81, 93

Flight controls, 60
  cockpit controls (control inceptors), 60
  control effectors, 60
  generic control effectors, 62
  primary, 60
  secondary, 60
  sign conventions, 61
  thrust control, 62
Fly by wire, 170
Flying qualities, 151
  aircraft classification, 156
  flight phase categories, 156
  handling qualities, 151
  HQR, 153
  illustrative use of standards, 156
  lateral–directional, 158
    coupled roll–spiral oscillation, 163
    Dutch roll, 158
    roll control effectiveness, 163
    roll mode, 159
    spiral stability, 161
  levels of flying qualities, 156
  longitudinal, 157
    flight path stability, 157
    phugoid stability, 157
    short period, 158
    static stability, 157
  metrics, 152
  pilot compensation, 151
  rating scale, 153
  simulators, 152
  specifications and standards, 155
    Background Information and User Guide, 155
  variable stability aircraft, 152
Forces and moments, 59
  flight controls, 60
  independent variables, 62
  non-dimensionalization, 62
  state rates, 60
  state variables, 60

Inertial accelerations, 43
  earth-centered moving reference frame, 46
  linear, 48
  mass, 47
  moment equations, 52
  point, 43
  rotational, 49
Initial value theorem, 220

Jacobian, 95

Lateral–directional modes
  coupled roll–spiral, 199
  coupled roll–spiral oscillation, 163, 198
  Dutch roll, 141, 142, 144, 145, 149, 150, 154, 158, 163, 172, 199
  lateral phugoid, 199
  roll mode, 141, 144, 159, 163, 170, 178, 179, 213
  spiral mode, 141, 147, 161, 163
Level flight, 87
Linearization, 93
  dimensional derivatives, 99
  force equation, 103
  kinematic equation, 99
  moment equation, 100
  non-dimensional, 100
  non-dimensional derivative, 100
  results
    dimensional lateral–directional equations, 106, 110
    dimensional longitudinal equations, 106, 109
    matrix forms, 108
    non-dimensional lateral–directional equations, 107, 110
    non-dimensional longitudinal equations, 107, 109
    scalar forms, 106
  scalar vs. matrix evaluation, 97
  taylor series, 94
Longitudinal Modes
  phugoid, 139
Longitudinal modes
  phugoid, 129, 130, 132, 136, 137, 150, 157, 188, 193, 207
  short period, 129, 130, 132–134, 138, 139, 149, 158, 166

M2-F2 Lifting Body, 198–207, 212–215, 243
MATLAB®, 2, 149, 231
  commands used, 273–282
  eigenvalues and eigenvectors, 129
  m-file, 248–249
  null space, 231
  plots, 183
  pseudo-inverse, 233
  random numbers, 228
  root locus, 186, 188
  singular value decomposition, 228
  transfer functions, 243
Moment equations, 52

Nonlinear differential equations
  linearization, 95
  systems of, 95

On-board model, 215

P-51 *Mustang*, 55
Phugoid, *see* Longitudinal modes
Principal axes, 11, 77, 102, 103, 105, 108, 111

Quaternions, 26

Reference frame
  atmospheric, 12
  body fixed, 10
    principal axes, 11
    stability, 11
    zero lift, 11
  earth centered, 8
  earth fixed, 8
  inertial, 7
  local horizontal, 8
  rotating, 31
  wind, 12
Roll mode, *see* Lateral–directional modes

Sensitivity analysis, 200
Short period, *see* Longitudinal modes
Singular value decomposition, 228
Spiral mode, *see* Lateral–directional modes
SR-71 *Blackbird*, 53
Stability and control derivatives, 63, 99
  dependencies, 63
    altitude, 64
    angle of attack, 64
    angular velocity, 68
    controls, 69
    sideslip, 66
    summary, 70
    velocity, 64
  linear, 71
  tabular data, 71
  wind and body axes, 63
Stability augmentation system, 169
State variables, 60

State-space equations, 96, 274
States, 53
Steady state, 139
   final value theorem, 138
Steady state, 81, 124, 126, 139
   conditions for, 82
   elevator response, 139
   Euler angles, 82
   final value theorem, 125, 138
   forces and moments, 82
   lateral–directional, 149
   mode approximation, 134, 147
   mode sensitivity, 123
   pitch rate, 188
   roll rate, 179, 183
   solutions, 81
   straight flight, 88
   throttle response, 139
Straight flight, 83
Symmetric flight, 57, 84

Taylor series, 94
Trim, 88
Turn and slip indicator, 84
Turning flight, 86

U-2 (ER-2), 60

Variable stability aircraft, 152
Vector
   cross product as matrix operator, 32
   notation, 13

Printed and bound by CPI Group (UK) Ltd, Croydon, CR0 4YY
09/06/2025
14685655-0001